Ernst Schering Research Foundation Workshop 32
The Role of Natural Products in Drug Discovery

Springer
Berlin
Heidelberg
New York
Barcelona
Hong Kong
London
Milan
Paris
Singapore
Tokyo

Ernst Schering Research Foundation
Workshop 32

The Role of Natural Products in Drug Discovery

J. Mulzer, R. Bohlmann
Editors

With 161 Figures and 23 Tables

Springer

Series Editors: G. Stock and M. Lessl

ISSN 0947-6075
ISBN 3-540-67540-X Springer-Verlag Berlin Heidelberg New York

CIP data applied for

Die Deutsche Bibliothek – CIP-Einheitsaufnahme
The role of natural products in drug discovery / Ernst Schering Research Foundation. J. Mulzer and R. Bohlmann, ed. - Berlin; Heidelberg; New York; Barcelona; Budapest; Hong Kong; London; Milan; Paris; Singapore; Tokyo: Springer, 2000
(Ernst Schering Research Foundation Workshop; 32)
ISBN 3-540-67540-X

Springer-Verlag Berlin Heidelberg New York
a member of BertelsmannSpringer Science+Business Media GmbH

Printed in Germany

Typesetting: Data conversion by Springer-Verlag
Printing and Binding: Druckhaus Beltz, Hemsbach
SPIN:10763341 21/3134/AG–5 4 3 2 1 0 – Printed on acid-free paper

Preface

Natural products are special. During their long evolution and selection they have acquired unique qualities, mostly in connection with biological functions in animal or plant organisms. Additionally, natural products are noted for their highly complex molecular architectures, and they show amazing arrangements of functional groups, strained ring systems, and other attractive structural attributes.

The reasons for interest in natural product chemistry are manifold. First, natural products may serve as lead compounds for new drugs. Second, they give us information on possible biomechanisms and thus on the molecular origin and basis of diseases. Third, their isolation has provoked novel analytical and spectroscopic instrumentation and techniques such as HPLC, NMR, and mass spectrometry. Fourth, natural products are a permanent challenge with respect to total synthesis and stimulate the development of new reagents and reactions; they have had a profound influence on the development of organic chemistry as a whole.

This workshop was designed to shed some light on new and encouraging developments in total synthesis and its applications in natural product research. Sophisticated preparative methods now permit the construction of rather complex structures. However, more economical and practical methods are still needed to improve their scope and efficiency. In particular, new chemical catalysts and biocatalysts, such as catalytic antibodies, or transgenic plants may be required to supplement the classical methodology.

A new facet of organic synthesis is the quickly growing field of combinatorial chemistry. The method promises to provide us with

The participants of the workshop

large collections of small molecules, based on similar reactions which were originally developed for the preparation of single molecules. As more and more target proteins for medical intervention are identified, massive parallel organic synthesis will be one key to meeting demand.

The demand for high molecular diversity is difficult to satisfy with a single method of production. Surely, microbiology will remain an important source of interesting new natural products as lead structures for drug development. Collections of natural products as well as their derivatives and analogues are valuable starting points for drug discovery projects.

The economic and ethical rewards are so huge that individuals and organizations are willing to go to great lengths to discover useful new drugs. Even if chances are low, considerable efforts are made to evaluate new and unconventional ideas. Whatever the costs, natural products and their derivatives are the major-league players that combine high potential with relatively low risk.

J. Mulzer, R. Bohlmann

Table of Contents

List of Editors and Contributors

Editors

J. Mulzer
Universität Wien, Institut für Organische Chemie, Währinger Str. 38,
1090 Wien, Austria

R. Bohlmann
Preclinical Drug Research, Schering AG, Müllerstr. 178,
13342 Berlin, Germany

Contributors

A.G.M. Barrett
Chemistry Department, Imperial College of Science Techology and Medicine,
Exhibition Road, South Kensington, London SW7 1AY, UK

O. Brackhagen
PH-R LSC-NP, Bayer AG, Postfach 101709, 42069 Wuppertal, Germany

R. Brunne
PH-R LSC-NP, Bayer AG, Postfach 101709, 42069 Wuppertal, Germany

I.P. Bucheler
Department of Chemistry, Texas A&M University, College Station,
Texas 77842–3012, USA

S.T. Cohn
Department of Chemistry, Texas A&M University, College Station,
Texas 77842–3012, USA

M.M. Cox
Department of Chemistry, Texas A&M University, College Station,
Texas 77842–3012, USA

S. Danishefsky
Department of Chemistry, Columbia University, 3000 Broadway, MC 3106,
New York, NY 10027, USA

W.W. Doubleday
Department of Chemistry, Colorado, State University, Fort Collins, Colorado,
80523, USA

A.B.Dounay
Department of Chemistry, University of Minnesota, 207 Pleasant Street S.E.,
Minneapolis, MN 55455–0431, USA

J. Ellmann
Department of Chemistry, Room 724 Latimer Hall, University of California,
Berkely, CA 94720–1460, USA

C.J. Forsyth
Department of Chemistry, University of Minnesota, 207 Pleasant Street S.E.,
Minneapolis, MN 55455–0431, USA

S. Grabley
Hans-Knöll-Institut für Naturstoff-Forschung e.V., Beutenbergstr. 11,
07745 Jena, Germany

T. Gross
Chemistry Department, Imperial College of Science, Technology and Medicine, Exhibition Road, South Kensington, London SW7 1AY, UK

D. Hamprecht
Chemistry Department, Imperial College of Science, Technology and Medicine, Office Room 733, Exhibition Road, South Kensington,
London SW7 1AY, UK

T. Henkel
PH-R LSC-NP, Bayer AG, Postfach 101709, 42069 Wuppertal, Germany

J.P. Henschke
Chemistry Department, Imperial College of Science, Technology and Medicine, Exhibition Road, South Kensington, London SW7 1AY, UK

D. Hilvert
Laboratory of Organic Chemistry, Swiss Federal Institute of Technology (ETH), Universitätsstrasse 16, 8092 Zürich, Switzerland

M. Inoue
Department of Chemistry, Columbia University, 3000 Broadway, MC 3106, New York, NY 10027, USA

R.A. James
Chemistry Department, Imperial College of Science, Technology and Medicine, Exhibition Road, South Kensington, London SW7 1AY, UK

K.D. Janda
Department of Chemistry, BCC582, The Scripps Research Insitute, 10550 N. Torrey Pines Road, La Jolla, CA 92037, USA

K. Kasdorf
Chemistry Department, Imperial College of Science Technology and Medicine, Exhibition Road, South Kensington, London SW7 1AY, UK

P. Kocienski
Department of Chemistry, Joseph Black Building, University of Glasgow, Glasgow G12 8QQ, UK

T.M. Kutchan
Leibniz Institut für Pflanzenbiochemie, Weinberg 3, 06120 Halle, Germany

J.M. Langenhan
Department of Chemistry, Texas A&M University, College Station, Texas 77842–3012, USA

N.B. Messerschmidt
Department of Chemistry, Texas A&M University, College Station, Texas 77842–3012, USA

H. Müller
H-R LSC-NP, Bayer AG, Postfach 101709, 42069 Wuppertal, Germany

R. Narquizian
Department of Chemistry, Joseph Black Building, University of Glasgow, Glasgow G12 8QQ, UK

M. Ohkubo
Chemistry Department, Imperial College of Science Technology and Medicine, Exhibition Road, South Kensington, London SW7 1AY, UK

K. Park
Department of Chemistry, Texas A&M University, College Station, Texas 77842–3012, USA

P.A. Procopiou
Glaxo Wellcome Research and Development Ltd., Medicinal Chemistry 1, Gunnels Wood Road, Stevenage, Hertfordshire SG1 2NY, UK

F. Reichel
PH-R LSC-NP, Bayer AG, Postfach 101709, 42069 Wuppertal, Germany

D. Romo
Department of Chemistry, Texas A&M University, College Station, Texas 77842–3012, USA

R.M. Rzasa
Department of Chemistry, Texas A&M University, College Station, Texas 77842–3012, USA

S.F. Sabes
Department of Chemistry, University of Minnesota, 207 Pleasant Street S.E., Minneapolis, MN 55455–0431, USA

I. Sattler
Hans-Knöll-Institut für Naturstoff-Forschung e.V., Beutenbergstr. 11, 07745 Jena, Germany

W.D. Schmitz
Department of Chemistry, Texas A&M University, College Station, Texas 77842–3012, USA

H.A. Shea
Department of Chemistry, Texas A&M University, College Station, Texas 77842–3012, USA

M. Shibasaki
Graduate School of Pharmaceutical Sciences, University of Tokyo, 7–3-1 Hongo, Bunkyo-ku, Tokyo 113–0033, Japan

R. Thiericke
Hans-Knöll-Institut für Naturstoff-Forschung e.V., Beutenbergstr. 11, 07745 Jena, Germany

D. Trauner
Department of Chemistry, Columbia University, 3000 Broadway, MC 3106, New York, NY 10027, USA

G.J. Tustin
Chemistry Department, Imperial College of Science, Technology and Medicine, Exhibition Road, South Kensington, London SW7 1AY, UK

R.A. Urbanek
Department of Chemistry, University of Minnesota, 207 Pleasant Street S.E., Minneapolis, MN 55455–0431, USA

A.J.P. White
Chemistry Department, Imperial College of Science, Technology and Medicine, Exhibition Road, South Kensington, London SW7 1AY, UK

D.J.Williams
Chemistry Department, Imperial College of Science, Techology and Medicine, Exhibition Road, South Kensington, London SW7 1AY, UK

J. Yang
Department of Chemistry, Texas A&M University, College Station, Texas 77842–3012, USA

1 Synthesis of Immunomodulatory Marine Natural Products

S.J. Danishefsky, M. Inoue, D. Trauner

1.1 Introduction

There has been a long-term fruitful synergy between the study of organic chemistry and the quest for the identification of novel and potentially valuable products from nature (Barton and Nakanishi 1999). Following the isolation of such naturally derived agents, chemistry has provided the indispensable intellectual framework for structure elucidation. In the early days, these assignments were accomplished through the medium of classical degradative chemistry. The validity of chemistry-based structure deduction was extremely dependent on the powerful logic which unifies the basic reactions of organic chemistry and allows for organization and interpretation of reaction sequences.

With the advent of more and more powerful spectroscopic, not to speak of crystallographic techniques, structure determination has become significantly less reliant on ingenious correlations of chemical transformations. Indeed, these days, it is not at all uncommon that the structure of even a complex naturally occurring agent can be fully assigned without the benefit of conducting a single chemical transformation on the target structure. Obviously, there are great advantages to our new-found capabilities, particularly as regards the speed with which the problems can be solved, the more accurate nature of the solutions, and, above all, the feasibility of achieving structure determination on minute quantities of the target systems. The advent of increasingly discriminating and well-focused biological screening protocols provides additional incentives for persisting in the isolation-structure assignment exercises. Thus increasing numbers of structures of compounds, identified through systematic bioassays, are assigned on trace amounts of material.

Accordingly, and unlike the situation in the classical era, the structures enter the literature unaccompanied by any established patterns of chemical personality. In many instances, the report of the isolation of a natural product, its structure assignment, and the statement as to early findings regarding biological activity become in effect the end point of the program. When the quantity of isolates is inadequate for future development, the projects are usually dropped in preference to repetition and scale-up of the isolation procedure. However, in some instances, the chemical novelty of the assigned structure and the promise of its potential biological activity serve to attract the interest of the community of synthetic organic chemists. Indeed, synthesis may become the only practical option for revitalizing such projects, for eventually producing enough of the agents for fuller evaluation of their biological profiles, and for probing their mechanisms of action.

Marine-derived natural products are often chemically novel and of difficult accessibility (Faulkner 1999). In this chapter we will describe two separate cases where the discovery of marine-derived natural agents prompted such follow-ups from the field of chemical synthesis. In each instance, chemical synthesis was indispensable for obtaining useful quantities of the agents, exploring their SAR profiles, and paving the way for deeper insights into the promise of the program for drug development. Herein, we focus on two goal systems of marine origin that elicited our interests as synthetic organic chemists. One was a family of compounds known as the frondosins and the other one is an alkaloid known as halichlorine. We will show how the confluence of biological promise, structural novelty, and difficult accessibility created opportunities and challenges which were met by the science of chemical synthesis.

1.2 Frondosin B, an IL-8 Receptor Antagonist

Interleukin-8 (IL-8), a chemoattractant for neutrophils, is produced by macrophages and endothelial cells (Hock and Schraustätter 1996). IL-8 has been implicated in a wide range of acute and chronic inflammatory disorders including psoriasis and rheumatoid arthritis (Scitz et al. 1991; Miller et al. 1992). Furthermore, various animal models of inflammation have established IL-8 as a principal chemotactic factor directing neutrophil recruitment to the inflammatory focus. Therefore, an IL-8 receptor antagonist represents a promising target for the development of novel pharmacological agents against autoimmune hyperactivity. Frondosins A–E were recently isolated from the sponge *Dysidea frondosa*. They each inhibited the binding of IL-8 to its receptor in the low micromolar range (Pateil et al. 1997). The structures of these compounds were determined mainly by NMR spectroscopy and feature bicyclo[5.4.0]undecane ring systems in the context of permuted linkages to various hydroquinone-based moieties.

A National Cancer Institute (NCI) team also isolated frondosin A and D from the HIV-inhibitory organic extract of the marine sponge (Hallock et al. 1998). It is worth noting that the frondosins exhibit opposite optical rotations with different absolute values. Thus the frondosins may occur as scalemic mixtures. It was anticipated that further biological studies of frondosins could spur fruitful investigations into IL-8 activity and present new possibilities for the development of novel anti-inflammatory agents. These considerations, as well as the novel structure of the frondosins, prompted us to undertake synthetic studies directed at these targets. We chose to focus on frondosin B (**1**) (Hallock et al. 1998), which contains an intriguing benzofuran ring system. In this paper, we describe a highly efficient total synthesis of this compound.

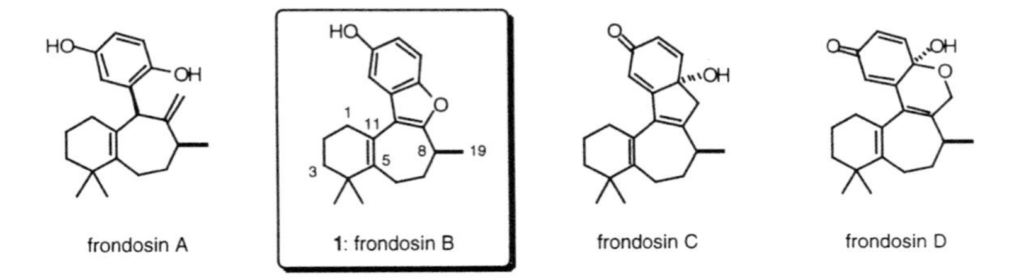

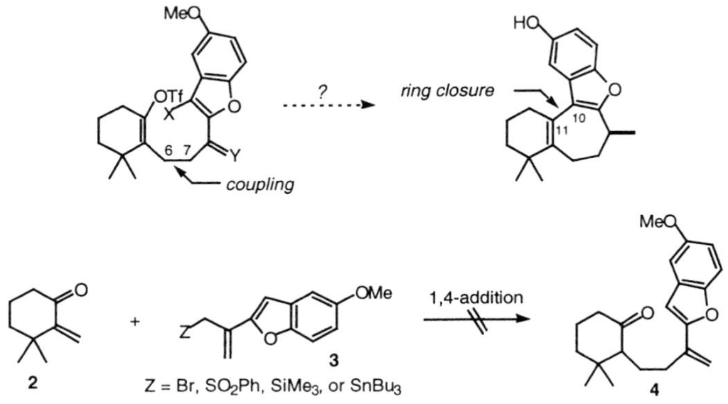

Scheme 1. First-generation strategy

Different synthetic disconnections were employed in pursuing various strategies. Initially, we planned to couple two fragments by forming a bond between C6 and C7 then closing the 7-membered ring via C10–11 bond formation (Scheme 1). Following preparation of the two

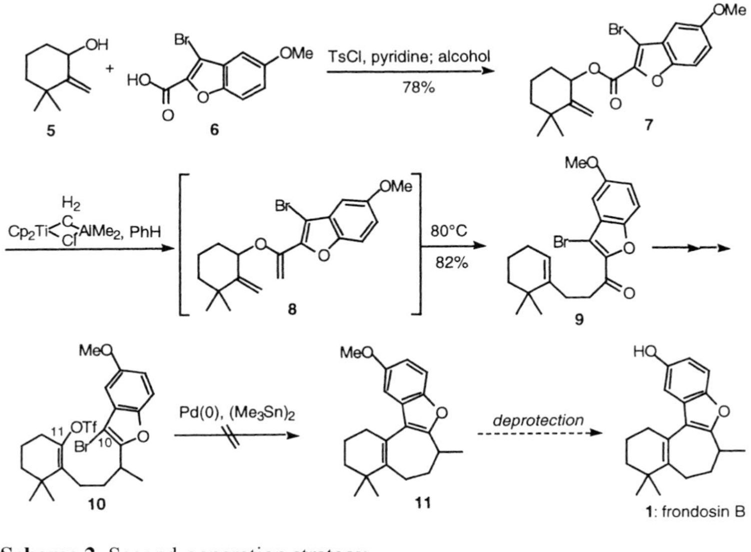

Scheme 2. Second-generation strategy

required segments, the first difficulty we encountered was that 1,4-addition of right-hand fragment **3** to α,β-unsaturated ketone **2** did not occur, although numerous reaction conditions were studied. Presumably, these difficulties reflect the steric demand of the *gem*-dimethyl group and the strictness associated with the trajectory for nucleophilic attack at the exocyclic methylene group.

Accordingly, we modified the coupling strategy to rely on an intramolecular delivery reaction (Scheme 2). Ester **7**, which was prepared from the corresponding alcohol **5** and carboxylic acid **6**, was transformed to a diene **8** using the Tebbe reagent (Tebbe et al. 1978). Subsequent heating cleanly gave the Claisen rearrangement product **9** in good yield (Kinney et al. 1985). Functional group processing of **9** led to substrate **10** in anticipation of the crucial ring closure. Unfortunately, various reaction conditions that were attempted failed to furnish the desired target **11** (Kelly et al. 1990; Grigg et al. 1991; Mori et al. 1991; Huang et al. 1999).

These disappointing results obliged us to change our approach. Our next plan was to alter the sequence by first coupling two segments at C10–11. This initial merger would be followed by bond formation between C8 and C9 (Scheme 3). In this way, the seven-membered ring would be constructed. Although the left- and right-hand pieces (**12**, **13**) could be synthesized in a straightforward manner, coupling with palladium(**0**) (Knight 1991) was low-yielding and unreliable, especially

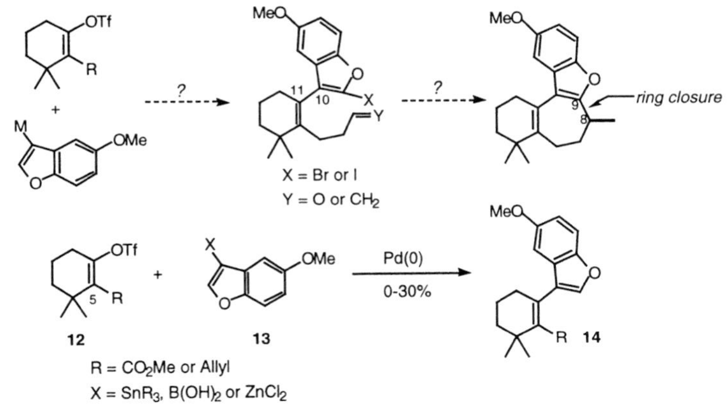

Scheme 3. Third-generation strategy

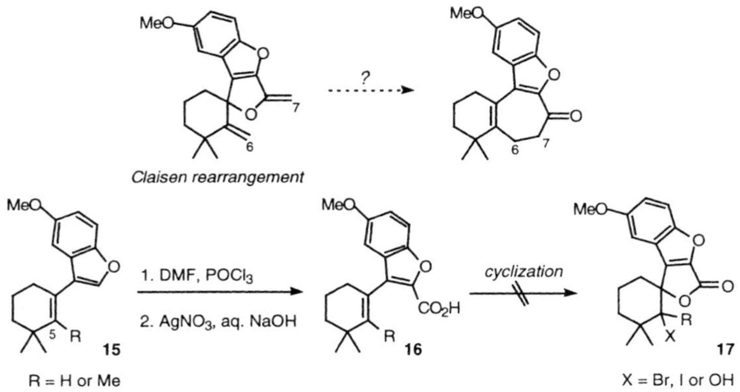

Scheme 4. Fourth-generation strategy

when we operated on scales larger than a few milligrams. Because of the inefficiency of these couplings, the subsequent investigation of seven-membered ring closure was severely hampered.

In the course of studying coupling reactions of this type, we found that the yields depended on the length of the chain attached to C5. Reactions could be performed in acceptable yield without a side chain at C5 or with a methyl in place (Scheme 4). Therefore, we decided to combine the second and third strategies, namely coupling at C10–11 followed by subsequent Claisen rearrangement to form the seven-membered ring at C6–7. Carboxylic acid **16** was prepared by palladium-mediated coupling, followed by a Vilsmeier reaction (Jones 1997) and oxidation (Sharmma and Rodrigues 1968). Unfortunately, spirolactone **17** was never obtained under a variety of conditions.

The series of negative results described above, augmented by many others (Frontier 1999), underscored two important points: (a) bond formation in spatial proximity to the dimethyl group promised to be extremely difficult, especially with advanced intermediates, and (b) attachment of tetra-substituted olefin derivatives to C10 of the furan by organometallic methods was unable to efficiently supply material for further steps.

With these considerations in mind, we developed a totally new strategy seeking to construct the final cyclohexane ring containing the *gem*-

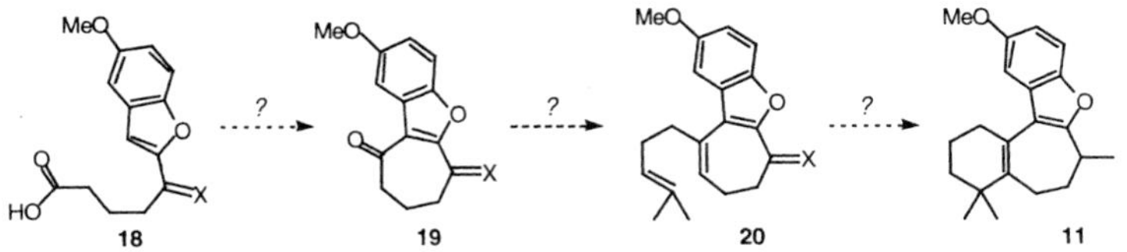

Scheme 5.

dimethyl group onto a pre-existing matrix wherein the seven-membered ring was already in place (Scheme 5). In this stepwise approach, the seven-membered ring diene **20** would be synthesized by addition of a homoprenyl group to the ketone **19**. This bond formation would set the stage for subsequent acid-induced cyclization, exploiting the activated benzofuran ring to promote and guide the sense of electrophilic attack on the cycloheptenyl double bond.

The synthesis started from the acetyl benzaldehyde **22**, which was prepared from 2-hydroxy-5-methoxy benzaldehyde **21** by a known protocol (Landelle et al. 1991), as shown in Scheme 6. Four-carbon extension to methyl ketone **22** by the Wittig reaction proceeded smoothly to give olefin **23** in 87% yield (Daheiser and Halgason 1994). The olefin in **23** was carefully hydrogenated using palladium on carbon, and subsequent saponification furnished carboxylic acid **24** in quantitative yield. In this hydrogenation, extended reaction times resulted in formation of significant quantities of dihydro benzofuran. Friedel-Crafts reaction to build the seven-membered ring was performed in a one-pot by stepwise addition of oxalyl chloride and tin chloride, giving ketone **25** in good yield. Nucleophilic attack of the cerium reagent (Fujiwara et al. 1993; Guevel and Hart 1996), prepared from 4-methyl-3-pentenyl magnesium bromide, led to the tertiary alcohol, which was readily eliminated in $CDCl_3$ to afford the diene **26** (endo:exo=6:1, 93%, two steps). This susceptibility to elimination by the tertiary alcohol indicated participation of the benzofuran in stabilizing the α-cation. To our delight, cationic cyclization was found to proceed under a variety of conditions in good yield. Acid-induced cyclization (HCOOH, H_3PO_4, $BF_3 \cdot OEt_2$) resulted in the formation of the six-membered ring in an approximately 2.5:1 mixture favoring the desired compound **11** and olefinic isomer mixture **27**. The mercury trifluoroacetate-mediated reaction (Kurbanov

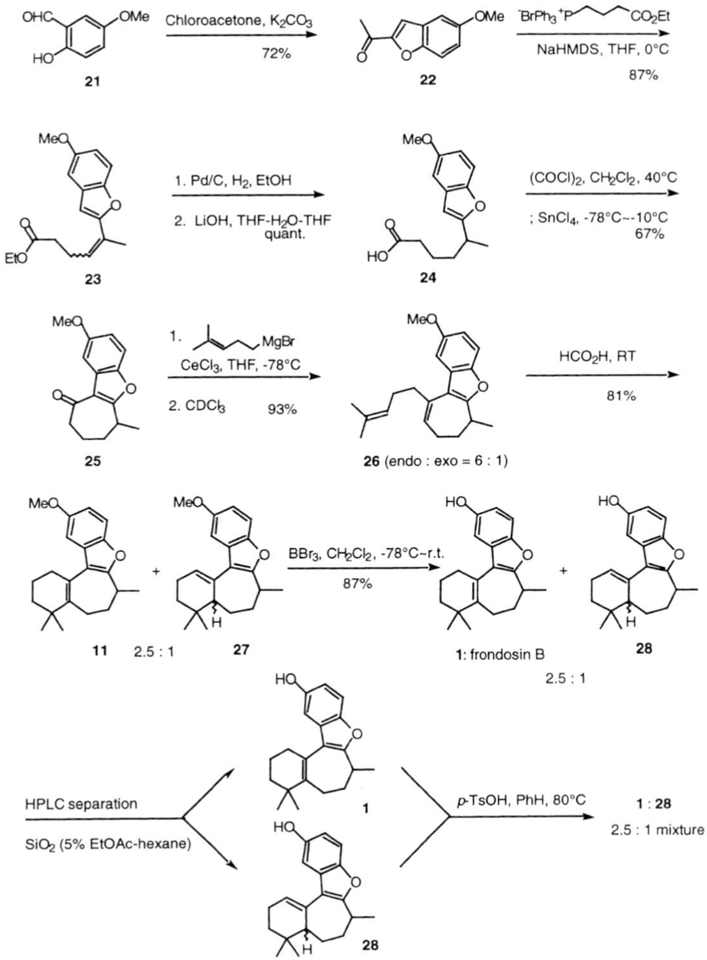

OHC OMe HO 21 Chloroacetone, K_2CO_3 72% 22 OMe $^-BrPh_3{}^+P$ CO_2Et NaHMDS, THF, 0°C 87%

MeO 23 EtO 1. Pd/C, H_2, EtOH 2. LiOH, THF-H_2O-THF quant. MeO 24 HO $(COCl)_2$, CH_2Cl_2, 40°C ; $SnCl_4$, -78°C~-10°C 67%

MeO 25 1. MgBr $CeCl_3$, THF, -78°C 2. $CDCl_3$ 93% MeO 26 (endo : exo = 6 : 1) HCO_2H, RT 81%

MeO 11 2.5 : 1 MeO 27 H BBr_3, CH_2Cl_2, -78°C~r.t. 87% HO 1: frondosin B HO 28 H 2.5 : 1

HPLC separation SiO_2 (5% EtOAc-hexane) HO 1 HO 28 H p-TsOH, PhH, 80°C 1 : 28 2.5 : 1 mixture

Scheme 6. Fifth-generation strategy

et al. 1972; Corey et al. 1980; Sato et al. 1982), in which olefin isomerization after the cyclization should be minimal, also proceeded smoothly but again gave the products as a mixture after the reduction of mercury. The similar isomeric ratios obtained in these reactions suggested that the product distribution reflected not only thermodynamic stability of the two isomers, but also the kinetic reaction rate of the cyclization-proton elimination sequence.

Though these isomers were inseparable by silica gel chromatography, we decided to complete the total synthesis (Scheme 6). Deprotection of the methyl ether was effected with tribromoborane to afford frondosin B **1** and its olefinic isomers **28** in 87% yield. These isomers were separated with HPLC (silica gel, 7% ethyl acetate-hexane), and synthetic frondosin B was identical to that isolated from natural sources in every respect (^{1}H, ^{13}C-NMR, and IR). Separate subjection of **1** and **28** to acidic condition led to formation of the thermodynamic mixture of isomers (ca. 2.5:1). Thus, we achieved the total synthesis of frondosin B in nine steps from commercially available 2-hydroxy 5-methoxy benzaldehyde. Although this synthesis is highly concise and assures quick access to a large amount of material, the problem of controlling the regiochemistry of olefin remained unsolved.

In order to develop a completely regiocontrolled synthesis, we decided to modify the synthetic route. We felt that it would be important to avoid acidic conditions, not only for the construction of a six-membered ring but also for subsequent synthetic manipulations. We thereby hoped to prevent the olefin migration. For this reason, a Diels-Alder reaction to form the C1–2 and C3–4 bonds was thought to be the best choice (Omodani and Shishido 1994). Construction of the corresponding diene (see **33**) was initiated via the aldol condensation of ketone **25** with acetone (Scheme 7). The zinc enolate of the ketone, prepared by transmetallation of the lithium enolate, reacted with acetone to furnish adduct **29** in 85% yield (Hagiwara et al. 1980; Witschel and Bestmann 1997). Not surprisingly, this compound was prone to retro-aldolization, especially with acidic media, as treatment of **29** with *p*-toluenesulfonic acid produced ketone **25** in 77% yield along with a trace amount of olefin **30**. After several experiments, dehydration of the tertiary alcohol was accomplished by treatment of **29** with mesyl chloride and triethylamine, leading to the mixture of olefinic isomers (ca. 1:1) in 88% yield. 16b Sodium methoxide in methanol effected the isomerization of the olefin

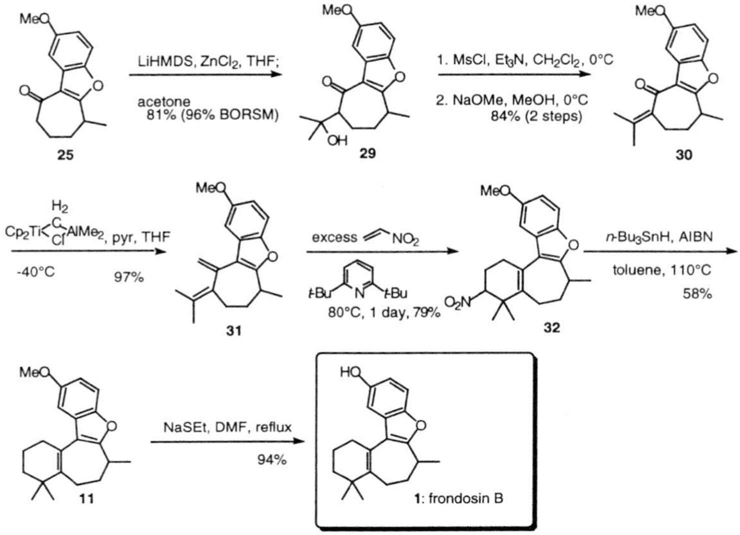

Scheme 7. Sixth-generation strategy

to give predominantly the desired compound **30** in 96% yield (Guerriero et al. 1990). Finally, diene **31** was prepared in 97% yield using Tebbe reagent buffered with pyridine.

Following preliminary experiments, this compound was found to be sensitive to acid and relatively unreactive. Reaction with maleic anhydride at 110°C for 2 days gave a Diels-Alder adduct, but only in 27% yield, along with some unidentified compound, presumably because of a trace amount of maleic acid. Considering the acid sensitivity of **31,** as well as the tendency of the expected product towards olefin migration, a Lewis-acid-promoted Diels-Alder reaction could not be utilized. Therefore, we chose to use nitroethylene as a reactive ethylene equivalent under acid-free conditions, expecting that the nitro group could be easily removed under radical conditions (Ranganathan et al. 1980; Corey and Myers 1985). In addition, it was anticipated that the nitro group would be beneficial for derivatization to the other functional groups, in order to synthesize structural variants for biological investigations or prepare an affinity column. Diels-Alder reaction of **31** with excess nitroethylene in the presence of di-*t*-butyl pyridine at 80°C finished in 24 h, giving adduct **32** in 79% yield. The nitro group was removed by radical reduction to afford methyl-protected frondosin B **11** in 58% yield (Ono and Kaji 1986). Finally, synthesis of geometrically pure frondosin B **1** was achieved by deprotection of the methyl group, this time with sodium ethyl thiolate (94%) (Feutrill and Mirrington 1972). By this alternative route, the total synthesis of frondosin B was realized in 12 steps, and this method has more potential for the synthesis of structural variants.

In summary, the total synthesis of frondosin B was achieved by two different routes, using a cationic cyclization and the Diels-Alder reaction as key steps (Inoue et al. 2000). These efficient syntheses not only provide a sufficient amount of the natural product for further biological studies, but also offer a general method for the preparation of a variety of structural derivatives possibly leading to promising anti-inflammatory agents. Further biological study of frondosin B and development of an asymmetric synthesis to determine the absolute configuration of frondosins are currently underway in this laboratory.

1.3 (-)-Halichlorine, an Inhibitor of VCAM-1 Expression

(-)-Halichlorine is a novel alkaloid recently isolated by the Uemura group from the marine sponge *Halichondria Okadai Kadota* (Kuramoto et al. 1996; Arimoto et al. 1998). The molecule was found to be a selective inhibitor of the expression of vascular cell adhesion molecule 1 (VCAM-1) and might therefore represent an important lead compound for the design of drugs against allergic inflammatory diseases.

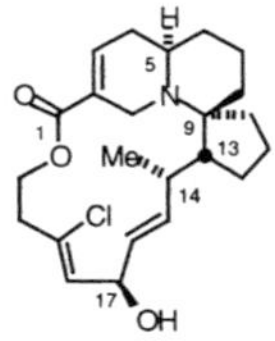

halichlorine (**33**)

VCAM-1 (=CD106), a member of the immunoglobulin superfamily, is expressed predominantly on the surface of vascular endothelial cells in response to various inflammatory stimuli (e.g., TNFα, IL-1; Barclay et al. 1997; Foster 1996; Springer 1994; Carlos and Harlan 1994).

Endothelial VCAM-1 contributes to the extravasation of certain leukocytes from blood vessels into inflamed tissue by interacting with the integrins α4β1 (very late antigen 4, VLA-4) and α4β7 found on these cells. The VCAM-1/VLA-4 interaction mediates both initial tethering and rolling of lymphocytes on endothelium and their subsequent firm adhesion and extravasation. Hence, compounds that interfere with this interaction or inhibit the expression of one of its participants should have an anti-inflammatory effect. Recently, it has also been demonstrated that FK506 (tacrolimus) inhibits extravasation of lymphoid cells by abrogating VLA-4/VCAM-1 mediated extravasation (Tsuzuki et al. 1998). FK506 might therefore exert its potent immunosuppressive action partially by interfering with leukocyte recruitment. Interestingly, a soluble form of VCAM-1 has been found to promote angiogenesis (Koch et al. 1995). The overall biological significance of VCAM-1 is further demonstrated by the observation that VCAM-1 knockout mice die during embryogenesis (Kwee 1995).

Apart from these interesting biological aspects, halichlorine would attract the interest of the synthetic organic chemist simply due to its intriguing structure. The most provocative feature of the molecule certainly consists of a dehydroquinolizidine ring system connected to a five-membered carbocycle through a unique azaspiro linkage. In addition to this amino-substituted quaternary center, the tetracyclic molecular framework contains four tertiary stereocenters at C5, C13, C14, and C17. The latter three stereogenic carbons are embedded in a 15-membered aza-macrolactone, which also features a highly unusual *Z*-vinyl chloride moiety. Retrosynthetically, halichlorine can be disconnected into two halves of unequal size and complexity by imagining saponification and cleavage of the C15–16 double bond (Scheme 8). This leads to a spiroquinolizidine subunit (**35**) and a carbon chain containing the vinyl chloride moiety and two oxygen-bearing functionalities (**34**). The central feature of the spiroquinolizidine around which every synthetic approach must revolve is its amino-substituted quaternary stereocenter at C9 – the spiro linkage. One of the most efficient methods of establishing such a stereocenter involves the addition of a suitable nucleophile to an imminium ion, or – even better – an *N*-acyl-imminium ion (Hiemstra and Speckamp 1991). Such an intermediate could be retrosynthetically conjured up by linking a C-15 carbonyl group (obtained through "ozonolytic" cleavage of the corresponding double bond) with

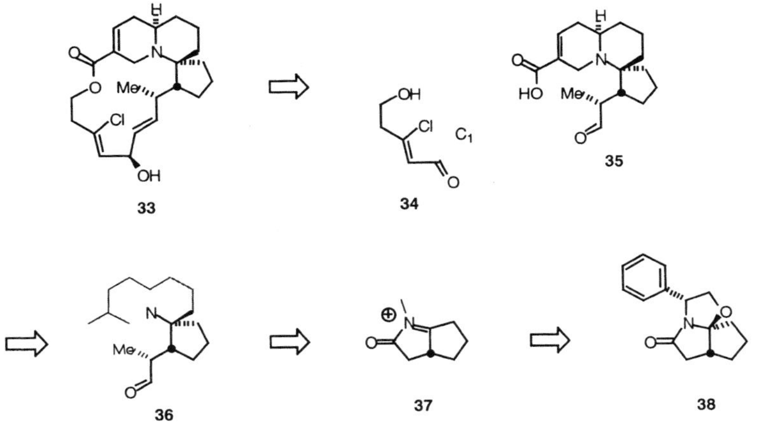

Scheme 8. Retrosynthetic analysis

the nitrogen after laying it bare of its surrounding carbons (see **35**→**36**). An additional advantage of this approach is the possibility of stereoselectively installing the methyl group in α-position to the C15-carbonyl through deprotonation/alkylation. The enantiomerically pure, bicyclic *N*-acyl-imminium ion **37** thus identified can then be traced back to its most common precursor, a Meyers lactam, e.g., compound **38**. Though the planning exercise would tend to focus on the establishment of the very interesting spiro linkage, solutions for other important synthetic issues would have to be found. For example, the selective installment of the C5-bridgehead stereocenter and the vinyl chloride moiety and the exact way in which the macrocyclic ring system would be closed. At this stage of our synthetic planning, however, we decided to defer these decisions to a later phase of our synthetic venture.

With such a general synthetic concept in mind, we embarked on the total synthesis of halichlorine (Scheme 9). Gratifyingly, the required Meyers lactam **38** turned out to be a known compound, which was obtained in excellent yield by condensation of the racemic ketocarboxylic acid **39** with D-phenylglycinol Interestingly, the entire racemic starting material is channeled into one enantiomer (and diastereomer) under the conditions shown (Ragan and Claffey 1995). Treatment of **38** with excess allyltrimethylsilane in the presence of titanium tetrachloride

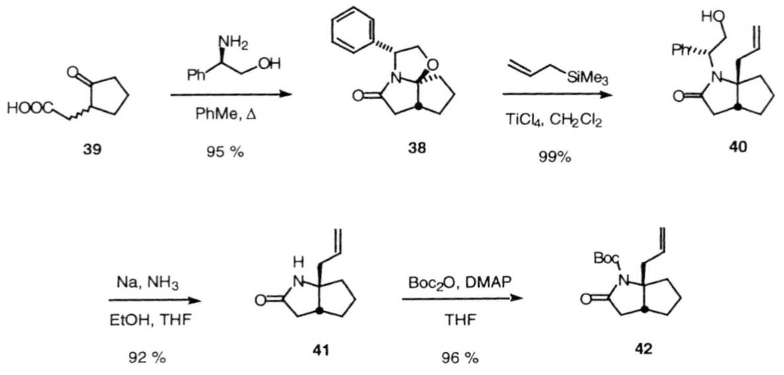

Scheme 9.

furnished bicyclolactam **40** in virtually quantitative yield (Burgess and Meyers 1991; Meyers et al. 1995; Meyers and Brengel 1997; Romo and Meyers 1991). The efficiency of this reaction is probably due to the high reactivity of the strained bicyclic *N*-acyl-imminium ion intermediate. Following reductive debenzylation of **40**, the resulting bicyclolactam **41** was transformed into the imide **42** by treatment with Boc-anhydride in the presence of one equivalent of DMAP, thus activating the system towards the ensuing alkylation.

The resulting bicyclic *N*-Boc lactam **42** was stereoselectively *C*-methylated, as shown, to afford compound **43** (Scheme 10). Since all efforts to reduce **43** directly to the desired primary alcohol **45** failed, the lactam ring was cleaved to provide the interesting enantiomerically pure γ-amino acid **44**. Following activation of the latter as the mixed anhydride, in situ reduction with sodium borohydride afforded the primary alcohol **45**. The hydroxyl group was protected as its *tert*-butyl-diphenylsilyl derivative **46**. Thus three of the five stereocenters of halichlorine had been installed in a systematic and completely stereoselective way.

Having found a pleasing solution for the quaternary center, and for arranging the two adjacent tertiary stereocenters, we set forth to close the first heterocyclic ring (a piperidine). We now faced the question of how the substitution of the cyclopentane ring might influence the stereochemical outcome of a heterocyclization event. In the course of this a reaction the C5-bridgehead stereocenter would be formed. After much

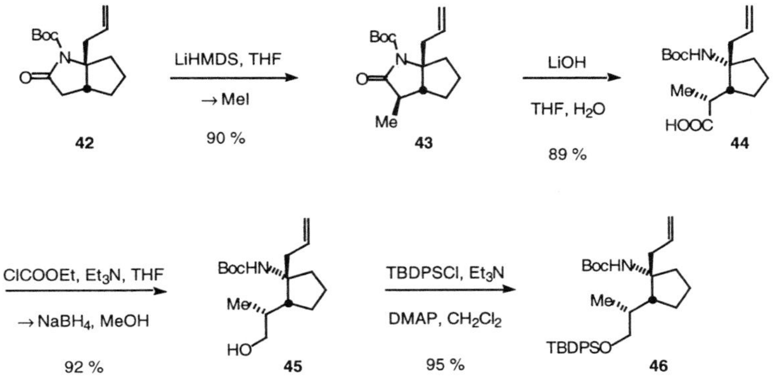

LiHMDS, THF → MeI 90 %; LiOH THF, H_2O 89 %; ClCOOEt, Et_3N, THF → $NaBH_4$, MeOH 92 %; TBDPSCl, Et_3N DMAP, CH_2Cl_2 95 %

Scheme 10.

deliberation, the following highly stereoselective sequence was devised (Scheme 11): Hydroboration of the allylic side chain of **46**, followed by palladium mediated Suzuki-coupling (Miyaura and Suzuki 1995; Balog et al. 1998; Sasaki et al. 1998; Fürstner and Konetzki 1998; Ohba et al. 1996; Johnson and Braun 1993), with methyl *Z*-3-iodoacrylate, afforded the crude unsaturated ester **47** which, upon deprotection of the amino function with TFA and subsequent basification, underwent intramolecu-

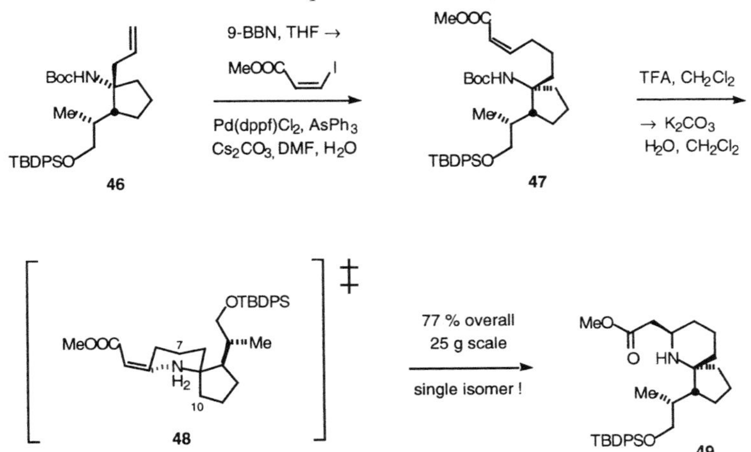

Scheme 11.

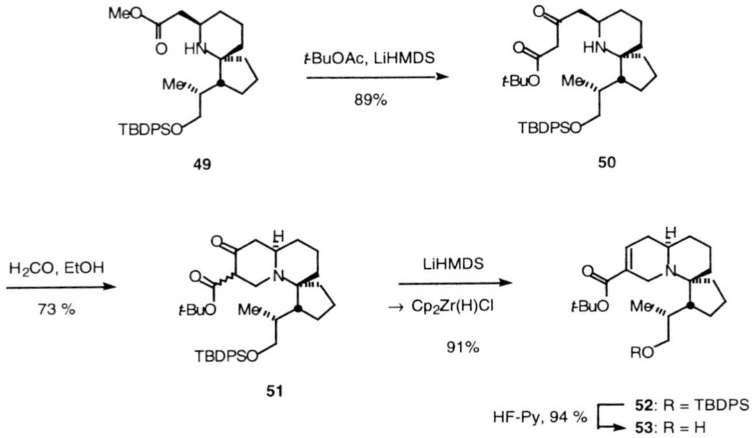

Scheme 12.

lar Michael addition to afford piperidine **49** *as the only isolated isomer* in good overall yield. The high stereoselectivity of this cyclization reaction can be explained in terms of a chair-shaped transition state **48**, wherein the larger substituents adopt *pseudo*-equatorial positions. The corresponding chair-shaped transition state leading to the undesired diastereomer, wherein the *cis*-α,β-unsaturated ester moiety is rotated downwards (i.e., resides in a *pseudo*-axial position) would suffer from severe unfavorable 1,3-diaxial interactions with the axial proton at C7 and the C10 methylene group. Not surprisingly, and in full accordance with this transition state model, the stereoselectivity of the conjugate cyclization was somewhat diminished when methyl-*E*-iodoacrylate was used instead of a geometrical isomer in an otherwise identical sequence (affording a 16:1 mixture of diastereomers in favor of **49**).

The transformation of the β-aminoester **49** into the dehydro-spiroquinolizidine ring system found in halichlorine is shown in Scheme 12. Due to the low reactivity of the extremely hindered secondary amine towards further conjugate additions we decided to pursue an intramolecular Mannich strategy. Crossed Claisen condensation of **49** with *tert*-butyl acetate afforded β-keto ester **50** in good yield. The quinolizidine ring system was then closed by a Mannich reaction to afford β-keto

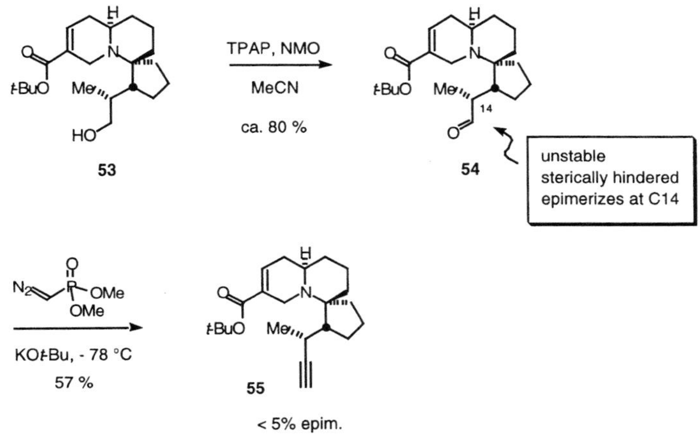

tBuO Me HO 53
TPAP, NMO
MeCN
ca. 80 %
tBuO Me 14 54
unstable
sterically hindered
epimerizes at C14

N_2 P OMe OMe
KOtBu, - 78 °C
57 %
tBuO Me 55
< 5% epim.

Scheme 13.

ester **51** as a mixture of diastereomers and tautomers. Efficient conversion of the β-keto ester moiety to the corresponding α,β-unsaturated ester was achieved in one synthetic step and in excellent yield using Ganem's protocol involving deprotonation followed by treatment with the Schwartz reagent (Godfrey and Ganem 1992). Interestingly, a more classical sequence involving reduction of the ketone, activation, and base-induced elimination failed in our hands. The resulting dehydroquinolizidine, **52**, was finally deprotected with HF-pyridine to afford the primary alcohol **53** (Trauner and Danishefsky 1999).

With the spiroquinolizidine subunit in hand, we investigated the elongation of the side chain as a prerequisite for the formation of the macrocyclic ring (Scheme 13). This required oxidation of a C15 primary alcohol to the corresponding aminoaldehyde, a transformation that proved to be quite difficult due to the instability of the resulting aminoaldehyde towards racemization at C14. Following extensive experimentation, it was found that this oxidation could be performed by prolonged treatment with a small amount of TPAP (<5 %) in the presence of excess NMO in acetonitrile (Ley et al. 1994). The crude material obtained under these conditions contained less than 5% of C14 epimer. Unfortunately, all attempts to olefinate the aldehyde **54** using one of the many available variants of the Wittig or the Horner-Wadsworth-Em-

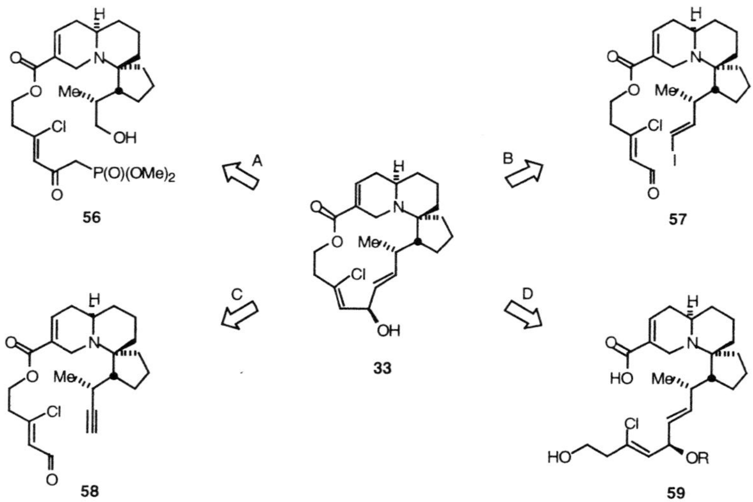

Scheme 14. Ring-closing strategies

mons reaction were either completely unsuccessful or led again to extensively epimerized material. After many failed attempts to install the C15–C16 double bond by other means (e.g., via the Takai reaction or Julia olefination) we found a reagent that was sufficiently small and nucleophilic to react with the sterically hindered aldehyde **54** at low temperature. Treatment of **54** with excess dimethyl diazophosphonate (Gilbert's reagent) (Gilbert and Weerasooriya 1979; Rao et al. 1997) and a base at low temperature afforded alkyne **55** in good overall yield and with little racemization at C14. Considering the plethora of useful reactions involving a terminal alkyne, confidence in our overall synthetic plan began to increase.

At this point we had to commit to a specific reaction for the formation of the macrocyclic ring. Four possible strategies which were under investigation at some point during our synthetic venture are shown in Scheme 14. Attempts to perform an intramolecular or Horner-Wadsworth-Emmons ring closure (A) failed due to our inability to oxidize hydroxyphosphonate **56** to the corresponding aldehyde. In view of our discouraging experiences with *inter*molecular Horner-

Wadsworth-Emmons-olefinations (see above), we decided against persistent pursuit of this option. Unfortunately, despite numerous attempts, a planned intramolecular Nozaki-Hiyama-Kishi ring closure (B) (Fürstner 1999) involving iodoalkene **57** (obtained in three steps from **55**) failed as well. Control experiments with model compounds showed that the β-chloro-α,β-unsaturated aldehyde moiety was extremely unstable towards the reducing power of chromium(II). A third option (C), based on an intramolecular nucleophilic addition of alkyne-aldehyde **58**, was also unsuccessful in our hands.

Hence we decided to focus on the seemingly most straightforward solution for the formation of the macrolactone: a macrolactonization (strategy D). We were fully aware, however, that such an approach would require careful attention to a protection/deprotection scheme. Furthermore, an efficient *inter*molecular coupling of the key alkyne **55** (or one of its derivatives) with a C_5-aldehyde representing the C17–C21 portion of halichlorine had to be developed. Ideally, the C17 stereocenter would be formed with a high degree of stereoselectivity in the course of this reaction. The required *Z*-β-chloro-2-alkenal **64** was prepared as shown in Scheme 15, using the known Weinreb amide **63** (Keen and Weinreb 1998) as a key intermediate.

Hydrozirconation of **55** with Schwartz reagent gave the vinyl zirconium intermediate **65**, which underwent metal exchange with dimethyl zinc at –65°C (Wipf and Ribe 1998; Wipf and Coish 1997; Wipf and Xu 1994). The resulting vinyl-zinc species **66** was then smoothly added to the sensitive aldehyde **64** in the presence of Soai's chiral aminoalcohol **67** (Soai and Niwa 1992; Soai and Takahashi 1994) to afford a 4:1 mixture of the desired divinylcarbinol **68** and its C17-epimer (not shown). In the absence of the chiral additive, this critical coupling proceeded at a greatly reduced rate and exhibited virtually no diastereoselection.

Having successfully assembled the entire carbon skeleton of halichlorine, we were in a position to investigate the formation of the 15-membered macrolactone ring (Scheme 17). This required deprotection of the carboxylic acid function and the primary alcohol in the presence of the very acid sensitive divinylcarbinol moiety. Treatment of *t*-butyl ester **68** (as an inseparable 4:1 mixture of diastereomers) with TBSOTf in the presence of 2,6-lutidine led to transesterification to the corresponding silyl ester with concomitant protection of the secondary C17-hydroxy group (Meng et al. 1997; Jones et al. 1990). The crude

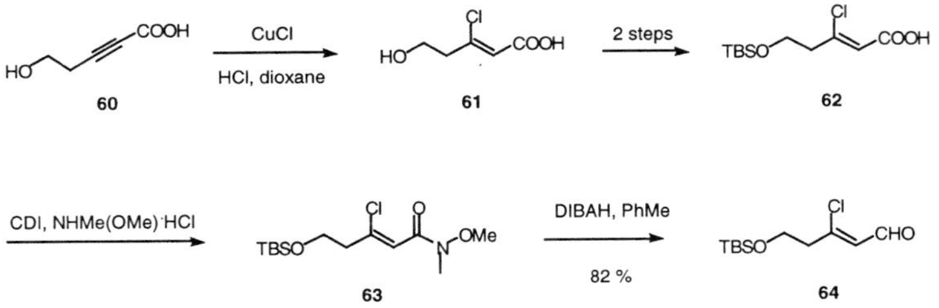

Scheme 15.

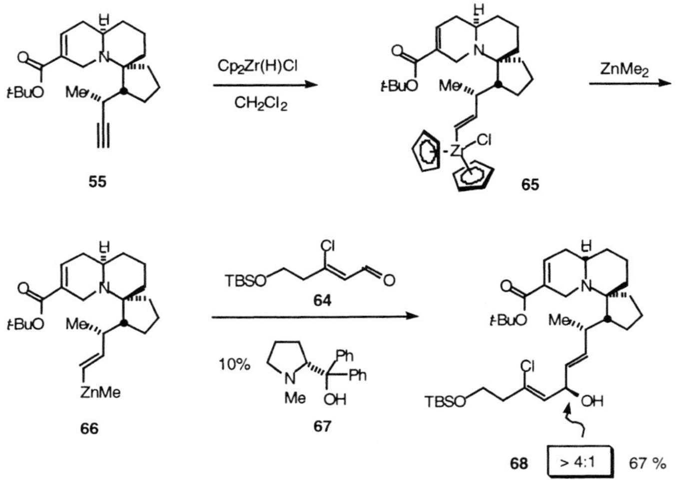

Scheme 16.

hydroxy amino acid **69** thus obtained was selectively deprotected under essentially neutral conditions upon exposure to ammonium fluoride in aqueous methanol (Zhang and Robins 1992; White et al. 1992). The valuable protecting group at the secondary alcohol function was retained under these conditions. Subsequent macrolactonization under Keck conditions (Boden and Keck 1985; Weber et al. 1992) produced 17-TBS halichlorine **70** as an 8:1 mixture of diastereomers, which were readily separated at this stage. Finally, deprotection of the major diastereomer at C17 with HF/pyridine afforded totally synthetic (+)-

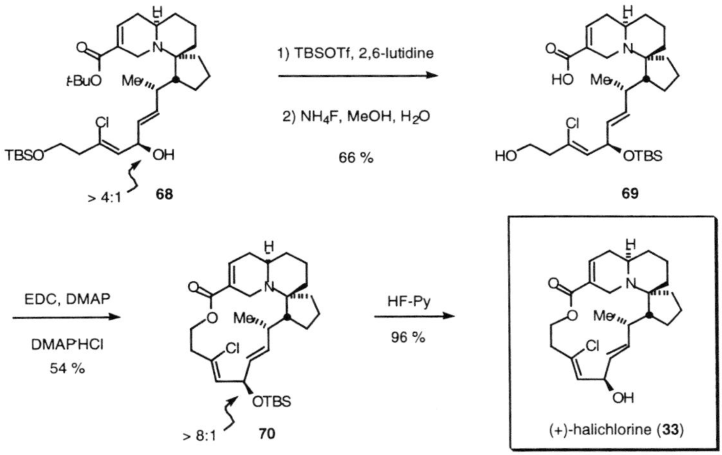

Scheme 17.

halichlorine (Trauner et al. 1999). Comparisons of the ^{1}H-NMR, ^{13}C-NMR, IR, and MS spectra, as well as chromatographic profiles of fully synthetic halichlorine with the corresponding properties of an authentic specimen sample provided by Professor Uemura, showed them to be identical and left no doubts about the correctness of the reported stereostructure (Kurbanov et al. 1972; Corey et al. 1980; Sato et al. 1982) Comparison of the optical rotations of synthetic halichlorine ($[\alpha]_D$=+234.9, c=1.04, MeOH) with that reported for the natural product ($[\alpha]^{25}_D$=+240.7, c=0.54, MeOH) further supported the previously reported absolute configuration of halichlorine.

In summary, a short (18 steps) and highly diastereoselective total synthesis of (-)-halichlorine has been achieved (Scheme 18). The synthesis features state-of-the-art organometallic chemistry in a complex setting, such as a *B*-alkyl Suzuki coupling (**46**→**47**), the novel zirconium-mediated deoxygenation (**51**→**52**), and the efficient addition of a highly functionalized organozinc compound to a sensitive aldehyde (**66**→**68**). More traditional transformations, e.g., a very stereoselective intramolecular Michael addition (**47**→**49**), a Mannich reaction (**50**→**51**), or a macrolactonization (**69**→**70**) were used to close the heterocyclic rings. Finally, Scheme 18 also displays the carbon building

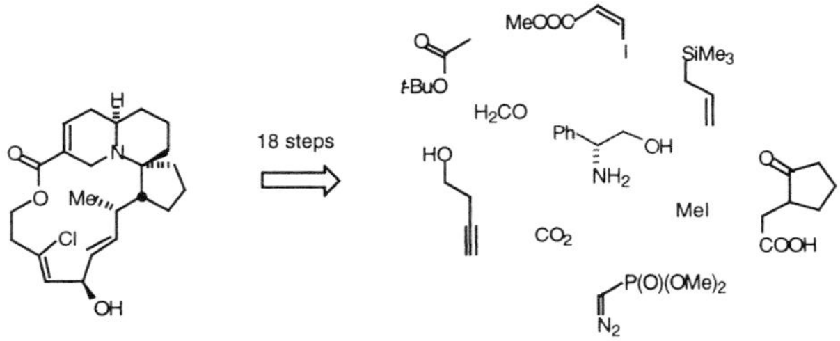

Scheme 18.

blocks used and indicates the source of the nitrogen atom and the chirality of our synthetic halichlorine: D-(-)-phenylglycinol.

In conclusion, it is seen that chemical synthesis proved itself to be equal to the challenge of producing amounts for biological investigations which are now being launched. We continue to believe that there is a particularly strong potential synergy between natural product isolation and chemical synthesis in advancing difficultly accessible marine-derived systems for development as drug candidates.

References

Arimoto H, Hayakawa I, Kuramoto M, Uemura D (1998) Tetrahedron Lett 39:861–862

Balog A, Harris C, Savin K, Zhang K-G, Chou T-C, Danishefsky SJ (1998) Angew Chem Int Ed Engl 37:2675–2678

Barclay AN, Brown MH, Law SKA, McKnight AJ, Tomlinson MG, Van der Merwe PA (1997) The leucocyte antigen facts book. Academic, Oxford

Barton DHR, Nakanishi K (eds) (1999) Comprehensive natural products chemistry. Elsevier, Amsterdam

Boden EP, Keck GE (1985) J Org Chem 50:2394–2395

Burgess LE, Meyers AI (1991) J Am Chem Soc 113:9858–9859

Carlos TM, Harlan JK (1994) Blood 84:2068–2101

Corey EJ, Myers AG (1985) J Am Chem Soc 107:5574–5576

Corey EJ, Tius MA, Das J (1980) J Am Chem Soc 102:1742–1744

Danheiser RL, Halgason AL (1994) J Am Chem Soc 116:9471–9479

Ennis MD, Hoffman RL, Ghazal NB, Old DW, Mooney PA (1996) J Org Chem 61:5813–5817

Faulkner DJ (1999) Nat Prod Rep 16:155–198

Feutrill GI, Mirrington RN (1972) Aust J Chem 25:1719–1729

Fürstner A (1999) Chem Rev 99:991–1045
Fürstner A, Konetzki I (1998) J Org Chem 63:3072–3080
Foster CA (1996) J Allergy Clin Immunol S272–S277
Frontier AJ (1999) PhD thesis, Columbia University
Fujiwara T, Iwasaki T, Takeda T (1993) Chem Lett 1321–1324
Gilbert JC, Weerasooriya U (1979) J Org Chem 44:4997–4998
Godfrey AG, Ganem B (1992) Tetrahedron Lett 33:7461–7464
Grigg R, Teasdale A, Sridharan V (1991) Tetrahedron Lett 32:3859–3862
Guerriero A, Cuomo V, Vanzanella F, Pietra F (1990) Helv Chim Acta 3:2090–2096
Guevel AC, Hart DJ (1996) J Org Chem 61:465–472
Hagiwara H, Uda H, Kodama T (1980) J Chem Soc Perkin Trans I:963–977
Hallock YF, Cardellina II JH, Boyd MR (1998) Nat Prod Lett 11:153–160
Hiemstra H, Speckamp WN (1991) In: Trost BM, Fleming I (eds) Comprehensive organic synthesis, vol 2, chap 4.5. Pergamon, Oxford
Hock RC, Schraustätter IU, Cochrane CG (1996) J Lab Clin Med 128:134–145
Huang AX, Xiong Z, Corey EJ (1999) J Am Chem Soc 121:9999–10003
Inoue M, Frontier AJ, Danishefsky SJ (2000) Angew Chem Int Ed (in press)
Johnson CR, Braun MP (1993) J Am Chem Soc 115:11014–11015
Jones AB, Villalobos A, Linde RG, Danishefsky SJ (1990) J Org Chem 55:2786–2797
Jones G (1997) Org React 49:1–330
Keen SP, Weinreb SM (1998) J Org Chem 63:6739–6741
Kelly TR, Li Q, Bhushan V (1990) Tetrahedron Lett 31:161–164
Kinney WA, Coghlan MJ, Paquette LA (1985) J Am Chem Soc 107:7352–7360
Knight DW (1991) In: Trost BM, Fleming I (eds) Comprehensive organic synthesis, vol 3. Pergamon, Oxford, pp 481–520
Koch AE, Halloran MM, Haskell CJ, Shah MR, Polverini PJ (1995) Nature 376:517–519
Kuramoto M, Tong C, Yamada K, Chiba T, Hayashi Y, Uemura D (1996) Tetrahedron Lett 37:3867–3870
Kurbanov M, Semenovsky AV, Smit WA, Shmelev LV, Kucherov VF (1972) Tetrahedron Lett 22:2175–2178
Kwee L (1995) Development 121:489–503
Landelle H, Godard AM, Laduree D, Chenu E, Robba M (1991) Chem Pharm Bull 39:3057–3060
Ley SV, Norman J, Griffith WP, Marsden SP (1994) Synthesis 639–666
Meng D, Bertinato P, Balog A, Su D-S, Kamenecka T, Sorenson E, Danishefsky SJ (1997) J Am Chem Soc 119:10073–10092
Meyers AI, Brengel GP (1997) J Chem Soc Chem Commun 1–8

Meyers AI, Tschantz MA, Brengel GP (1995) J Org Chem 60:4359–4362
Miller E, Cohen A, Nagos S, Griffith D, Maunder T, Martin T, Weiner-Kronish J, Sticherling M, Christophers E, Matthay M (1992) Am Rev Respir Dis 146:427–432
Miyaura N, Suzuki A (1995) Chem Rev 95:2457–2483
Mori M, Kaneta N, Shibasaki M (1991) J Org Chem 56:3486–3493
Ohba M, Kawase N, Fujii T (1996) J Am Chem Soc 118:8250–8257
Omodani T, Shishido K (1994) Chem Commun 2781–2782
Ono N, Kaji A (1986) Synthesis 693–704
Patil AD, Freyer AJ, Killmer L, Offen P, Carte B, Jurewicz AJ, Johnson RK (1997) Tetrahedron 53:5047–5060
Ragan JA, Claffey MC (1995) Heterocycles 41:57–70
Ranganathan D, Rao CB, Ranganathan S, Mehrotra AK, Iyengar R (1980) J Org Chem 45:1185–1189
Rao M, McGuigan MA, Zhang X, Shaked Z, Kinney WA, Bulliard M, Laboue B, Lee NE (1997) J Org Chem 62:4541–4545
Romo D., Meyers AI (1991) Tetrahedron 47:9503–9569
Sasaki M, Fuwa H, Inoue M, Tachibana K (1998) Tetrahedron Lett 39:9027–9030
Sato C, Ikeda S, Shirahama H, Matsumoto T (1982) Tetrahedron Lett 23:2099–2102
Scitz M, Dewald B, Gerber N, Baggiolini M (1991) J Clin Invest 87:463–469
Sharmma M, Rodrigues HR (1968) Tetrahedron 24:6583–6589
Soai K, Niwa S (1992) Chem Rev 92:833–856
Soai K, Takahashi K (1994) J Chem Soc Perkin Trans 1:1257–1258
Springer TA (1994) Cell 76:301–314
Tebbe FN, Marshall GW, Reddy GS (1978) J Am Chem Soc 100:3611–3613
Trauner D, Danishefsky SJ (1999) Tetrahedron Lett 40:6513–6516
Trauner D, Schwarz J, Danishefsky SJ (1999) Angew Chem Int Ed 38:3542–3545
Tsuzuki S, Toyama-Sorimachi N, Kitamura F, Tobita Y, Miyasaka M (1998) FEBS Lett 430:414–418
Weber AE, Steiner MG, Krieter PA, Coletti AE, Tata JR, Halgren TA, Ball RG, Doyle JJ, Schorn TW, Stearns RA, Miller RR, Siegl PKS, Greenlee WJ, Patchett AA (1992) J Med Chem 35:3755–3773
White JD, Amedio JC, Gut S, Ohira S, Jayasinghe LR (1992) J Org Chem 57:2270–2284
Wipf P, Coish PDG (1997) Tetrahedron Lett 38:5073–5076
Wipf P, Ribe S (1998) J Org Chem 63:6454–6455
Wipf P, Xu W (1994) Tetrahedron Lett 35:5197–5200
Witschel MC, Bestmann HJ (1997) Synthesis 107–112
Zhang W, Robins MJ (1992) Tetrahedron Lett 33:1177–1180

2 The Pederin Family of Antitumor Agents: Structures, Synthesis and Biological Activity

R. Narquizian, P.J. Kocienski

2.1 Structures

In 1775 the Danish entomologist Johann Christian Fabricius (1745–1808) first described the genus *Paederus*, which at that time included only two species. In the ensuing two centuries, over 600 species have been identified, including *Paederus fuscipes*, whose natural history deserves some mention (Frank and Kanamitsu 1987). *Paederus fuscipes* is about 8 mm long with a black head and abdominal apex, an orange thorax and abdominal base and iridescent blue elytra (wing case). It inhabits riverbanks, marshes, and irrigated fields, where it feeds mainly on insects, mites, soil nematodes, and decaying vegetable matter. Like most members of the genus, *Paederus fuscipes* is a predator of the fly population, but it is also a pest to man. The insect does not sting or bite, but a toxin in its hemolymph causes severe dermatitis when it is crushed on the skin, and the eyes are particularly sensitive though the palms of the hands and the soles of the feet are resistant. In addition to the lesions, severe symptoms such as fever, edema, neuralgia, arthralgia, and vomiting are observed with erythema persisting for several months. It has been suggested (Frank and Kanamitsu 1987) that both the affliction and its causative agent were known to Chinese medicine over 1200 years earlier. An insect called *ch'ing yao ch'ung* was described by Ch'en in 739 A.D.: "It contains a strong poison and when it touches the skin it causes the skin to swell up. It will take the skin off one's face and remove tattoo marks completely. It is used as a caustic for toxic boils, nasal polypi, and ringworm."

The active chemical agent responsible for the dermatitis was first isolated in crystalline form by Netolitzky in 1919 (Netolitzky 1919). The research which eventually led to the correct structure began in 1952 with Pavan and Bo (1953), who named the toxic agent pederin and determined its melting point (112°C) on a sample derived from 25 million specimens (ca. 100 kg). The correct molecular formula ($C_{25}H_{45}O_9N$) established by Quilico et al. (1961) in 1961 led to detailed study of the chemical constitution of pederin and a structure, devoid of stereochemical definition, was proposed in 1965 (Cardani et al. 1965). An independent investigation by Matsumoto's team at Sapporo gave corroborating evidence (Matsumoto et al. 1964, 1968). With one minor exception (see below) all the conclusions drawn from the degradation and ^{1}H NMR studies of Quilico et al. (1966) and Matsumoto were later confirmed by an X-ray crystallographic analysis of pederin di-*p*-bromobenzoate (Furusaki et al. 1968; Bonamartini Corradi et al. 1971), which also established the absolute and relative stereochemistry.

Single *Paederus riparius* specimens reared from the egg and kept for prolonged periods of time show that only the females are able to biosynthesize pederin (Kellner 1998; Kellner and Dettner 1995, 1996). Preimaginal stages efficiently store pederin transferred by the females into their eggs, and the males' pederin content decreases slowly over time. Only males with access to eggs containing the substance moderately increase their pederin load. The females begin to accumulate the toxin a few weeks after imaginal eclosion and build up reserves for the egg-laying period within 60 days.

For many years, pederin (**1**), pseudopederin (**2**) (Cardani et al. 1965), and pederone (**3**) (Cardani et al. 1967) (also isolated from *Paederus fuscipes*) were structurally unique in the realm of natural products (Scheme 1). However, in 1988 routine screening for antiviral agents identified two marine natural products which bore a close structural

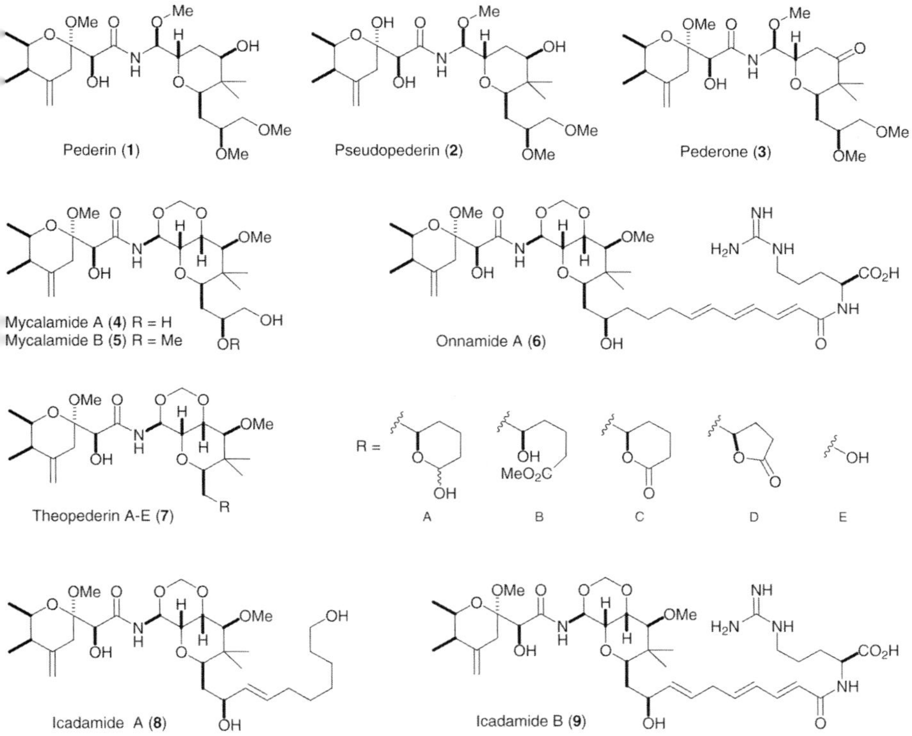

Pederin (**1**) Pseudopederin (**2**) Pederone (**3**)

Mycalamide A (**4**) R = H
Mycalamide B (**5**) R = Me

Onnamide A (**6**)

Theopederin A-E (**7**)

Icadamide A (**8**) Icadamide B (**9**)

Scheme 1.

resemblance to pederin. Mycalamide A (**4**) was isolated from a sponge of the genus *Mycale*, found in the Otago harbor off New Zealand (Perry et al. 1988, 1990), while onnamide A (**6**) was isolated from a sponge of the genus *Theonella* found in Okinawan waters (Sakemi et al. 1988). The pederin family had grown to 24 members by 1995 with the isolation of mycalamide B (**5**) (Perry et al. 1990), a further 11 onnamides (Matsunaga et al. 1992; Kobayashi et al. 1993), and theopederins A–E (**7**) (Fusetani et al. 1992). The most recent additions to the pederin family are icadamides A (**8**) and B (**9**), isolated from a sponge of the genus *Leiosella* (Clardy and He 1995). The significance of the mycalamides, onnamides, theopederins, and icadamides in sponge physiology is unclear, although it has been suggested that the occurrence of closely related compounds in such taxonomically remote animals as sponges and terrestrial beetles may indicate connection by a common producer, possibly a symbiotic micro-organism (Fusetani and Sugawara 1993).

2.2 Synthetic Studies

All members of the pederin family are rare, difficult to isolate, and comparatively frail; many of them have potent and potentially useful activity as antiviral and antitumor agents (see below); this has stimulated considerable interest in their total synthesis. Total syntheses of pederin (Matsuda et al. 1982, 1983, 1988; Nakata et al. 1985a,b; Willson et al. 1987; Jarowicki et al. 1990; Kocienski et al. 1991, 1998a) , mycalamide A (Hong and Kishi 1990; Nakata et al. 1994, 1996), mycalamide B (Hong and Kishi 1990; Kocienski et al. 1998b), onnamide A (Hong and Kishi 1991), and theopederin D (Kocienski et al. 1998a) have been reported, as have significant syntheses of various fragments (Nakata et al. 1994; Isaac and Kocienski 1982; Isaac et al. 1983; Kocienski and Willson 1984; Matsumoto et al. 1984;Willson et al. 1990a,b; Hoffmann and Schlapbach 1992, 1993; Roush and Marron 1993; Toyota et al. 1995, 1998a,b; Marron and Roush 1995; Roush et al. 1997; Roush et al. 1998; Breitfelder et al. 1998; Trotter et al. 1999). The parent member of the family, pederin itself, is also the simplest. It is composed of two tetrahydropyran rings (a left fragment and a right fragment) connected by an *N*-acyl aminal bridge, as depicted in structure **10** (Scheme 2). The left fragment of all members of the pederin family is identical but all

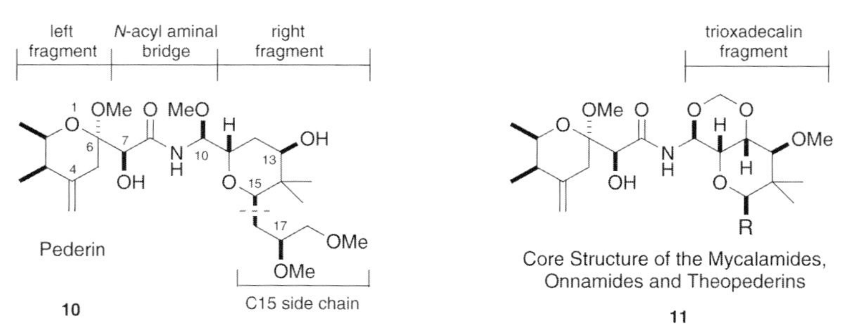

Scheme 2.

remaining members of the family (the mycalamides, onnamides, theopederins, and icadamides) have a trioxadecalin right fragment as shown in structure **11**. The principal site of structural variation is the side chain attached to C15. The common left fragment also imposes a common problem: the high acid lability associated with the alkene at C4, which activates the acetal function at C6. A comprehensive discussion of the syntheses of members of the pederin family is beyond the scope of this review, so we will focus on one of the most demanding challenges: the construction of the *N*-acyl aminal bridge linking the left and right fragments. The following discussion is arranged according to the bond formed in the fragment linkage strategy.

2.2.1 The C8–N9 Connection via Imidate Ester Acylation

Matsumoto and his team reported the first total synthesis of pederin in 1982, in which the left and right fragments were connected by the *N*-acylation of an imidate ester (Scheme 3) (Matsuda et al. 1982;

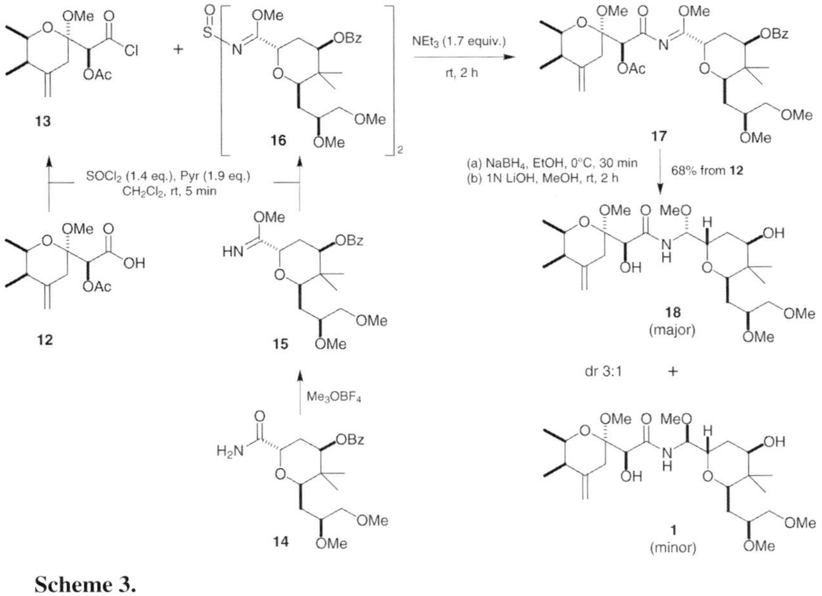

OMe O OAc Cl 13 + OMe OBz OMe OMe 16 2 NEt_3 (1.7 equiv.) rt, 2 h OMe O OMe OAc OBz OMe OMe 17

$SOCl_2$ (1.4 eq.), Pyr (1.9 eq.) CH_2Cl_2, rt, 5 min

(a) $NaBH_4$, EtOH, 0°C, 30 min (b) 1N LiOH, MeOH, rt, 2 h 68% from **12**

OMe O OH OAc 12 OMe HN OBz OMe 15 OMe OMe O MeO H OH N H OH OMe 18 (major) OMe

Me_3OBF_4 dr 3:1 +

O H_2N OBz OMe 14 OMe OMe O MeO H OH N H OH OMe 1 (minor) OMe

Scheme 3.

Yangiya et al. 1982). The crucial coupling reaction was accomplished by *brief* treatment of (+)-acetylpederic acid (**12**) with thionyl chloride in the presence of pyridine in dichloromethane, followed by addition of methyl pedimidate (**15**). The reaction time had to be minimized owing to the instability of the acid chloride intermediate **13**. Unfortunately, the highly hindered carboxylic acid was converted slowly to the acid chloride **13** under the reaction conditions, thereby allowing time for the reaction of imidate ester **15** with thionyl chloride to give *N,N′*-sulfinyl-bis(methyl pedimidate) (**16**), which can also serve as a substrate in the acylation of acid chloride **13**. Thus, the formation of *N*-acyl imidate intermediate **17** occurred by two different paths concurrently. The *N*-acyl aminal bridge was then constructed by reduction of **17** with sodium borohydride. The synthesis was completed by hydrolysis of the benzoate and acetate ester protecting groups to give a mixture of pederin (**1**) and 10-*epi*-pederin (**18**) in a ratio of 1:3 [68% overall from (+)-benzoylpedamide (**14**)], with pederin being the minor product.

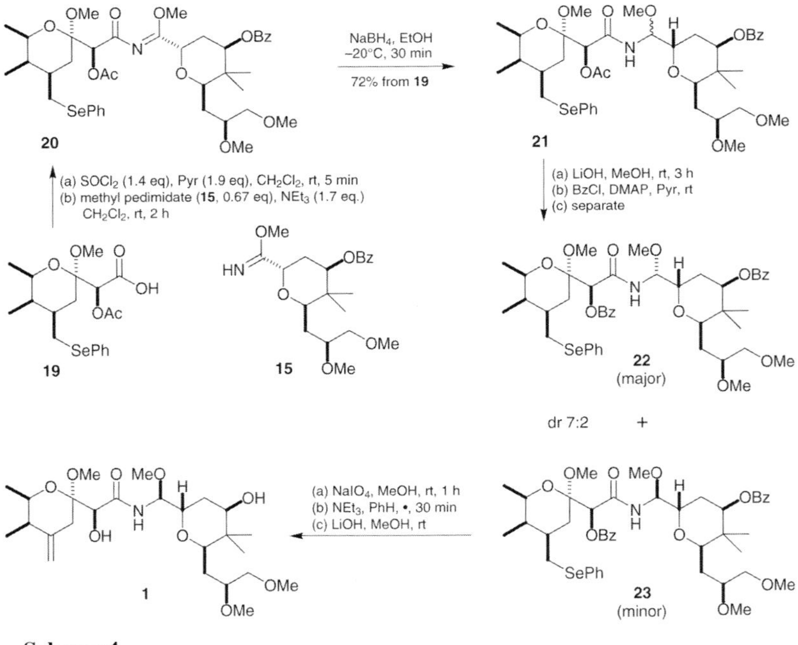

$NaBH_4$, EtOH
−20°C, 30 min
72% from **19**

(a) $SOCl_2$ (1.4 eq), Pyr (1.9 eq), CH_2Cl_2, rt, 5 min
(b) methyl pedimidate (**15**, 0.67 eq), NEt_3 (1.7 eq.) CH_2Cl_2, rt, 2 h

(a) LiOH, MeOH, rt, 3 h
(b) BzCl, DMAP, Pyr, rt
(c) separate

20 **21** **19** **15** **22** (major) dr 7:2 + **23** (minor) **1**

(a) $NaIO_4$, MeOH, rt, 1 h
(b) NEt_3, PhH, •, 30 min
(c) LiOH, MeOH, rt

Scheme 4.

The high acid sensitivity of the homoallylic acetal in (+)-acetylpederic acid (**12**) was a major obstacle to Matsumoto's first synthesis. In a subsequent refinement, the C4 alkene was carried through the synthesis in latent form, with the troublesome alkene being introduced in the penultimate step of the synthesis as shown in Scheme 4 (Matsuda et al. 1983). However, the reduction of *N*-acyl imidate **20** once again afforded an unfavorable mixture of *N*-acyl aminals **21** in 72% overall yield (dr 7:2). After chromatographic separation, the requisite alkene was introduced by brief thermolysis of the selenoxide derived from oxidation of the minor selenoether **23**, whereupon hydrolysis of the C7 and C13 benzoates afforded pederin (**1**).

The poor stereoselectivity in the reduction of the *N*-acyl imidates **17** and **20** in the foregoing studies was beyond repair, and so isomerization of the *N*-acyl aminal in the 10-*epi*-pederin series was investigated. Transacetalization of 10-*epi*-pederin derivative **22** with acetyl chloride in methanol (rt, 3 h) gave an equilibrium mixture (**22**:**23** = 3:1), showing that the undesired 10-*epi* series was the thermodynamic product. The possibility of selective conversion of the 10-*epi*-pederin derivative **22** into pederin derivative **23** under *kinetically* controlled conditions was next examined, by taking into account the acceleration effect of the alkoxy-exchange reaction in methanol by a large alkoxy group (Matsuda et al. 1988). Thus, treatment of **22** with acetyl chloride (Scheme 5) in isopropanol gave **25** selectively after 7 days through initial formation of the kinetically controlled product **24**. Compound **25** was unstable and

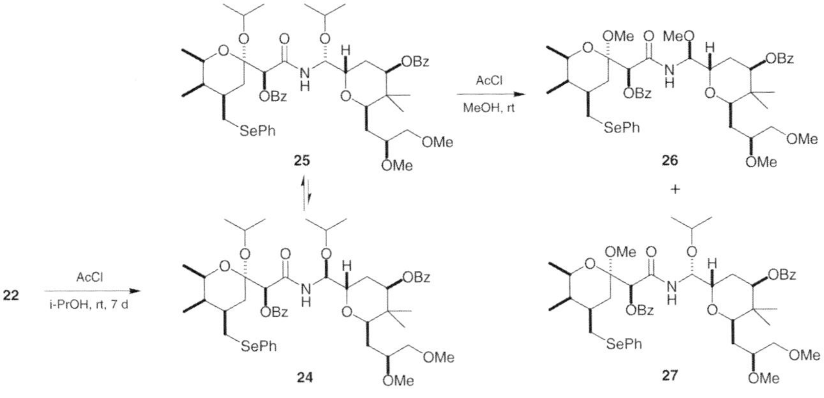

Scheme 5.

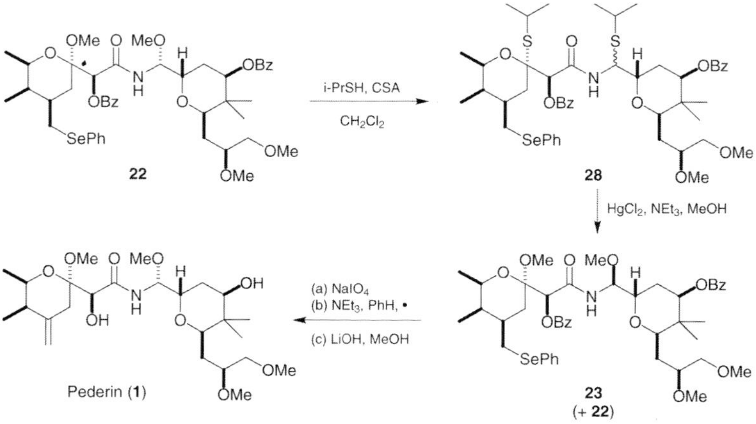

Scheme 6.

could not be isolated in a pure state, but kinetically controlled transacetalization of **25** with acetyl chloride in methanol (rt, 4.5 h, 50% conversion) proceeded in a stereoselective manner to give a 60% isolated yield (based on consumed **25**) of **22** and a 14% yield of **26** (**26:22** = 4:1). The 10α-isopropoxy compound was recovered in 42% yield in the form of a 6α-methoxy compound (**27**).

Nakata and his associates (1985a,b) discovered that reduction of the 7-*O*-benzoyl analogue of *N*-acyl imidate **20** with $NaBH_4$ in a mixture of isopropanol and CH_2Cl_2 gave **23** (28%) and **22** (30%) (Scheme 6). The 10-*epi* derivative **22** was converted into **23** by reaction of **22** with 2-propanethiol and camphorsulfonic acid (CSA) in CH_2Cl_2 to give thioacetal **28**, which was then treated with $HgCl_2$ in MeOH in the presence of NEt_3 to give **23** in 47% yield (from **22**) along with epimer **22** (36%). The completion of the synthesis was performed as previously described by Matsumoto.

2.2.2 The C8–N9 Connection via Aminal Acylation

In the Matsumoto-Nakata approaches to pederin, the *N*-acyl aminal bridge was constructed by first condensing an activated carboxylic acid

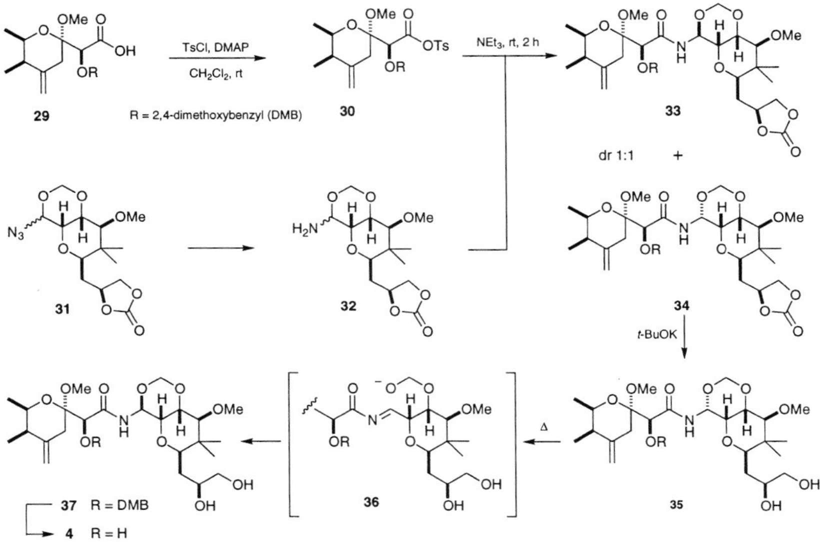

Scheme 7.

(e.g., acid chloride **13**) with an imidate ester to give an *N*-acyl imidate, which was reduced with sodium borohydride to generate the *N*-acyl aminal – a reaction which gave poor stereoselectivity. In the mycalamides, onnamides, and theopederins, the incorporation of the oxygen atom of the aminal into a dioxane ring makes the Matsumoto-Nakata protocol less attractive, owing to the difficulty associated with the synthesis of the appropriate imidate ester. Therefore, the first syntheses of mycalamide A, mycalamide B (Hong and Kishi 1990), and onnamide A (Hong and Kishi 1991) by Kishi and Hong adapted the Matsumoto-Nakata protocol by condensing an aminal **32** with the activated carboxylic acid derivative **30**, as illustrated in Scheme 7. The C10 aminal unit of **32** is configurationally unstable under acidic, basic, and neutral conditions, and consequently a 1:1 mixture of adducts **33** and **34** was obtained. However, treatment of **34** with potassium *tert*-butoxide at room temperature first accomplished the transesterification of the carbonate group of **34** to the corresponding diol **35**, which then epimerized on heating to yield **37**. Owing to competing decomposition, the reaction

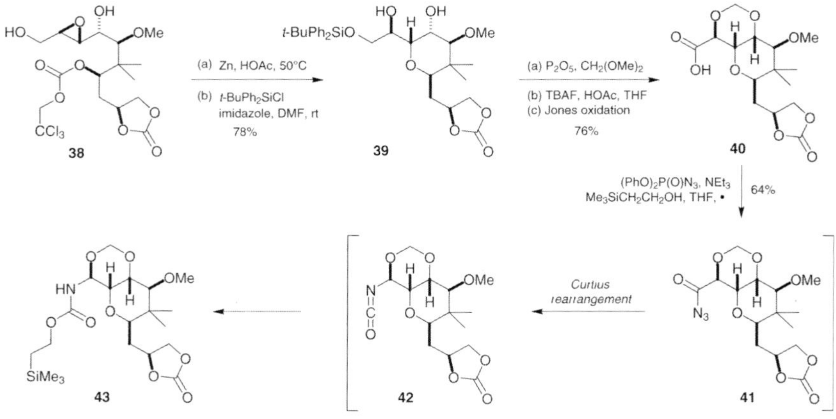

Scheme 8.

was stopped at approximately 60% completion to yield the epimerized natural diastereoisomer **37** in 42% yield along with the unnatural diastereoisomer **35** (33% yield). Nakata has also described a synthesis of mycalamide A in which the *N*-acyl aminal bridge is constructed by the Kishi procedure (Nakata et al. 1996).

In 1993 Roush and Marron described a route to the mycalamides which incorporates two significant advances. First, the stereochemistry of the aminal was assured by a Curtius rearrangement, and second, the configuration of the aminal center was stabilized as a temporary carbamate appendage. The steps leading up to the Curtius rearrangement are noteworthy. Reductive cleavage of the 2,2,2-trichloroethyl carbonate **38** (Scheme 8) with Zn released an alkoxide that performed a nucleophilic attack on the neighboring oxirane to give a cyclic diol, which was selectively protected as its mono-*tert*-butyldiphenylsilyl ether **39** (Marron and Roush 1995). After closure of the 1,3-dioxane ring, the *tert*-butyldiphenylsilyl ether was cleaved and the resultant alcohol oxidized to the corresponding carboxylic acid **40**. The key Curtius rearrangement was performed by treatment of acid **40** with diphenylphosphoryl azide at 65°C. The intermediate acyl azide **41** underwent the desired Curtius rearrangement with clean retention of configuration to the isocyanate **42**, which was trapped with 2-trimethylsilylethanol to give the car-

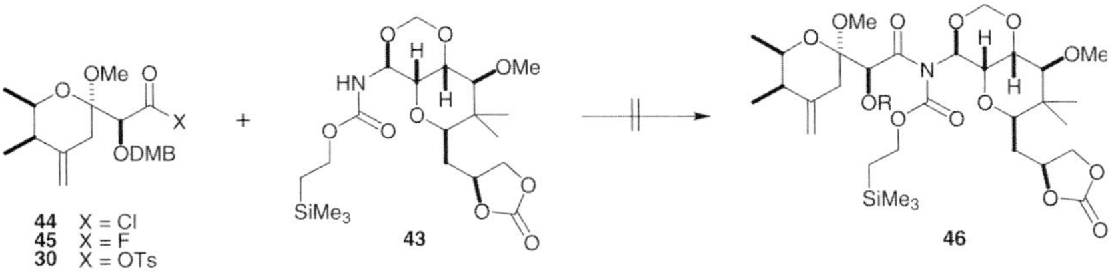

Scheme 9.

bamate **43** as a single diastereoisomer. Hoffmann published an approach to the trioxadecalin ring system of the mycalamides in 1993; it was similar to that of Roush, in that it included a Curtius rearrangement to form the C10-aminal diastereoselectively (Hoffmann and Schlapbach 1993).

Roush and co-workers later published a very brief route to the pederic acid derivatives **44**, **45**, and **30** and they achieved some success in the construction of *N*-acyl aminal bridges with model systems, but the high steric hindrance of the carboxyl group in the left fragments once again thwarted condensation with the carbamate **43** (Scheme 9) under a wide variety of conditions, and none of the desired adduct **46** was observed (Roush et al. 1997, 1998; Roush and Pfeifer 1998).

2.2.3 The C7–C8 Connection

In 1996 Hoffmann and co-workers proposed an original strategy to the construction of the *N*-acyl aminal bridge of mycalamide B based on union of the ester **47** with the acyl anion **48** to forge the C7–C8 bond (Scheme 10). Once again, the stereochemistry of the aminal is to be controlled by a Curtius rearrangement.

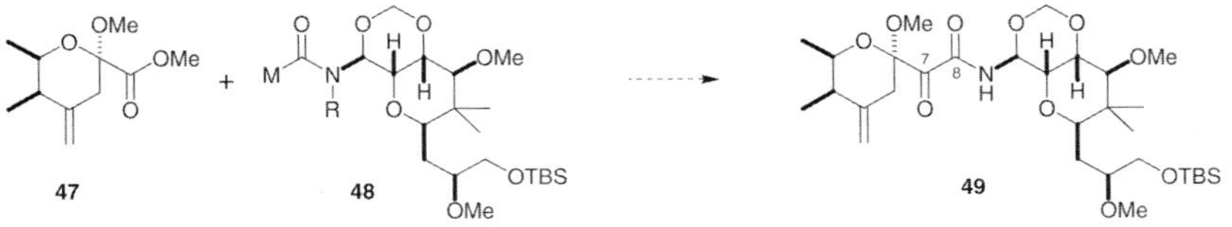

Scheme 10.

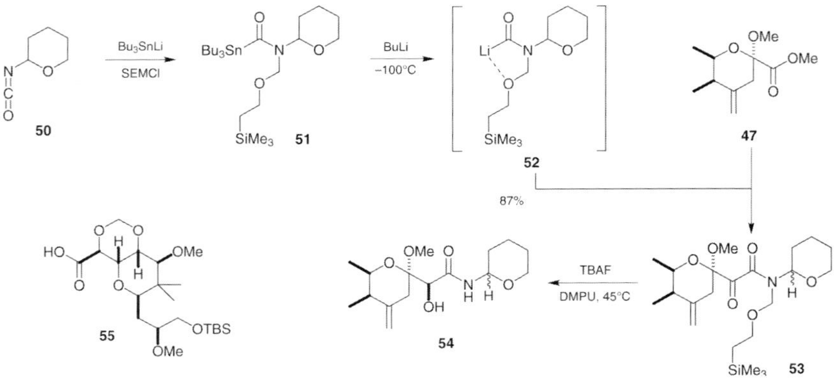

Scheme 11.

Model studies for the novel coupling step were carried out with the commercially available isocyanate **50** as a surrogate for the right fragment. The acylstannane **51** (Scheme 11), protected as its 2-(trimethylsilyl)ethoxymethoxy (SEM) derivative, was easily prepared by addition of Bu_3SnLi to the isocyanate **50** followed by *N*-alkylation with 2-(trimethylsilyl)ethoxymethyl chloride. The use of SEM protection for the N-atom was a judicious choice, the purpose being to stabilize the highly labile acyllithium **52** which was generated and trapped in situ by transmetallation with BuLi at −100°C. The desired model adduct **53** was obtained in 87% yield. A fortuitous and unprecedented reduction of the α-oxo ester accompanied the deprotection of the SEM group in adduct **53** with TBAF in *N,N′*-dimethylpropyleneurea (DMPU) to give the desired α-hydroxyamide **54** as a mixture of diastereoisomers (dr 3:1). The origin of the hydride for the reduction of the keto function is unknown. Progress towards implementation of the strategy has been reported: a synthesis of **55** from D-arabinose has been accomplished (Breitfelder et al. 1998), but union of the fragments remains elusive.

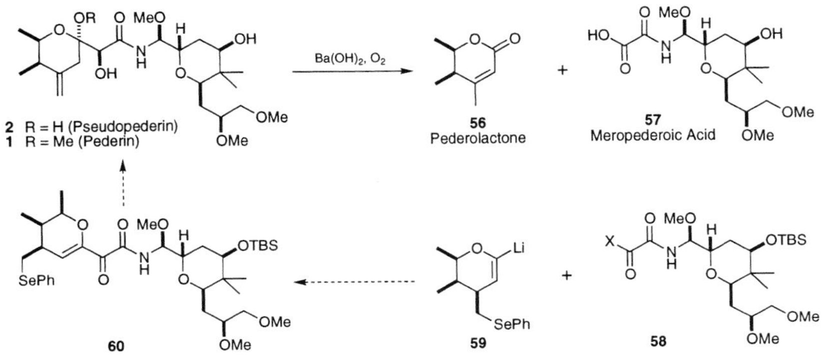

Scheme 12.

2.2.4 The C6–C7 Connection: The Metallated Dihydropyran Approach

In our early approaches to the synthesis of pederin, we chose the well-used Matsumoto-Nakata protocol to construct the *N*-acyl aminal bridge (Willson et al. 1987; Isaac and Kocienski 1982; Isaac et al. 1983; Kocienski and Willson 1984; Willson et al. 1990a,b). We were similarly chastened by the difficulty of reconciling slow reactions with unstable reagents, which has been a persistent feature of nearly all of the accounts reported to date. The problem was compounded by the poor stereoselectivity of the subsequent reaction of the *N*-acyl imidate adducts. In the mid 1980s we turned to a new strategy which attempted to alleviate the problems associated with the high steric hindrance surrounding the acetal center at C6. Our new strategy was inspired by some of the elegant degradation studies conducted by Quilico and co-workers during their structure elucidation of pederin (Cardani et al. 1966). Pseudopederin (**2**, Scheme 12), the hydrolysis product of pederin, undergoes an easy retroaldol reaction on heating in base in the presence of air to give pederolactone (**56**) and meropederoic acid (**57**), wherein the *N*-acyl aminal group is still intact. These transformations suggested an alternative disconnection between C6 and C7 that circumvented the cascade of problems which beset the closing stages of the previous syntheses. The new strategy required a metallated dihydropyran **59**

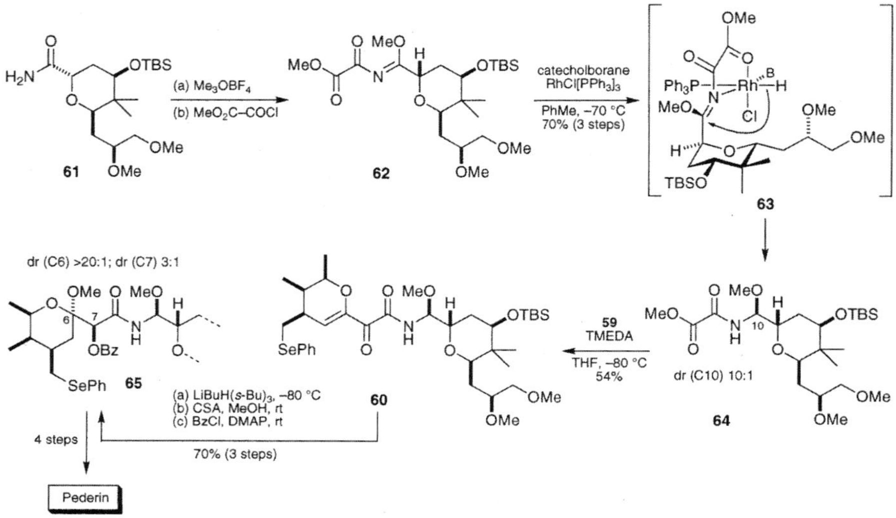

Scheme 13.

functioning as an acyl anion equivalent in reaction with a suitably activated meropederoic acid derivative **58** (Jarowicki et al. 1990; Kocienski et al. 1991).

Amide **61** was converted to the *N*-acylimidate **62** in two steps using standard transformations. The imidate ester intermediate **62** was prone to hydrolysis, but good yields were obtained by working fast and with minimal purification. Reduction of the *N*-acylimidate **62** was achieved by using an unprecedented reaction – reduction with catecholborane in the presence of a catalytic amount of $[Ph_3P]_3RhCl$. Under these conditions a 70% yield of a mixture of diastereoisomeric *N*-acyl aminals was obtained in which the desired isomer **64** predominated (10:1). Thus, for the first time, a metal hydride reduction of an *N*-acyl imidate in the pederin series afforded appreciable selectivity in favor of the desired diastereoisomer at C10. The stereochemistry of the reduction was interpreted in terms of an intermediate **63**, in which an octahedral Rh complex delivers a hydride intramolecularly as indicated in Scheme 13.

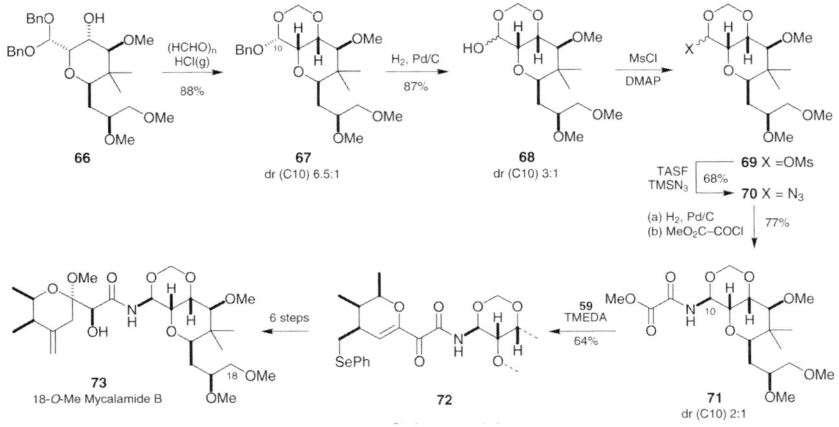

Scheme 14.

The key reaction of the sequence entailed addition of the metallated dihydropyran **59** to the methyl ester **64** in the presence of TMEDA at low temperature to give a 54% yield of the adduct **60**. With the bulk of the pederin skeleton now constructed, completion of the synthesis required merely the introduction of the two adjacent stereogenic centers at C6 and C7 and a few functional group transformations. The stereogenic center at C7 was introduced by metal hydride reduction of the enone function in **60**. Use of the bulky reducing agent LiBH(*s*-Bu)3 afforded the desired diastereoisomer in modest diastereoselectivity (3:1). Addition of methanol to the dihydropyran occurred with excellent diastereoselectivity (dr=20:1). Completion of the synthesis required four further steps which were well precedented.

The success of the metallated dihydropyran approach in the synthesis of pederin suggested an easy adaptation of the strategy to syntheses in the mycalamide-theopederin series. An initial foray published in 1996 was superseded by tactical improvements, depicted in Scheme 14, for the synthesis of 18-*O*-methyl mycalamide B (Kocienski et al. 1996), which had been identified as the most potent of the mycalamide derivatives in assays against a series of human tumors (Richter et al. 1997; Thompson et al. 1992) (see below). The construction of the 1,3-dioxane ring was accomplished by treatment of **66** with paraformaldehyde in the

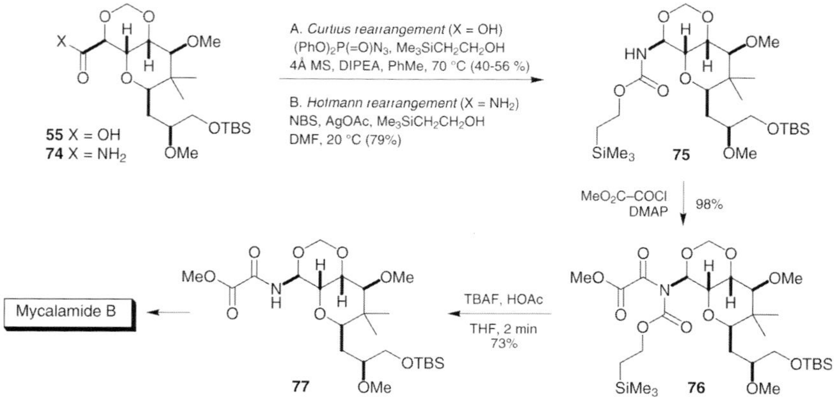

Scheme 15.

presence of HCl to give a mixture of diastereoisomeric 1,3-dioxane acetals **67** (dr=6.5:1) in 88% yield. Separation of acetals **67** by column chromatography was possible but useless, since hydrogenolysis of the benzyl group gave the same mixture of hemiacetals **68** (dr 3:1).

Replacement of the hydroxyl group in **68** by an azido group via substitution of the mesylate **69** by Bu_4NN_3 had been reported by Hong and Kishi (1990), but in our hands the yields ranged from 20% (typically) to 72% (rarely). We therefore developed a new method which, to our knowledge, is novel: the crude mesylate **69** derived from the mixture of hemiacetals **68** was treated with trimethylsilyl azide in the presence of tris(dimethylamino)sulfonium difluorotrimethylsilicate (TASF) to give the azides **70** as a mixture of diastereoisomers in the ratio 1:1 to 2:1, depending on the reaction conditions. The isomers could be separated for purposes of characterization but in practice it was best to carry the mixture of azides forward to the next stage of the synthesis. Catalytic reduction of the azides **71** gave a sensitive mixture of aminals which were acylated with methyl oxalyl chloride in the presence of DMAP to afford the diastereoisomeric methyl oxalamides (1:2) in 77% yield. The diastereoisomers were separated by column chromatography and the minor crystalline diastereoisomer **71**, having the correct stereochemistry at C10, was added to a solution of the dihydro-2*H*-pyranyllithium reagent **59** in the presence of excess TMEDA to give the acylated dihydro-

2*H*-pyran derivative **72** in 64% yield. The remainder of the synthesis was conducted as described above for the synthesis of pederin (Kocienski et al. 1991).

The successful implementation of the metallated dihydropyran approach to 18-*O*-methyl mycalamide B (Scheme 14) was gratifying, but the unfavorable stereochemistry at the C10 aminal center was a blemish which eluded correction. Therefore, we examined the Roush-Hoffmann-Curtius protocol as part of a synthesis of mycalamide B (Kocienski et al. 1998b). Treatment of the carboxylic acid **55** (Scheme 15) with diphenylphosphoryl azide afforded an acyl azide intermediate that rearranged to an isocyanate, which was trapped by 2-(trimethylsilyl)ethanol to give the carbamate **75**. However, the elevated temperatures (70°C) required for the rearrangement resulted in some decomposition with an overall reduction in yield to 56% at best, with typical yields being more like 40%. A much better alternative was a classical Hofmann rearrangement using Ag(I)-assisted rearrangement of the *N*-bromoamide derived from amide **74** (Kocienski et al. 2000). The reaction occurred at room temperature with clean retention of configuration to give the carbamate **75** in 79% overall yield. The remaining 2-carbon fragment was installed by reaction of carbamate **75** with methyl oxalyl chloride in the presence of DMAP to yield the imide derivative **76**. To complete the sequence, the carbamate function was expunged using TBAF buffered with acetic acid to give the *N*-acyl aminal intermediate **77** of mycalamide B. A similar strategy was used in the first synthesis of a member of the theopederin family, theopederin D (**7D**) Kocienski et al. 1998a).

2.3 Biological Activity

2.3.1 The Natural Products

Pederin is a very weak antibacterial agent but it is highly toxic to eukaryotic cells. Ingestion can cause severe internal damage and intravenous injection causes death at levels which suggest that it is more potent than cobra venom. The toxicity of pederin appears to be related to its inhibition of protein biosynthesis and cell division. Using human tonsil ribosomes, Vazquez (1979; Carrasco et al. 1976) showed that pederin binds irreversibly to the ribosome, preventing translation of mRNA. Inhibition occurred at the translocation step during the elongation cycle. The irreversible binding of pederin to ribosomes and its vesicant activity suggest it may function as an alkylating agent with the homoallylic acetal or *N*-acyl aminal as sites of potential reactivity. The biochemical and pharmacological activities are probably not related, however, since the hydrogenated derivative dihydropederin is not a vesicant though it remains a potent inhibitor of protein biosynthesis. Unfortunately, there have been very few attempts to evaluate the clinical potential of pederin. Pavan (1982) demonstrated that elderly patients with chronic necrotic and purulent sores completely recovered in some cases after treatment with minute amounts of pederin. A Russian study has also revealed a therapeutic effect on eczema and neurodermatitis with no complications (Mikhailova 1967). Potential use as an anticancer agent has been suggested based on the ability of pederin to block mitosis in normal and tumor cells at doses of 1 ng/ml and on reports that it inhibited sarcoma-180 tumors in mice (Soldati 1966).

Like pederin, the mycalamides and their derivatives induce severe dermatitis. Mycalamides A and B reveal potent in vitro cytotoxicity and in vivo antitumor efficacy against several leukemia and solid tumor model systems, as well as antiviral activity. They inhibit in vitro replication of murine lymphoma P388 cells (IC_{50} 3.0±1.3 and 0.7±0.3 ng/ml, respectively, and human promyelocytic (HL-60), colon (HT-29), and lung (A549) cells (IC_{50} <5 n*M*) (Burres and Clement 1989). Mycalamide A was also active against B16 melanoma, Lewis lung carcinoma, M5076 ovarian carcinoma, colon 26 carcinoma, and the human MX-1 (mammary), CX-1 (colon), LX-1 (lung), and Burkitt's lymphoma tumor xenografts (Burres and Clement 1989). Mode of action studies confirm that the mycalamides, like pederin, are protein synthesis inhibitors. Mycalamide A also disrupts DNA metabolism but does not intercalate into DNA itself. A correlation between their relative ability to inhibit protein synthesis, their cytotoxicity, and their in vivo efficacy suggests that their antitumor activity is a consequence of protein synthesis inhibition (Burres and Clement 1989).

Antiviral assays on mycalamide A by the Munro group (Perry et al. 1988, 1990) indicated that the minimum dose that inhibited the cytopathic effect of the test viruses (herpes simplex type-1 and polio type-1 viruses) over a whole (17 mm) well was 5 ng/disk. No in vivo antiviral results on pure mycalamide A were available from the initial screen, but

Table 1. IC_{50} values (ng/ml) in vitro for members of the pederin family against murine P388 leukemia cells

Compound[a]	Munro[b]	Fusetani[c]	Fusetani[d]	Kobayashi[e]
Pederin (**1**)	0.07		–	–
Mycalamide A (**4**)	0.5		–	–
Mycalamide B (**5**)	0.1	–	–	–
Theopederin A (**7A**)	–	0.05	–	–
Theopederin B (**7B**)	–	0.1	–	–
Theopederin C (**7C**)	–	0.7	–	–
Theopederin D (**7D**)	–	1.0	–	–
Theopederin E (**7E**)	–	9.0	–	–
Onnamide A (**6**)	0.4	–	10	2.0
13-des-*O*-Methyl-onnamide A (**78**)	–	–	150	–
21,22-Dihydo-onnamide A (**79**)	–	–	40	4.6
Pseudo-onnamide A (**80**)	–	–	130	–
Onnamide B (**81**)	–	–	130	–
17-Oxo-onnamide B (**82**)	–	–	100	–
Onnamide C (**83**)	–	–	70	–
Onnamide D (**84**)	–	–	20	–
Onnamide E (**85**)	–	–	inactive	–
21,22-Dihydro-17-oxo-onnamide A (**86**)	–	–		16.0
17-Oxo-onnamide A (**87**)	–	–		9.2
4(Z)-Onnamide A (**88**)	–	–		1.5

[a]For structures see Schemes 1 and 16.
[b]Perry et al. 1990; Thompson et al. 1992.
[c]Fusetani et al. 1992.
[d]Matsunaga et al. 1992.
[e]Kobayashi et al. 1993. Murine L1210 lymphoma cells were used.

in vitro assays showed that it was responsible for the in vitro activity of the crude sponge extract and thus probably the in vivo activity as well. A crude extract (ca. 2% mycalamide A) was tested in mice infected with A59 coronavirus. Four mice dosed with virus and the extract at 0.1 mg/kg survived 14 days, while eight mice dosed with the virus only all died within 8 days. For further data relevant to the antiviral activity of mycalamide A see Sect. 3.4.

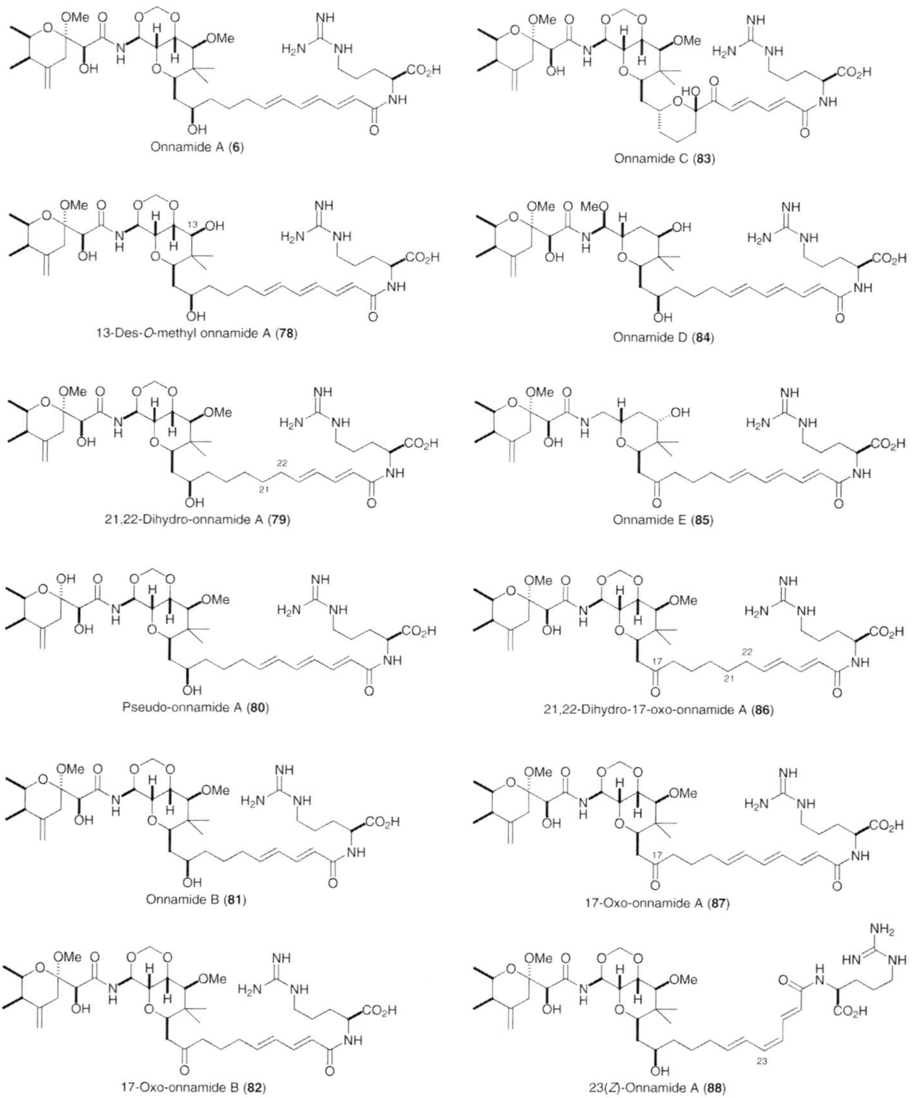

Onnamide A (**6**)

Onnamide C (**83**)

13-Des-*O*-methyl onnamide A (**78**)

Onnamide D (**84**)

21,22-Dihydro-onnamide A (**79**)

Onnamide E (**85**)

Pseudo-onnamide A (**80**)

21,22-Dihydro-17-oxo-onnamide A (**86**)

Onnamide B (**81**)

17-Oxo-onnamide A (**87**)

17-Oxo-onnamide B (**82**)

23(*Z*)-Onnamide A (**88**)

Scheme 16.

Onnamide A is approximately equivalent in potency to mycalamides A and B against murine P388 leukemia cells (IC_{50} 2.4±0.3 n*M*) in vitro but it was inactive against P388 cells in vivo (15% increase in life span at 40 μg/kg). Onnamide A is about 70 times less active against HL-60, HT-29, and A549, in line with its reduced potency as an inhibitor of protein synthesis (Burres and Clement 1989). Biological data for the remaining members of the onnamide subfamily are limited to in vitro studies against P388, and the relevant data are collected in Table 1. Onnamide A and 23(*Z*)-onnamide A (**88**, Scheme 16) are the most potent members of the family, while onnamide E (**85**), lacking the *N*-acyl aminal functionality, is inactive. All members of the onnamide subfamily were vesicants (Matsunaga et al. 1992). There are also very few published data regarding the antiviral activity of the onnamides. In their original report on the isolation and structural determination of onnamide A, Sakemi et al. (1988) claimed potent antiviral activity against herpes simplex type-1, vesicular stomatitis virus, and coronavirus A-59.

Theopederins A–E (**7A–E**) were markedly cytotoxic against P388 leukemia cells (see Table 1; Fusetani et al. 1992). Theopederins A and B also showed promising antitumor activity against P388 (i.p.): T/C = 205% (0.1 mg/kg/day, treated on days 1, 2, and 4–6, i.p.) and T/C = 173% (0.4 mg/kg/day, treated on days 1, 2, and 4–6, i.p.), respectively.

2.3.2 Simple Derivatives of the Natural Mycalamides

In 1992, Thompson et al. (1992) prepared 34 simple acyl, alkyl, and silyl derivatives of the C7, C17, and C18 hydroxyl groups and the *N*-amido group of mycalamides A and B and their relative potency was assayed in vitro against P388 cells. The most noteworthy conclusions from this study are (a) methylation of the amide nitrogen together with the C7 hydroxyl group causes at least a 10^3-fold reduction of activity; (b) derivatization of the C7 hydroxyl group causes a 10-to 100-fold reduction in activity; and (c) methylation of both the C17 and C18 hydroxyl groups (as found in pederin) renders the mycalamides as active as pederin. From these observations, Munro concluded that the intact *N*-acyl aminal bridge is vitally important for the biological activity of the mycalamides.

In 1997 we studied the biological activities of 18-*O*-methyl mycalamide B (**73**), 10-*epi*-18-*O*-methyl mycalamide B, and pederin, all prepared by total synthesis (Richter et al. 1997). The activities of 18-*O*-methyl mycalamide B and pederin were virtually indistinguishable when evaluated in DNA or protein synthesis assays and in cytotoxicity assays using human carcinoma cell lines (IC_{50}s 0.2–0.6 n*M*). In the assays, 10-*epi*-18-*O*-methyl mycalamide B was 10^3 times less toxic than its diastereoisomer, demonstrating that the cytotoxicity of 18-*O*-methyl mycalamide B is inseparable from its ability to inhibit protein synthesis. Short-term exposure of squamous carcinoma cells to 18-*O*-methyl mycalamide B or pederin caused an irreversible inhibition of cellular proliferation and induced cellular necrosis. In contrast, the antiproliferative effects of the compounds on human fibroblasts were reversible and there was no evidence of necrosis.

2.3.3 Degradation Products of the Natural Mycalamides

An extensive program on the chemistry of the mycalamides by the New Zealand group of Munro and Blunt involved treatment of mycalamide A and some of its alkyl derivatives with alkoxide, hydroxide, oxide bases, sodium borohydride, and azide (Thompson et al. 1994). The study was extended to acid-catalyzed degradations, acetal exchange reactions, catalytic hydrogenation, epoxidation, and oxidation reactions (Thompson et al. 1995). The numerous derivatives and degradation products were then tested in vitro against P388 cells. The IC_{50} values in ng/ml are given together with the relevant structures in Scheme 17.

Oxazolidinones **89–91** and the 7-*O-benzyl* derivatives **92** and **93** displayed poor activity, as was expected from the previous studies which had established the need for a free hydroxyl group at C7. The relative inactivity of the cleavage fragments **97–103** demonstrates further that both segments of the mycalamide structure are essential to the biological activity. Only the reduction product **96** (40-fold deactivation) has a significant biological activity, showing that the C10 configuration was crucial for activity to be displayed. The moderate activity of the reduction product **96** is surprising, since the aminal function is generally considered a crucial structural motif for biological activity. The importance of the C6 acetal is shown by the retention of activity with a C6

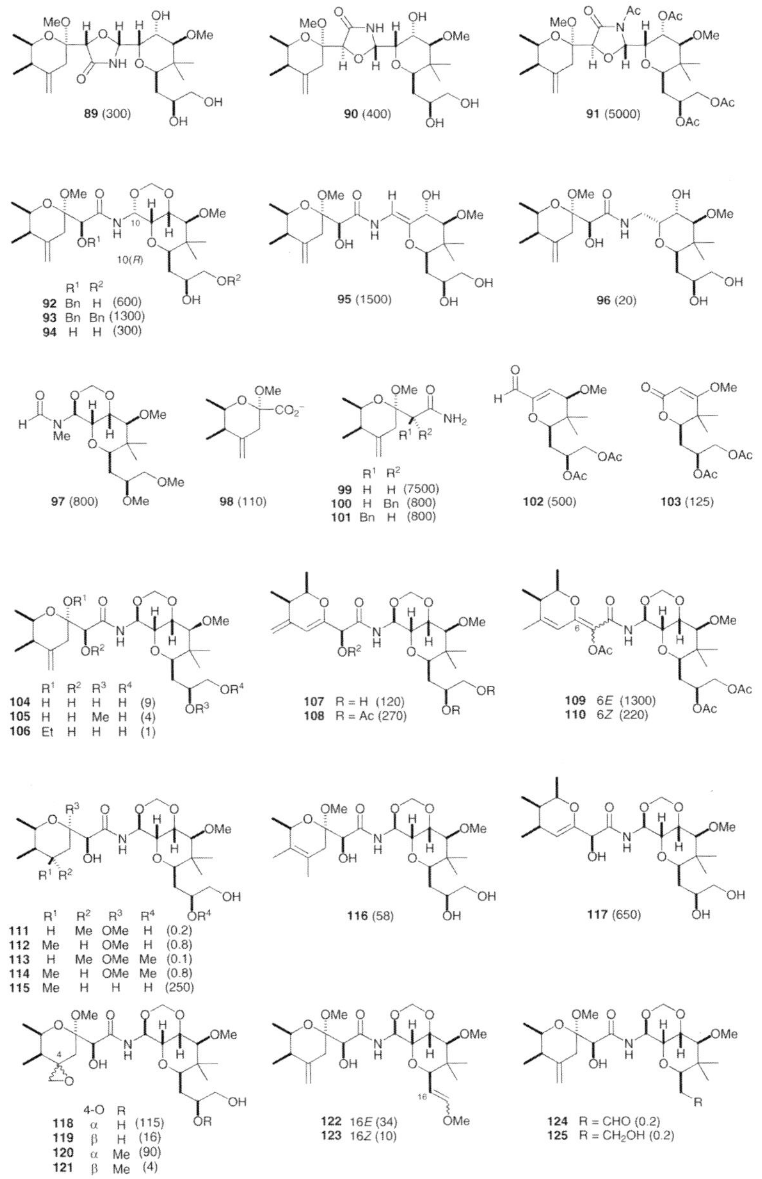

89 (300)

90 (400)

91 (5000)

	R[1]	R[2]	
92	Bn	H	(600)
93	Bn	Bn	(1300)
94	H	H	(300)

10(*R*)

95 (1500)

96 (20)

97 (800)

98 (110)

	R[1]	R[2]	
99	H	H	(7500)
100	H	Bn	(800)
101	Bn	H	(800)

102 (500)

103 (125)

	R[1]	R[2]	R[3]	R[4]	
104	H	H	H	H	(9)
105	H	H	Me	H	(4)
106	Et	H	H	H	(1)

107 R = H (120)
108 R = Ac (270)

109 6*E* (1300)
110 6*Z* (220)

	R[1]	R[2]	R[3]	R[4]	
111	H	Me	OMe	H	(0.2)
112	Me	H	OMe	H	(0.8)
113	H	Me	OMe	Me	(0.1)
114	Me	H	OMe	Me	(0.8)
115	Me	H	H	H	(250)

116 (58)

117 (650)

	4-O	R	
118	α	H	(115)
119	β	H	(16)
120	α	Me	(90)
121	β	Me	(4)

122 16*E* (34)
123 16*Z* (10)

124 R = CHO (0.2)
125 R = CH_2OH (0.2)

Scheme 17.

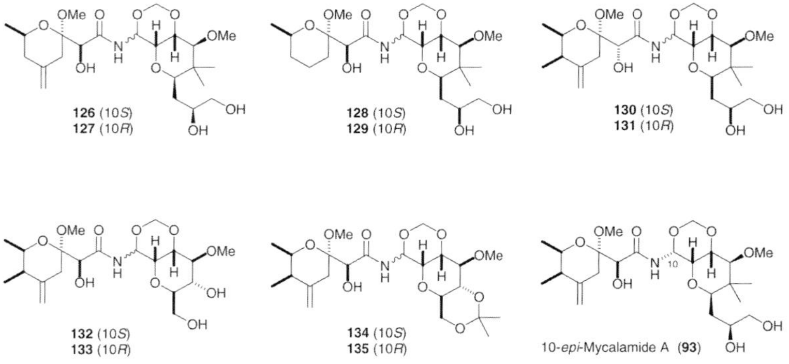

Scheme 18.

Table 2. Biological activity of mycalamide A analogues[a,b]

Compound	Cytotoxicity against HeLa cells	Antiviral activity against HSV-1		Antiviral activity against VZV	
	IC_{50}	MIC[c]	IC_{50}[d]	MIC[e]	IC_{50}[f]
5-Fluorouracil	3.0	–	–	–	–
Acyclovir	–	1.6	1.6	6.25	>50.0
Mycalamide A (**4**)	<0.03	<0.4	<0.4	<0.4	<0.4
10-epi-Mycalamide A (**93**)	3.0	12.5	12.5	1.6	12.5
126	<0.03	<0.4	<0.4	<0.4	<0.4
127	3.0	50.0	50.0	1.6	>50.0
128	0.03	3.1	3.9	<0.4	1.6
129	>10.0	50.0	>50.0	3.1	>50.0
130	>10.0	50.0	12.5	<0.4	12.5
131	>10.0	50.0	>50.0	12.5	>50.0
132	1.0	25.0	25.0	3.1	>50.0
133	>10.0	>50.0	>50.0	25.0	>50.0
134	3.0	50.0	>50.0	12.0	>50.0
135	10.0	50.0	50.0	12.5	>50.0

[a]Data taken from Fukui et al. 1997.
[b]IC_{50} (μg/ml), MIC (μg/ml).
[c]Against HSV-1 cells.
[d]Against vero cells.
[e]Against VZV.
[f]Against HEL cells.

ethoxy derivative **106**, a 20-to 40-fold drop in activity for a C6 hydroxy substituent as in compounds **104** and **105**, and further losses (>100-fold) with elimination or hydrogenolysis at C6 (compounds **107–110**, **115**, **117**). Compounds involving derivatization or transposition of the exocyclic double bond retained significant activity: thus, 4β-dihydromycalamide A **111** was significantly more active than mycalamide A but 4β-dihydromycalamide B **113** had the same activity as mycalamide B and Δ^3-mycalamide A **116** was 100-fold less active than mycalamide A. The epoxide derivatives **118–121** were also much less active, although there was an even more pronounced isomer effect.

Modifications at the C16 side chain were expected to be well tolerated, so it was surprising that both Δ^{16}-normycalamide B isomers (**122** and **123**) were approximately 100 times less active than mycalamide B. However, the C17 aldehyde and alcohol derivatives (**124** and **125**) were significantly more active than both mycalamide A and theopederin E (IC_{50} = 9 ng/ml against P388).

2.3.4 Advanced Synthetic Analogues of Mycalamide A

In 1997 the Nakata group (Fukui et al. 1997) synthesized a set of ten advanced analogues of mycalamide A (Scheme 18) in order to probe the minimum structural requirements of the left fragment for biological activity and to explore the possibility of replacing the right fragment with glucose derivatives. The cytotoxicity against HeLa cells and antiviral activity against herpes simplex type 1 (HSV-1) and varicella-zoster virus (VZV) were tested in vitro, along with 5-fluorouracil and acyclovir as standards. The data are summarized in Table 2. For comparison, parallel tests were run on mycalamide A itself and its 10-epimer (**95**).

Mycalamide A, **125** and **127**, showed very potent cytotoxicity against HeLa cells but their corresponding 10-epimers (i.e., 10-*epi*-mycalamide A, **127** and **129**) were 100 times less active, suggesting that the C10 configuration is a crucial determinant for high cytotoxicity. As expected from the results described in the previous section, the high activity of compound **127** verifies that the presence of the C4-*exo*-methylene and C3-methyl groups is not an important factor for the potent cytotoxicity. The unnatural C7 hydroxyl isomer **130** and its C10-epimer **131** decreased the activity, which suggests that the configuration of the C7-hydroxy group is also essential for potent cytotoxicity. It is noteworthy that **132** and **134**, in which the right fragment is replaced by glucose derivatives, showed nearly the same activity as 5-fluorouracil.

A compound can be judged to have significant antiviral activity if its therapeutic ratio (TR = IC_{50}/MIC) is higher than that of acyclovir. The antiviral activity of mycalamide A **126** and **128** against HSV-1 is very strong. However, their cytotoxicity (IC_{50}) against vero cells is also very strong: TRs of all synthetic compounds tested are less than 1 (TR of acyclovir = 32). Although mycalamide A, **44** and **46** showed strong activity against VZV, their potent cytotoxicity (IC_{50}) against HEL cells was also observed. 10-*epi*-Mycalamide A (**94**), **127**, **129**, **130**, and **132** showed potent antiviral activity against VZV and low cytotoxicity against HEL cells: TRs of 10-*epi*-mycalamide A (**94**), **127**, **129**, **130,** and **132** are 8, >32, >16, >32, and >16, respectively (cf. TR of acyclovir >8). Thus 7- or 10-epimeric compounds showed significant antiviral activity against VZV.

2.3.5 Simple Synthetic Derivatives

None of the biological data presented thus far identify the *minimum* structural requirements for cytotoxicity or antiviral activity. In 1997, Abell et al. (1997) evaluated the cytotoxicity of simple analogues (Table 3) of the *N*-acyl aminal bridge against the P388 leukemia cell line in vitro. In general, compounds **136a–i** with a (1'*R*,2*S*) configuration (equivalent to C7 and C10 in the natural products) show significantly greater in vitro cytotoxicity than the corresponding (1'*S*,2*S*) derivatives **137a–i**. Notable exceptions were the parent natural products (pederin, mycalamides A and B) for which the equivalent C10 position is *S*. A preference for a (1'*R*) configuration over a (1'*S*) configuration does not seem to be evident within the cyclic oxazolidinone series **138–143** (Scheme 19), where the (1'*S*) and (1'*R*) compounds show similar in vitro cytotoxicity.

A variety of R^1 groups appear to be accommodated for the induction of in vitro cytotoxicity. For example, the corresponding acetates of **136a** and **137a**, compounds **136e** and **137e**, show comparable activity. By comparison, acylation of the C7-hydroxyl group of mycalamide A and

Table 3. IC_{50} values of **136** and **137** against P388 cells

	R^1	R^2	IC_{50} (µg/ml) **136** (1'*R*,2*S*)	**137** (1'*S*,2*S*)
a	OH	Ph	52	>340
b	OH	Me	>125	>188[a]
c	NHZ	Ph	14	>188[b]
d	NH-Ala-Z	Ph	36	>125
e	OAc	Ph	101	>313
f	OH	Et	176	43
g	NHZ	Me	>375	>375
h	$OCOC_6H_4Br$	i-Pr	105[c]	105
i	O-camphanyl	Ph	42[d]	78[d]

[a] 1:1 mixture of epimers.
[b] 3:1 mixture of epimers.
[c] 17:3 mixture of epimers.
[d] 9:1 mixture of epimers.

136 (1'*R*,2*S*)

137 (1'*S*,2*S*)

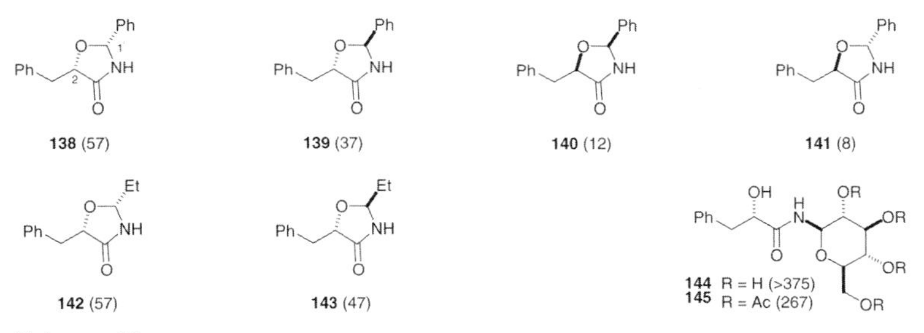

Scheme 19.

B (analogous to C2 in **136/137**) results in compounds with significantly decreased activity. The (1')-epimeric pairs **136c,d,g** and **137c,d,g** were designed to give the derivatives more peptide character. The most active compounds in this series, compounds **136c,d** show activities comparable to, or better than, that of **136a**. Again a preference for a (1'*R*) configuration is noted (**136c/137c** and **136d/137d**). A change from R^2=Ph to Et appears to be tolerated, although in this case, contrary to the other compounds, a (1'*S*) configuration seems to give the most potent in vitro bioactivity (**136f >137f**). It should be noted that **137f** and the acetal contributes to high activity but not essential

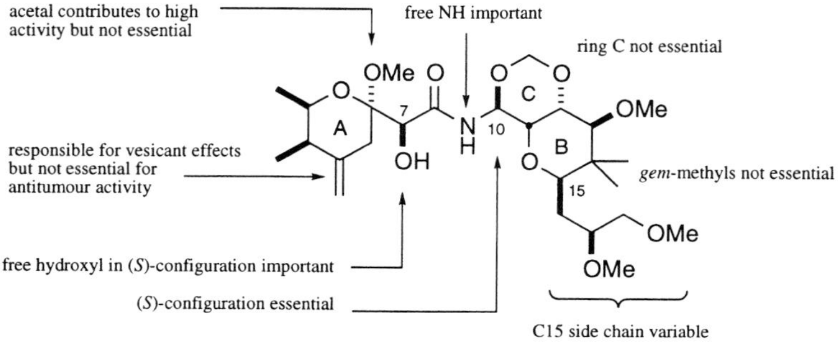

free NH important

ring C not essential

responsible for vesicant effects but not essential for antitumour activity

gem-methyls not essential

free hydroxyl in (*S*)-configuration important

(*S*)-configuration essential

C15 side chain variable

Scheme 20.

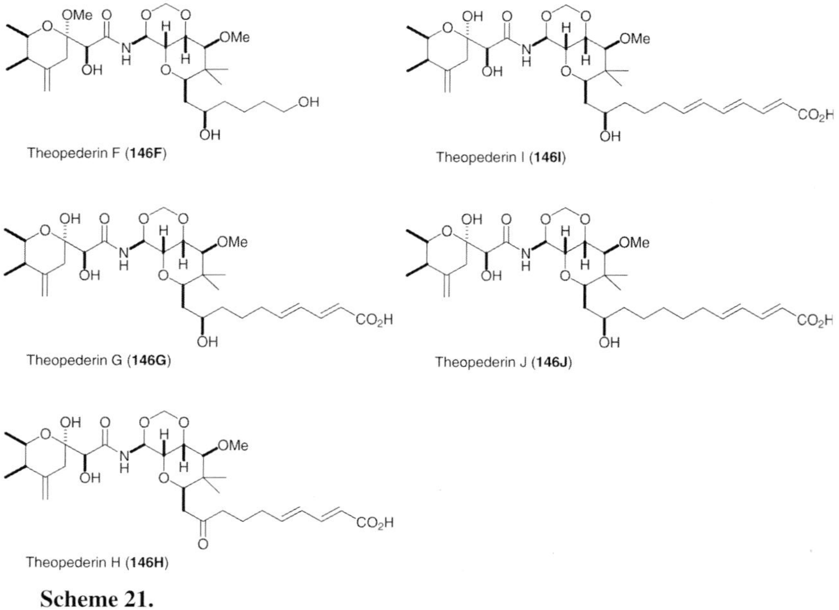

Theopederin F (**146F**)

Theopederin I (**146I**)

Theopederin G (**146G**)

Theopederin J (**146J**)

Theopederin H (**146H**)

Scheme 21.

parent natural products possess the same relative configuration at this center (i.e., *S*). The introduction of a methyl group at the R^2 position resulted in compounds with significantly reduced activity (see compounds **136b,g**, **137b,g**). Finally, the glucosyl derivatives **143** and **144** show less activity than the corresponding acyclic analogues **136** and **137**, where R^2=Et and Ph.

In conclusion, the foregoing structure-activity studies on the pederin family of antitumor agents established that the *N*-acyl aminal bridge is the pharmacophore. The homoallylic acetal array encompassing C4–C6, which is responsible for the acid lability of the natural products as well as their vesicant effects, is not necessary for antitumor or antiviral activity. The C6 acetal function contributes to the high activity of the natural products, though simpler analogue studies reveal that it is not essential. The presence of a free hydroxyl group at C7 with the (*S*) configuration is important for high activity. The configuration of the aminal center is also very important, with the (*S*) configuration at C10 being significantly more active as an antitumor agent than the (*R*) epimer; however, compounds with the (*R*) configuration at C10 remain potent antiviral agents. The complex trioxadecalin ring system characteristic of the mycalamides, onnamides, and theopederins is not essential for high activity since pederin, with its simpler monocyclic right half, is one of the most active of the natural products, followed closely by 18-*O*-methyl-mycalamide B, a simple synthetic derivative of natural mycalamide B. Finally, the side chain at C15 tolerates considerable variation with comparatively little impact on activity. A summary of the SAR data is given in Scheme 20.

Addendum. Theopederins F–J (**145F–J**, Scheme 21) have recently been isolated from *Theonella swinhoei* (Tsukamoto et al. 1999). Theopederin F was antifungal against an *erg*6 mutant of *Saccharomyces cerevisiae* deficient in (*S*)-adenosylmethionine-Δ^{24}-methyltransferase involved in the biosynthesis of ergosterol. Theopederin F was also cytotoxic against murine P388 cells with an IC_{50} of 0.15 ng/ml. No biological data were reported for theopederins G–J. The close structural kinship between theopederins G–J and the onnamides suggests that the theopederins may be biosynthetic precursors of the onnamides.

Acknowledgements. We thank the EPSRC, Pfizer Central Research (Sandwich), and AstraZeneca Pharmaceuticals (Macclesfield) for their generous financial support. We also thank Dr. F.T. Boyle (AstraZeneca) for many helpful discussions.

References

Abell AD, Blunt JW, Foulds GJ, Munro MHG (1997) J Chem Soc Perkin Trans 1:1647

Bonamartini Corradi A, Mangia A, Nardelli N, Pelizzi G (1971) Gazz Chim Ital 101:591

Breitfelder S, Schlapbach A, Hoffmann RW (1998) Synthesis 468

Burres NS, Clement JJ (1989) Cancer Res 49:2935

Cardani C, Ghiringhelli D, Mondelli R, Quilico A (1965) Tetrahedron Lett 2537

Cardani C, Ghiringhelli D, Mondelli R, Quilico A (1966) Gazz Chim Ital 96:3

Cardani C, Ghiringhelli D, Quilico A, Selva A (1967) Tetrahedron Lett 41:4023

Carrasco L, Fernandez-Puentes C, Vasquez D (1976) Mol Cell Biochem 10:97

Clardy J, He H (1995) In: Biologically active amides containing a bicyclo moiety. Cornell Research Foundation, Inc, USA, US 547953

Frank JH, Kanamitsu K (1987) J Med Entomol 24:155

Fukui H, Tsuchiya Y, Fujita K, Nakagawa T, Koshino H, Nakata T (1997) Bioorg Med Chem Lett 7:2081

Furusaki A, Watanabe T, Matsumoto T, Yanagiya M (1968) Tetrahedron Lett 6301

Fusetani N, Sugawara T (1993) Chem Rev 93:1793

Fusetani N, Sugawara T, Matsunaga S (1992) J Org Chem 57:3828

Hoffmann RW, Schlapbach A (1992) Tetrahedron 48:1959

Hoffmann RW, Schlapbach A (1993) Tetrahedron Lett 34:7903

Hoffmann RW, Breitfelder S, Schlapbach A (1996) Helv Chim Acta 79:346

Hong CY, Kishi Y (1990) J Org Chem 55:4242

Hong CY, Kishi Y (1991) J Am Chem Soc 113:9693

Isaac K, Kocienski P (1982) J Chem Soc Chem Commun 460

Isaac K, Kocienski P, Campbell S (1983) J Chem Soc Chem Commun 249

Jarowicki K, Kocienski P, Marczak S, Willson T (1990) Tetrahedron Lett 31:3433

Kellner RLL (1998) Z Naturforsch C 53:1081

Kellner RLL, Dettner K (1995) J Chem Ecol 21:1719

Kellner RLL, Dettner K (1996) Oecologia 107:293

Kobayashi J, Itagaki F, Shigemori H, Sasaki T (1993) J Nat Prod 56:976

Kocienski P, Jarowicki K, Marczak S (1991) Synthesis 1191

Kocienski P, Willson TM (1984) J Chem Soc Chem Commun 1011
Kocienski P, Raubo P, Davis JK, Boyle FT, Davies DE, Richter A(1996) J Chem Soc Perkin Trans 1:1797
Kocienski P. J, Narquizian R, Raubo C, Smith C, Farugia LF, Muir K (2000) J Chem Soc Perkin Trans 1 (in press)
Kocienski PJ, Narquizian R, Raubo P, Smith C, Boyle FT (1998a) Synlett 1432
Kocienski PJ, Narquizian R, Raubo P, Smith C, Boyle FT (1998b) Synlett 869
Marron TG, Roush, WR (1995) Tetrahedron Lett 36:1581
Matsumoto T, Tsutsui S, Yanagiya M, Tasuda S, Maeno S, Kawashima J, Uete A, Murakami M (1964) Bull Chem Soc Jpn 37:1892
Matsumoto T, Yanagiya M, Maeno S, Yasuda S (1968) Tetrahedron Lett 6297
Matsuda F, Yanagiya M, Matsumoto T (1982) Tetrahedron Lett 23:4043
Matsuda F, Tomiyoshi N, Yanagiya M, Matsumoto T (1983) Tetrahedron Lett 24:1277
Matsumoto T, Matsude F, Hasegawa K, Yanagiya M (1984) Tetrahedron 40:2337
Matsuda F, Tomiyoshi N, Yanagiya M, Matsumoto T (1988) Tetrahedron 44:7063
Matsunaga S, Fusetani N, Nakao Y (1992) Tetrahedron 48:8369
Mikhailova LA (1967) Vestn Dermatol Venerol 41:67
Nakata T, Nagao S, Mori N, Oishi T (1985a) Tetrahedron Lett 26:6461
Nakata T, Nagao S, Oishi T (1985b) Tetrahedron Lett 26:6465
Nakata T, Matsukura H, Jian DL, Nagashima H (1994) Tetrahedron Lett 35:8229
Nakata T, Fukui H, Nakagawa T, Matsukura H (1996) Heterocycles 42:159
Netolitzky F (1919) Koleopterol Rundsch 7:121
Pavan M (1982) Summary of the present data on pederin. Pubblicazioni dell 'Instituto di Entomologia dell' Universita di Pavia
Pavan M, Bo G (1953) Physiol Comp Oecol 3:307
Perry NB, Blunt JW, Munro MHG, Pannell LK (1988) J Am Chem Soc 110:4850
Perry NB, Blunt JW, Munro MHG, Thompson AM (1990) J Org Chem 55:223
Quilico A, Cardani C, Mondelli R, Quilico A (1961) Chem Ind (Milan) 43:1434
Richter A, Kocienski P, Raubo P, Davies DE (1997) Anti Cancer Drug Design 12:217
Roush WR, Marron TG (1993) Tetrahedron Lett 34:5421
Roush WR, Pfeifer LA (1998) J Org Chem 63:2062
Roush WR, Marron TG, Pfeifer LA (1997) J Org Chem 62:474
Roush WR, Pfeifer LA, Marron TG (1998) J Org Chem 63:2064
Sakemi S, Ichiba T, Kohmoto S, Saucy G, Higa T (1988) J Am Chem Soc 110:4851

Soldati M (1966) Experientia 22:176
Thompson AM, Blunt JW, Munro MHG, Perry NB, Pannell LK (1992) J Chem Soc Perkin Trans 1:1335
Thompson AM, Blunt JW, Munro MHG, Clark BM (1994) J Org Chem Perkin Trans 1:1025
Thompson AM, Blunt JW, Munro MHG, Perry NB (1995) J Chem Soc Perkin Trans 1:1233
Toyota M, Yamamoto N, Nishikawa Y, Fukumoto K (1995) Heterocycles 40:115
Toyota M, Hirota M, Nishikawa Y, Fukumoto K, Ihara M (1998a) J Org Chem 63:5895
Toyota M, Nishikawa Y, Fukumoto K (1998b) Heterocycles 47:675
Trotter NS, Takahashi S, Nakata T (1999) Org Lett 1:957
Tsukamoto S, Matsunaga S, Fusetani N, Toh-e A (1999) Tetrahedron 55:13697
Vazquez D (1979) Inhibitors of protein biosynthesis. Springer, Berlin Heidelberg New York
Willson T, Kocienski P, Faller A, Campbell S (1987) J Chem Soc Chem Commun 106
Willson TM, Kocienski P, Jarowicki K, Isaac K, Faller A, Campbell SF, Bordner J (1990a) Tetrahedron 46:1757
Willson TM, Kocienski P, Jarowicki K, Isaac K, Hitchcock PM, Faller A, Campbell SF (1990b) Tetrahedron 46:1767
Yanagiya M, Matsuda F, Hasegawa K, Matsumoto T (1982) Tetrahedron Lett 23:4039

3 Biotherapeutic Potential and Synthesis of Okadaic Acid

C.J. Forsyth, A.B. Dounay, S.F. Sabes, R.A. Urbanek

3.1 Introduction

Nature has long provided an array of secondary metabolites that have been utilized by man or humans for a large variety of therapeutic applications. The polyether marine natural product okadaic acid (OA, Fig. 1) represents a particularly interesting example, with a rich modern history whose drama is continuing to unfold (Scheuer 1995). Originally isolated from extracts of the marine sponge *Halichondria okadai* as a potential anticancer agent (Tachibana et al. 1981), OA was subsequently found to have cancerous tumor-promoting activity in the two-stage model of carcinogenesis on mouse skin (Suganuma et al. 1988). Moreover, these seemingly paradoxical responses to OA exposure have subsequently led to the now wide-spread recognition of the central roles that protein serine/threonine phosphatases (PPases), the primary cellular

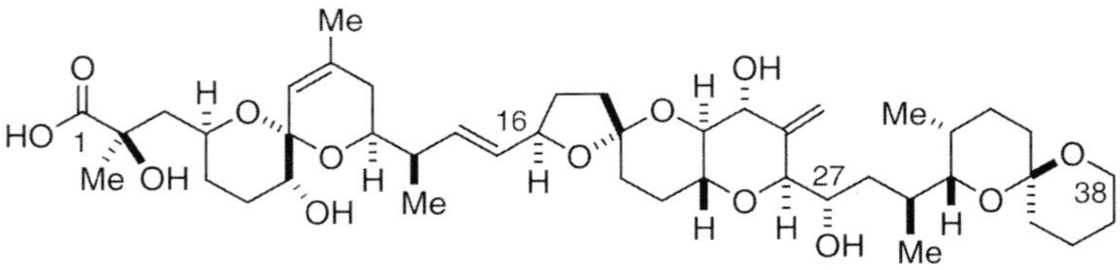

Okadaic Acid (1)

Fig. 1. Okadaic acid

targets of OA, play in the regulation of many essential cellular processes, including metabolism, growth, division, and death (Cohen and Cohen 1989). Thus, OA has emerged as a key laboratory tool for identifying and studying the myriad of events associated with the inhibition of select PPases (Cohen et al. 1990; Schonthal 1992). It has also spawned an entire class of remarkably distinct secondary metabolites from such disparate organisms as yeast, bacteria, blue-green algae, dinoflagellates, and even insects that together comprise the "okadaic acid class of tumor promoters" (Fujiki et al. 1989).

Today, OA is being employed in basic studies directed towards understanding such diverse human disease-related processes as cancer, AIDS, inflammation, osteoporosis, Alzheimer's, and diabetes, among others. However, the therapeutic potential of OA is presently limited by several factors. First, OA lacks sufficient specificity in its inhibition of OA-sensitive PPases. Hence, relatively indiscriminate phosphatase inhibition may simultaneously affect a variety of important cellular processes, not only the targeted ones (Schonthal 1992). Second, the current commercial source of OA is primarily isolation in small quantities from cultured dinoflagellates, single-celled marine micro-organisms (Dickey et al. 1990; Murakami et al. 1982). This provides only research quantities at appreciably high costs. However, alternative sources of OA and structurally related compounds have been emerging in recent years via total laboratory synthesis (Isobe et al. 1986; Forsyth et al. 1997b; Ley et al. 1998). Although an industrial-scale synthesis of OA itself may not yet be feasible in terms of cost and safety, total synthesis offers unique opportunities to tailor the details of the OA architecture, with the aim of developing an empirical understanding of the relationship between structure, function, and specificity.

In this chapter, an overview of the unifying impact that OA has had on the fields of natural products chemistry, cellular biology, and modern synthetic organic chemistry will be provided. Emphasis will be placed on the present and projected synthetic access to OA and its non-natural derivatives, the latter of which may prove to be particularly valuable for capitalizing on the biotherapeutic potential of potent and specific inhibitors of the various subtypes of okadaic acid-sensitive protein phosphatases.

3.2 Okadaic Acid

3.2.1 A Potential Anticancer Agent from Sponges

In the 1970s, researchers at the Fujisawa pharmaceutical company recognized that crude extracts of the black sponge *Halichondria okadai*, isolated off the coast of Japan, bore remarkable cytotoxic activity. From these extracts was isolated a principle with moderate anti-KB activity in vitro in 1975, which was named "halichondrine A" (Scheuer 1995). It was not until after the carboxylic acid of this isolate was derivatized as an *o*-bromophenacyl ester and its name changed to the more appropriate "okadaic acid" that its complete structure was determined by X-ray crystallography by Van Engen and Clardy at Cornell University in 1981 (Tachibana et al. 1981). Although it would soon become one of the world's most widely used marine natural products in biological research, purified okadaic acid did not seem to maintain the high level of cytotoxic activity associated with the crude sponge extract. Instead, additional polyethers called the "halichondrins" were subsequently isolated from *Halichondria okadai* and found to represent the phenomenal levels of antimitotic activity associated with the crude sponge extract (Uemura et al. 1985; Aicher et al. 1992).

The X-ray analysis of the *o*-bromophenacyl ester derivative of OA revealed an intriguing array of functionality and structural features. Of the 38 contiguous carbons of its backbone, 23 are functionalized and 17 are stereogenic carbons. OA also contains three spiroketal moieties, a *trans*-alkene, an exocyclic alkene, an α-methyl, α-hydroxy carboxylic acid, and three secondary alcohols. Much of this functionality is believed to contribute to the cyclic conformation observed in the crystal

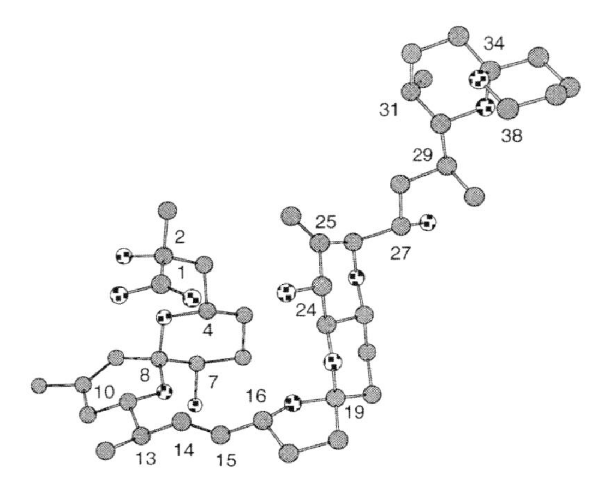

Fig. 2. Cyclic conformation of okadaic acid

structure (Fig. 2). Notably, a very similar cyclic conformation has been assigned to the parent carboxylic acid in solution on the basis of NMR studies (Norte et al. 1991; Matsumori et al. 1995). This pseudomacrolide conformation allows the hydroxyl group at C24 to form an intramolecular hydrogen bond with the C1 carboxylate moiety of the natural product. Additional structural features that may contribute to the cyclic conformation are the opportunity for another hydrogen bond between the C2 hydroxyl and the C4 oxygen, thermodynamically enforced anomeric configurations at C8 and C19, the C14-C15 *trans*-alkene, and the C19-C26 *trans*-dioxo-decalin system (Uemura and Hirata 1988). Projecting from the pseudomacrolide core is a C28–C38 lipophilic domain characterized by a spiroketal terminus. Also isolated from marine sources have been a number of compounds very closely related to OA, including acanthifolicin (Schmitz et al. 1981), the dinophysistoxins (Windust et al. 1997; Fujiki et al. 1988), and 7-deoxyokadaic acid (Fujiki et al. 1988).

Although purified OA displayed relatively modest in vitro cytotoxic activity in initial assays (Tachibana et al. 1981), its unique structure was subsequently found to elicit an amazing array of cellular responses. Of acute health concern as an environmental contaminant, OA was characterized as a causative agent of diarrhetic shellfish poisoning (Yasumoto et al. 1985), and a nonphorbol ester type of tumor promoter (Suganuma et al. 1988). The latter activity refers to the promotion of cancerous tumor growth in the two-stage model of carcinogenesis on mouse skin after preliminary treatment with mutagens, but without direct interaction with the phorbol ester receptor (Hecker 1978). Hence, a new mechanistic pathway of cancerous tumor promotion was invoked (Herschman et al. 1989; Sassa et al. 1989).

3.2.2 A Nonphorbol Type of Tumor Promoter

In 1980, Nishizuka identified the enzyme protein kinase C (PKC) as the receptor of the phorbol ester tumor promoters, thus implicating the potential role of protein phosphorylation in cellular processes involved in cancerous tumor promotion (Nishizuk 1984; Blumberg 1988). It was hypothesized that persistent activation of PKC by the phorbol esters leads to hyperphosphorylation of key proteins involved in signaling cellular growth. But if OA did not interact directly with PKC, then how could it elicit cellular responses similar to those of the phorbol esters? The answer came in the same year in which OA was characterized as a nonphorbol ester tumor promoter. Bialojan and Takai reported in 1988 that OA potently inhibits protein serine/threonine phosphatases 1 and 2A (Bialojan and Takai 1988), (PP1 and PP2A, respectively), cellular enzymes that may have complementary activity to PKC. K_i values for inhibition of PP1 and PP2A by OA have been reported to be 145 n*M* and 32 p*M*, respectively, but there are variations in the data available depending upon the source of the enzymes and the nature of the assays. In contrast to its subnanomolar inhibition of PP2A, concentrations of OA above 10 n*M* are required for inhibition of PP1, the immunomodulator calcineurin (PP2B), and the aortic PCM phosphatases. Other phosphatases, including type 2C, phosphotyrosyl, inositol triphosphate, and acid and alkaline phosphatases, are not affected by OA at concentrations up to 10 μ*M*.

Hyperphosphorylation caused by inhibition of the PPases was hypothesized to be functionally equivalent to that induced by activation of PKC (Fujiki 1992). Several structurally diverse natural products were subsequently characterized as additional members of the "okadaic acid class of tumor promoters" by virtue of their inhibition of OA-sensitive phosphatases. These compounds include acanthifolicin (Holmes et al. 1990), and the related dinophysistoxins (Ohuchi et al. 1989); calyculin A (Ishihara et al. 1989); microcystins (MacKintosh et al. 1990) and nodularin (Matsushima et al. 1990); tautomycin (MacKintosh and Klumpp 1990); cantharidin (Li and Casida 1992); thyrsiferyl 23-acetate (Matsuzawa et al. 1994); and fostriecin (Roberge et al. 1994). Competitive binding studies indicate that all of these inhibitors share binding domains at the active site of the phosphatases. Inhibition of okadaic acid-sensitive phosphatases may be responsible for many, if not all, of the observed cellular responses to the OA class of tumor promoters (Sheppeck et al. 1997). Therefore, in order to fully explore the biotherapeutic potential of OA, a detailed understanding of the structure and function of their primary cellular targets is required.

3.3 The Okadaic Acid Receptor

3.3.1 Protein Serine/Threonine Phosphatases 1 and 2A

The recognition that the naturally occurring tumor promoter okadaic acid was a potent inhibitor of serine/threonine phosphatases helped to bring these enzymes to the forefront of modern biomedical and cellular research. It had been appreciated for some time that protein kinases and phosphatases are two classes of enzymes that perform complementary roles in maintaining levels of protein phosphorylation. However, the emergence of okadaic acid and other naturally occurring inhibitors of the phosphatases provided valuable new tools for their study, analogous to the previously known exogenous activators of protein kinase C.

Protein serine/threonine phosphatases 1 (PP1) and 2A (PP2A), 2B (PP2B), and 2C (PP2C) are the four principle threonine/serine-specific protein phosphatases in the cytosol and nucleus of eukaryotic cells (Cohen and Cohen 1989). PP1 and PP2A-like enzymes are particularly significant to human health because they are intimately involved in the regulation of a multitude of important cellular processes, including transcriptional regulation of oncogenes and control of the cell cycle (Parsons 1998; Cohen 1989; Mumby and Walter 1993; Depaoli-Roach et al. 1994). For example, PP1 and PP2A are involved in the expression and activation of the oncogenic transcription factor AP-1 (Schonthal et al. 1991; Thevenin et al. 1991; Park et al. 1992; Ainbinder et al. 1997; Peng et al. 1997; Rosenberger and Bowden 1996), the regulation of the mitosis-promoting factor p34cdc2 kinase (Borgne and Meijer 1996), and replication of SV40 DNA (Virshup and Kelly 1989; Virshup et al. 1989; Moens et al. 1997). PP2A also down-regulates the mitogen-activated protein kinase (MAPK) cascade by dephosphorylating the activating sites in MAPK and MEK.

Members of the closely related PP1 and PP2A families share approximately 50% sequence homology in their catalytic subunits (Cohen and Cohen 1989). They have broad substrate specificity but are distinguished by their specificity towards phosphorylase kinase. PP1 dephosphorylates the phosphorylase kinase β-subunit, whereas PP2A dephosphorylates the α-subunit (Cohen 1989). Three endogenous mammalian proteins, inhibitor 1, DARP-32, and inhibitor 2, regulate the catalytic activity of cytosolic PP1 but do not affect PP2A (Cohen 1989). Levels of inhibitor 2 oscillate over the cell cycle in apparent concert with the hyperphosphorylated human retinoblastoma (Rb) gene product (Mihara et al. 1989). The association of hyperphosphorylation of the Rb gene product with oncogenesis suggests that inhibition of PP1 by inhibitor 2 is an important step in this process. The protein DARP-32 occurs in dopamine-innervated portions of the brain and is phosphorylated in response to dopamine (Hemmings et al. 1990). When phosphorylated at threonine-34 by cAMP-dependent protein kinase, DARP-32 inhibits PP1 (Hemmings et al. 1984). PP1 does not appreciably dephosphorylate phospho-DARP-32 in vitro, but both PP2B and PP2A are able to do so (Hemmings et al. 1990).

The in vivo composition and regulation of the PP1 and PP2A holoenzymes is complex. Each catalytic subunit associates with various noncatalytic subunits that direct the intracellular localization and modulate the activity of the enzymes. PP1 forms heterodimers with subunits such as G_M and G_L, which target the catalytic activity to striated muscle glycogen particles and liver glycogen, respectively (Armstrong et al. 1997; Brady et al. 1997; Printen et al. 1997). Other targeting subunits include the p53-binding protein p53BP2 (Helps et al. 1995) and the retinoblastoma gene product (Durfee et al. 1993). A recent X-ray crystallographic study of PP1 complexed with a 13-residue peptide corresponding to amino acids 63–75 of G_M revealed a specific PP1-binding motif present in many PP1-binding subunits, as well as a regulatory domain binding site on the catalytic subunit (Egloff et al. 1997). The subunit binding domain resides on the side opposite to the active site on the elliptical catalytic subunit. Whereas the natural product inhibitors of PP1 may interact primarily at the active site, phosphopeptide inhibitors 1, 2, and DARP-32 are thought to simultaneously occupy both the active site and the distal recognition site (Goldberg et al. 1995). Predictably, the specific residues on PP1 that make contact with the recognition motif are absent in PP2A.

PP2A generally occurs as a heterotrimeric complex comprised of a 36-to 38-kD catalytic subunit (C), a 61-kD subunit (A), and a variable B subunit (Mumby and Walter 1993; Csortos et al. 1996). Three major forms occur in mammalian tissue: $PP2A_0$, $PP2A_1$, and $PP2A_2$, with A and C subunits of approximately 65 kD and 36 kD, respectively. Several distinct B subunits of varying apparent mass (54–130 kD) are complexed with the $PP2A_0$ and $PP2A_1$ core dimers, whereas $PP2A_2$ generally occurs as an AC dimer. Two slightly different forms of each of the A and C subunits have been found. However, the amino acid sequences of the PP2A catalytic subunit is thought to be among the most highly conserved of any known enzyme (Orgad et al. 1990). Two selective, noncompetitive, nanomolar protein inhibitors of PP2A, termed I_1^{PP2A} and I_2^{PP2A} (apparent mol. wt. 30 kD and 20 kD, respectively) have been isolated from bovine kidney (Li et al. 1996). I_1^{PP2A} inhibits the PP2A catalytic subunit by 50% at 4 n*M*, but does not affect PP1 or other major protein phosphatases. I_2^{PP2A} has been identified as a truncated form of the myeloid leukemia-associated protein SET (PHAP-II or TAF). Phosphorylation of the PP2A catalytic subunits deactivates catalytic activity. Another putative regulatory mechanism for PP2A is carboxy terminus methylation; methyl esterification and de-esterification are mediated by a PP2A-specific methyl transferase (Lee et al. 1996; Floer and Stock 1994) and an OA-sensitive esterase (Lee et al. 1996), respectively. The carboxy terminus of the PP2A catalytic subunit is important for binding the B subunit but is not required for binding the polyoma virus middle tumor antigen (Ogris et al. 1997).

Numerous studies have demonstrated that subunit-subunit interactions play an important role in regulating the activity and specificity of the catalytic subunit (Kamibayashi et al. 1991; Mayer-Jaekel and Hemmings 1994; McCright et al. 1996; Turowski et al. 1997; Zolnierowicz et al. 1996). For PP2A, the B subunit binds directly to the A subunit of an AC complex and modulates the substrate affinity of the catalytic subunit, the rate of substrate dephosphorylation, and cellular localization. Transforming viral antigens replace the B subunit of PP2A in infected cells and thereby attenuate phosphatase activity towards several substrates (Yang et al. 1991). A dimeric form of the PP2A AC complex has been found to associate with tumor antigens in cells infected with the simian virus 40 small-t antigen and the small and medium t antigens of polyoma virus (Pallas et al. 1990; Walter et al. 1990). The phosphatase-complexed small-t antigen inhibits dephosphorylation of tumor suppressor protein p53 and large T antigen, and may thereby promote transformation (Yang et al. 1991; Sontag et al. 1993, 1997).

A growing number of other important cellular and disease-related proteins have also been found to associate with PP2A. These include translation termination factor eRF1 (Andjelkovic et al. 1996), cyclin G in a p53-dependent manner (Okamoto et al. 1996), an adenovirus E4 gene product (E4orf4) (Kleinberger and Shenk 1993), leukemia-associated proteins HRX and SET (Adler et al. 1997), HOX11 (Kawabe et al. 1997), and casein kinase 2α (Heriche et al. 1997). Many of these proteins are required for oncogenic proliferation. Because the leukemia-associated protein SET (I_2^{PP2A}) is known to be a potent inhibitor of PP2A, it has been postulated that the HRX/SET/PP2A complex may contribute to leukemogenesis by overriding a G2/M cell cycle checkpoint in human T-cell leukemias, as may HOX11. Microtubule-associated PP2A has been shown to specifically dephosphorylate Ser 202 and Thr 205 of hyperphosphorylated tau, the major building block of Alzheimer's disease neurofibrillary tangles (Merrick et al. 1996; Sontag et al. 1996). These heteroprotein-PP1/PP2A interactions represent a general theme of in vivo modulation of PP2A phosphatase activity by a wide variety of important gene products, often with dramatic and deleterious results. Nonetheless, the mechanisms or structural basis by which PP2A-associated proteins may regulate the catalytic activity are not known.

3.3.2 Structure of the PP1 Active Site

In 1995, separate X-ray crystallographic studies revealed the three-dimensional structures of the catalytic subunit of PP1 (derived from rabbit muscle cDNA) covalently linked to the natural product microcystin-LR at 2.1 Å resolution (Goldberg et al. 1995), of human $PP1_{\gamma 1}$ at 2.1 Å resolution (Egloff et al. 1995) and of PP2B with and without FK506/FKBP. Analogous crystallographic studies of PP1 with additional ligands other than the G_M-related peptides, or of PP2A with or without bound ligands have not been successful. The X-ray structures of PP1 reveal a bi-metallic (Mn^{2+}/Fe^{2+}) (Egloff et al. 1995) active site at the bottom of a shallow pocket that is at the junction of three distinct surface grooves, termed acidic, lipophilic, and carboxy terminal. Models for the mechanism of phosphate hydrolysis involve binding of two phosphate oxygens to the two metals followed by in-line attack by a metal-activated water molecule. A transition-state phosphate may be stabilized by coordination to active site Arg 96, Arg 221, and Asn 124. General acid catalysis would then liberate inorganic phosphate and the serine or threonine product (Goldberg et al. 1995).

The natural product microcystin-LR, a member of the OA class of inhibitors, was found to simultaneously occupy the active site, the hydrophobic groove, and part of the C-terminal groove in the co-crystal (Goldberg et al. 1995). The carboxylate side chain of the natural product's D-glutamate residue was hydrogen bonded to a metal coordinated water. The carboxylate of the α-methyl aspartate residue of microcystin was hydrogen bonded with Arg 96 and Tyr 134, and the hydrophobic Adda side chain fit into the enzyme's lipophilic surface groove. Microcystin-LR was also covalently linked to Cys 273 of the enzyme's β12-β13 loop via the *N*-methyldehydroalanine residue, although such covalent modification is not necessary for potent inhibition (Moorhead et al. 1996). The overall conformation of microcystin-LR in the co-crystal structure is very similar to that determined in solution by NMR spectroscopy. Similarly, the Barford X-ray structure of PP1 without microcystin is conformationally very similar to the Kuriyan co-crystal structure (Goldberg et al. 1995).

Based upon the Kuriyan structure, phosphothreonine 34 of DARP-32 is postulated to occupy the bi-metallic active site, while the extended peptide chain may reside in hydrophobic and acidic grooves emanating from the active site (Goldberg et al. 1995), as well as the distal regulatory binding site (Egloff et al. 1997). Phosphorylation of Thr 320, which is seen to reside in the C-terminal groove, is known to inhibit PP1 (Dohadwala et al. 1994). Although the catalytic domains of PP1 and PP2A share considerable sequence homology, there are important differences in structure between the two enzymes, as manifested in the differential binding of ligands and substrates. The active sites are likely to be similarly constructed, due to the fact that every amino acid residue involved in metal binding and implicated in catalysis is strictly conserved. The hydrophobicity or hydrophilicity of the buried side chains in PP1 is largely conserved in PP2A, suggesting that the enzymes have similar tertiary structures. The lipophilic side chains lining the hydrophobic surface groove proximal to the active site of PP1 are also largely conserved in PP2A. However, many of the acidic residues found on the surface of PP1 are absent from PP2A; Asp 197, Glu 256, and Glu 275 of PP1 are replaced by His, Lys, and Arg, respectively, in PP2A. Also, Glu 218, Asp 220, Glu 252, and Asp 277 of PP1 are represented by neutral side chains in PP2A. Therefore, the surface electrostatic potential of PP2A is significantly less negative than that of PP1, consistent with the observation that PP2A dephosphorylates the acidic protein phospho-casein, while PP1 does not. Differences in substrate, subunit, and inhibitor recognition by PP2A may be rationalized on the basis of such key variations in surface functionalization and local topology.

There is evidence that the β12-β13 loop observed at the intersection of the surface grooves and proximal to the active site in the X-ray structures of PP1 (residues 262–290 of PP1) may contribute to inhibitor specificity. Replacement of the PP1 274–277 GEFD sequence with the corresponding residues of PP2A (YRCG) confers enhanced sensitivity of a mutant enzyme to inhibition by OA (Zhang and Lee 1997). Specific residues that may be important in this regard are Glu 275, Phe 276, Cys 273, and Tyr 272. Glu 275 of PP1 is represented by an Arg in the corresponding position in PP2A. Mutation of Phe 276 of PP1 to Cys causes a 40-fold increase in OA affinity but a slight decrease in the affinity to inhibitor 2, microcystin, nodularin, and calyculin A (Zhang et al. 1994, 1996). Similarly, mutation of Cys 269 of PP2A (corresponding to Phe 276 of $PP1_{\gamma 1}$) to Gly results in decreased sensitivity towards OA (Shima et al. 1994). Tyr 272, which interacts with microcystin, has been shown via mutagenesis experiments to be important for toxin sensitivity

(Zhang et al. 1996). Cys 127 and Cys 202, which occupy the lipophilic groove leading to the active site in PP1, are absent from PP2A. Combined X-ray, mutagenesis, and molecular-modeling (Gauss et al. 1997; Frydrychowski et al. 1998; Forsyth et al. 1997a; Bagu et al. 1997; Lindvall et al. 1997) studies suggest that small molecule inhibitors interact with a number of active site residues, including residues in the β12-β13 loop. In summary, while mutagenesis and crystallographic studies have provided detailed information regarding the plausible modes of regulation of PP1 and PP2B, much less is known about the structure and function of PP2A. This is due largely to the limitations in obtaining active recombinant enzyme and X-ray crystallographic data for PP2A.

3.4 The Okadaic Acid Class of Phosphatase Inhibitors

It is reasonable to assume that each of the structurally diverse OA-type inhibitors must be able to present their common receptor with a similar three-dimensional array of functionally equivalent structural moieties when bound. Reliable identification of the pharmacophore will require knowledge of both the ligand conformations and their essential structural moieties. OA is the most potent inhibitor of PP2A, and considerable structure-activity information exists; therefore, it is uniquely suited as the starting point for defining the structural basis for selective phosphatase inhibition. Previous assays of a variety of naturally occurring and semi-synthetic analogs of OA have provided data that indicate the importance of a number of OA's structural features for activity, and they are consistent with the hypothesis that the biologically active conformation of OA is similar to its solution (Norte et al. 1991; Matsumori et al. 1995; Uemura and Hirata 1988), structure and the solid state (Tachibana et al. 1981) structure of the ester (Fig. 2) (Takai et al. 1992, 1995; Nishiwaki et al. 1990). Many of the data have come from studies by Takai et al. (Takai et al. 1992, 1995) and Nishiwaki et al. (1990), who assayed the inhibitory effect of OA and several related compounds on purified rabbit skeletal muscle PP2A and a partially purified phosphatase preparation from mouse brain, respectively (Table 1).

The absence of X-ray data for OA-phosphatase complexes and limitations in the variety of structural analogs available from natural or

Table 1. PP2A inhibition data for okadaic acid and analogs (Takai et al. 1992, 1995; Nishiwaki et al. 1990)

Inhibitor tested	K_i	% inhibition (1 μg/ml)
Okadaic acid	30	100
35-Methyl-OA (DTX_1)	19	100
OA –9,10-episulfide	47	100
7-Deoxy-OA	69	–
14,15-Dihydro-OA	315	–
2-Deoxy-OA	899	–
7-*O*-Docosahexaenoyl-OA	$>10^5$	64
7-*O*-Palmitoyl-OA	$>10^5$	40
7-*O*-Palmitoyl-DTX_1	$>10^5$	–
Methyl okadaate	$>10^5$	0
2-Oxo-decarboxyl okaOA	$>10^5$	4.8
OA tetramethyl ether	–	2.2
OA –9,10-episulfide methyl ester	–	0
OA-glycine amide	–	51
1-Decarboxyl-1-amino-OA	–	17
1-Decarboxyl-1-hydroxyl-OA	–	41
Nor-okadanone	–	4.8
Nor-okadanol	–	14
C1-C14 fragment of OA	–	99
C15-C38 fragment of OA	–	99

synthetic sources have prevented the definitive identification of the bound conformation of OA and the importance of all of the natural product's various functionalities. Thus, neither a systematic examination of essential structural features nor positive identification of relevant conformations has been made. In principle, the cyclic conformational hypothesis is testable using synthetic conformationally constrained ligands based upon the OA architecture. Combined with a systematic empirical correlation of the effects of functional group modifications on phosphatase inhibition, this approach may help to define the OA pharmacophore. Such information would be of great value for the de novo design of selective small molecule modulators of phosphatase activity. *Herein may lie the greatest biotherapeutic potential of OA.*

3.5 Total Synthesis of Okadaic Acid

A flexible and efficient total synthesis of OA was designed that would provide access to the natural product and rationally designed analogs, in advance of the detailed structural information provided by the X-ray structures of PP1 that were published in 1995. This de novo entry into the OA architecture was specifically designed to overcome the limitations inherent in the previous total synthesis (Isobe et al. 1986) and reliance upon semi-synthetic or naturally occurring compounds. The original synthesis of OA by Isobe and co-workers (1986) spanned a total of 106 steps, with 54 steps in the longest linear sequence, and provided an overall yield of approximately 0.01%. Further, the necessity of deferring installation of several sensitive functional groups until after coupling of major fragments limited the potential of the synthesis to provide efficiently a variety of designed analogs. Therefore, an alternative assembly was designed with the goals of reducing the total number of linear steps by one half and increasing the overall yield by two orders of magnitude. This was initially accomplished in 1996 (Forsyth et al. 1997a), and substantial improvements have been made since then. In addition to offering an abbreviated entry into the OA system, this synthetic work is notable for its adaptability to construct designed structural variants. At the end of 1998, Prof. Steven Ley reported a third total synthesis of OA (Ley et al. 1998). Notably, nearly all of the key transformations in their synthesis highlighted the application of modern synthetic methods developed in the Ley group in recent years. Publication of further details of this synthesis are awaited.

3.5.1 Overall Strategy

Our strategy for a facile assembly of OA relied upon several key elements. Chief among these was to incorporate a maximal degree of functionalization into each advanced synthetic intermediate so as to minimize the extent of post-coupling transformations required to complete the synthesis. It was anticipated that the (19*R*)-configuration of the central C19-C23 spiroketal could be established near the end of the synthesis by an acid-triggered spiroketalization of a C16 hydroxyl upon a masked ketone at C19. Thus, the first retrosynthetic disconnection was

Scheme 1. Major retrosynthetic disconnections of okadaic acid

to open the C16-C19 tetrahydrofuran of OA to yield the γ-hydroxy mixed methyl acetal **2** (Scheme 1). Only acetal and benzyl ether protecting groups were chosen to differentiate **2** from OA. Thus, the end-game strategy relied upon a stereoselective spiroketalization of **2**, followed by final hydroxyl and carboxylate deblocking.

The *trans*-C14-C15 alkene of **2** was to be obtained by coupling **3** (Urbanek et al. 1998) and **4** (Sabes et al. 1998) under conditions sufficiently mild to allow the C1 carboxylate to be incorporated in **4**. The expected (*E*)-enone product would then serve as a substrate for diastereoselective ketone reduction to provide the (16*R*)-configuration of the natural product in **2**. Because the C16 carbonyl of **5** would be in an acyclic region of the carbon skeleton and insulated from the nearest stereogenic center by several carbons, reagent-based control would be required. Subsequent intramolecular ketalization under equilibrating conditions was then expected to provide predominately the required (19*R*)-configuration in the 1,6-dioxaspiro[4.5]decane ring system of **6**. The chair-chair conformation of the C19-C26 *trans*-dioxadecalin sys-

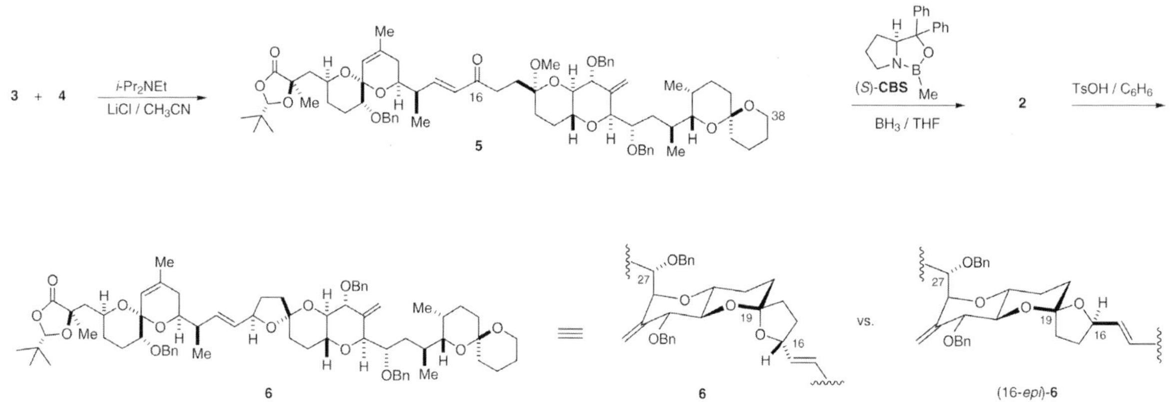

Scheme 2. Late-stage formation of the C16-C19 spiroketal of okadaic acid

tem, in concert with the anomeric stabilization afforded by an axial ketal oxygen at C19, should favor the formation of **2** over the alternative (19*S*)-epimer.

Dissection of intermediate **3** at the C27-C28 carbon-carbon bond and retrosynthetic simplification of the β-ketophosphonate moiety yields the C27 aldehyde **7** and C28 bromide **8** (Scheme 2). Ideally, a direct coupling of aldehyde **7** with an unstablized primary carbanion derived from **8** would generate the C27 carbinol with a useful degree of stereoselectivity and without the previous necessity of post-coupling removal of a carbanion stabilizing group (Isobe et al. 1986, 1987). The β,γ-alkenyl aldehyde **7** was expected to be susceptible to base-induced enolization-conjugation to give the corresponding α,β-unsaturated aldehyde (Ichikawa et al. 1984, 1987b). Nevertheless, we anticipated that, if successful, the convergency of adding a primary organometallic species derived from **8** directly to aldehyde **7** to generate the C27 carbinol would provide a substantial gain in synthetic efficiency. Ideally, only protection of the C27 hydroxyl and elaboration of the C16 terminus to the ketophosphonate would then be required to complete the assembly of **3**. The synthesis of OA thus began with the preparation of aldehyde **7** and bromide **8**, and an examination of their direct and convergent coupling.

3.5.2 Preparation of a C15-C38 β-Ketophosphonate (3)

3.5.2.1 Synthesis of the C16-C27 Fragment (**7**)

Examination of the central C22-C26 ring of OA immediately suggested that a d-altropyranoside could provide much of the stereochemistry (C23, C24, and C26) and functionality present in this portion of the natural product (Scheme 3). Conversion of D-altropyranose **9** into intermediate **7** would entail methylenation at C25, α-selective *C*-glycosidation to form the C21-C22 bond, annulation to construct the functionalized C19-C23 oxane ring, and final functional group transformation to provide the C27 aldehyde. While commercially available, the high costs of D-altrose and its derivatives precluded the selection of any of these sugars as an actual starting point for the synthesis of **4**. However, it was recognized that inversion of both the C2 and C3 configurations of inexpensive D-glucose would provide D-altrohexose with the correct

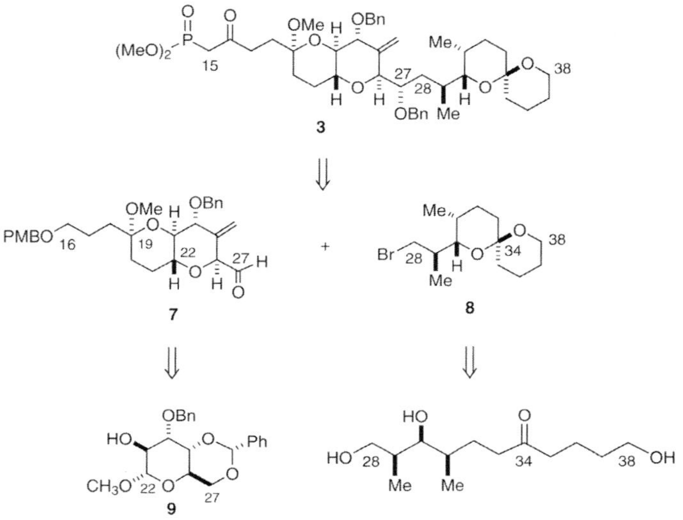

Scheme 3. Retrosynthetic analysis of the C15-C38 portion of okadaic acid

stereochemistry for C23 and C24, respectively. Therefore, a two-step protocol based upon conversion of 4,6-di-*O*-benzylidene methyl-α-d-glucopyranoside into the corresponding 2,3-anhydromannopyranoside (Hicks and Fraser-Reid 1974), followed by regioselective opening of the oxirane at C3 with sodium benzoxide was adopted to provide the requisite altropyranoside (Kunz and Weissmueller 1984).

The one-pot procedure of Bennek and Gray (1987) for *C*-glycosidation of methyl glycosides with allyl trimethylsilane provided a convenient and efficient method for installation of carbons 19–21. As a prelude, acidic methanolysis removed the benzylidene group from **9** (Scheme 4). The resultant triol **10** was treated with *N*,*O*-bis-trimethylsilyltrifluoroacetamide to generate the tris-TMS ether derivative. Without isolation, the tris-TMS ether was treated with trimethylsilyltrifluoromethanesulfonate and allyltrimethylsilane to produce, after aqueous work up, propenylated product **11** as the major component of an approximately 10:1 mixture of α:β epimers. Conversion of crude **11** into the corresponding anisylidene derivative yielded the C19-C27 intermediate **12**. Before carbons C16-C18 were installed, the free C23 hy-

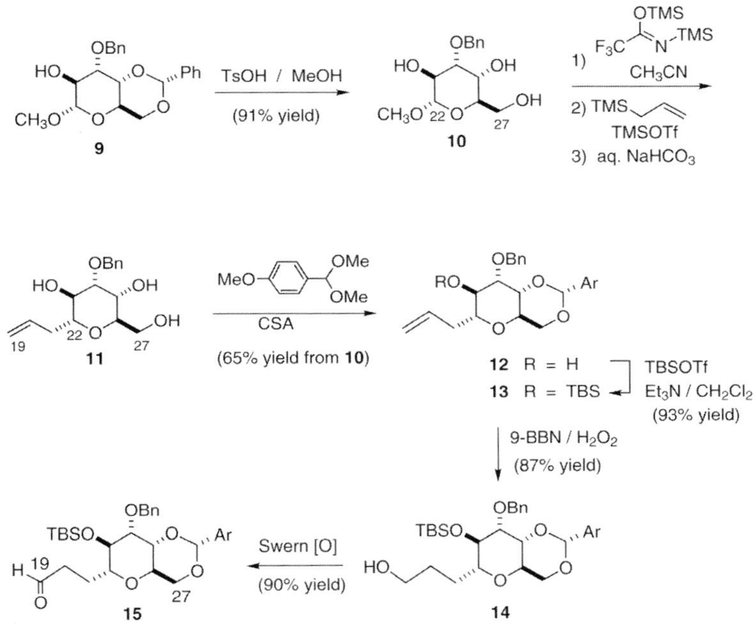

Scheme 4. Synthesis of the C19-C27 okadaic acid intermediate from altropyranoside **9**

droxyl group of **12** was capped as a silyl ether to give **13**. A routine hydroboration-oxidation sequence then converted alkene **13** into aldehyde **15**.

The remaining three carbons of the side chain were derived from 1,3-propane diol via bromide **18** (Scheme 5). Conversion of **18** into the corresponding organolithium followed by addition to aldehyde **15** gave a 1:1 diastereomeric mixture of C19 carbinols **19**. Treatment of **19** with Dess-Martin periodinane (Dess and Martin 1991) provided **20** with C19 at the ketone oxidation state required for OA. The C19 carbon of **20** was to remain at this oxidation state throughout the duration of the synthesis. However, this necessitated masking the carbonyl from nucleophilic attack or premature spiroketalization. A mixed methyl-ketal seemed an ideal choice for this. In addition to providing requisite protection of the

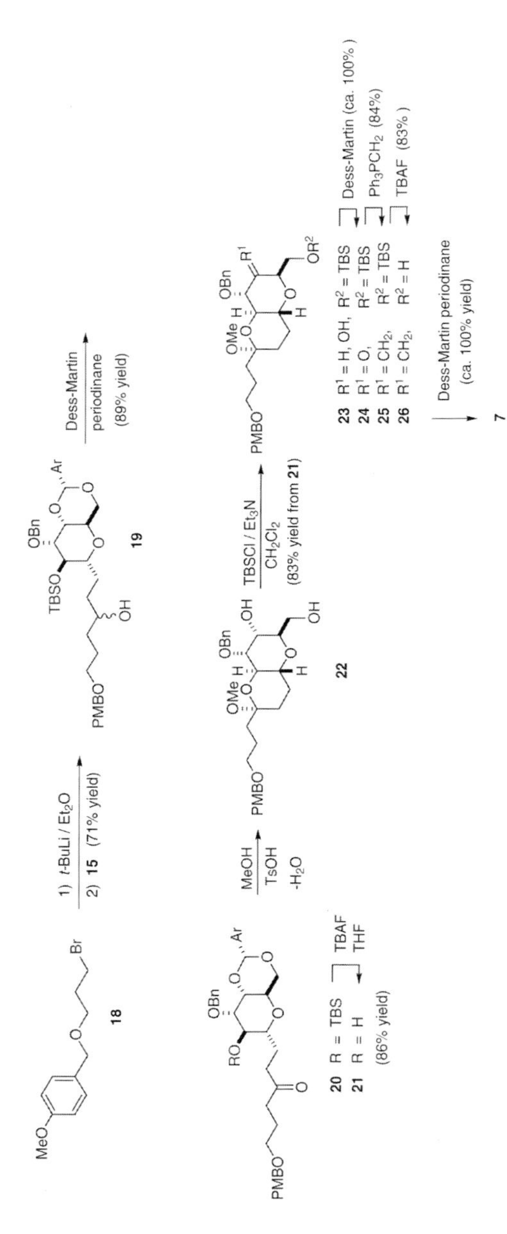

Scheme 5. Completion of the C16-C27 okadaic acid intermediate **7**

ketone, intramolecular mixed ketal formation was expected to provide a rigid and stable *trans*-dioxadecalin system with a well-defined single anomeric configuration. Further, a mixed methyl-acetal would be expected to participate readily in an acid-triggered C16-C23 spiroketalization at the appropriate juncture late in the total synthesis. The targeted mixed methyl-acetal (**22**) was obtained in a two-step process involving cleavage of the silyl ether of **20** by treatment with TBAF, followed by ketalization of the resultant keto-alcohol **21** in acidic methanol.

In the course of converting keto-alcohol **21** into methyl-acetal **22**, the central pyran ring is allowed to undergo a chair-chair ring flip concomitant with loss of the anisylidene group. The chair-chair conformation of **22** is buttressed by anomeric stabilization at C19 and equatorial deployment of the C19 alkyl side chain and C24 benzyloxy group, but it places the C27 hydroxymethyl substituent axial with respect to the *trans*-dioxadecalin system. Thus, enolization of a derived C27 aldehyde may cause epimerization or endocyclic migration of the alkene into conjugation.

Selective silylation of the primary alcohol of **22** yielded secondary alcohol **23** (Scheme 5). Treatment of **23** with Dess-Martin periodinane gave ketone **24**, which was methylenated according to Miljkovic and Glisin (1975). TBAF-induced removal of the silyl group from **25** provided primary alcohol **26**, which was converted into the corresponding aldehyde **7** with Dess-Martin periodinane. The β-methylenated aldehyde **7** was quite prone to isomerization to the α,β-conjugated carboxaldehyde. It was therefore best to prepare **7** from **26** immediately prior to use. Aldehyde **7** could thus prepared in 14–16 steps from methyl-3-*O*-benzyl-α-D-altropyranoside **10**.

3.5.2.2 Synthesis of the C28-C38 Fragment (**8**)

The C28-C38 domain of OA contains four stereogenic centers, including the C34 spiroketal carbon. The two chair perhydropyran rings of the C30-C38 spiroketal benefit from mutual anomeric stabilization, while the bulkier C29 substituent adopts an equatorial position in preference to the axially disposed C32 methyl branch. Therefore, the natural product configuration at C34 was expected to result preferentially from simple spiroketalization-thermodynamic equilibration of the corresponding δ,δ'-dihydroxy ketone (Scheme 6) (Tsuboi et al. 1997; Ichikawa et al. 1987a). In contrast to the original synthesis of OA where

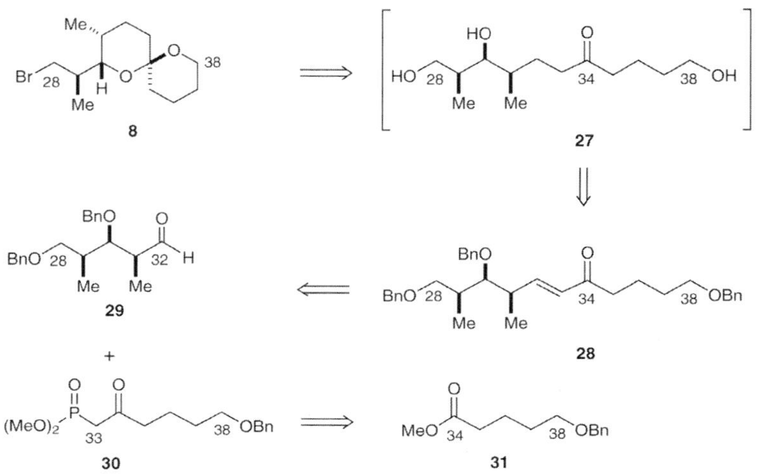

Scheme 6. Retrosynthesis of the C28-C38 domain of okadaic acid

a phenylsulfonyl group was installed at C28 to stabilize a carbanion for C27-C28 bond formation (Isobe et al. 1986), an unstabilized primary carbanion at C28 was planned here to simplify the overall synthetic sequence. Thus, the C28 bromide **8** was selected as a precursor to a range of potential organometallic species that could be used to explore formation of the C27-C28 bond. Because the bromide functionality could be elaborated from the corresponding primary alcohol, acid-induced spiroketalization-equilibration of an acyclic tri-hydroxy ketone (**27**) was planned. Keto-triol **27** would derive from the corresponding tri-benzyloxy α,β-unsaturated ketone (**28**), which, in turn, could be obtained by Horner-Emmons coupling of a C33-C38 β-ketophosphonate (**30**) and a C28-C32 aldehyde (**29**). All of the stereochemistry of **8** would derive from the latter. This general approach could also provide synthetic access to the related dinophysistoxin natural products (Hu et al. 1992; Sakai and Rinehart 1995) by simply varying the methyl substitution at C31 and C35 in the aldehyde and ketophosphonate.

Carbons 34–38 were derived conveniently from methyl ester **31** (Scheme 6). Addition of the lithium anion of dimethyl methyl phosphonate to **31** provided β-ketophosphonate **30**. The stereochemical triad

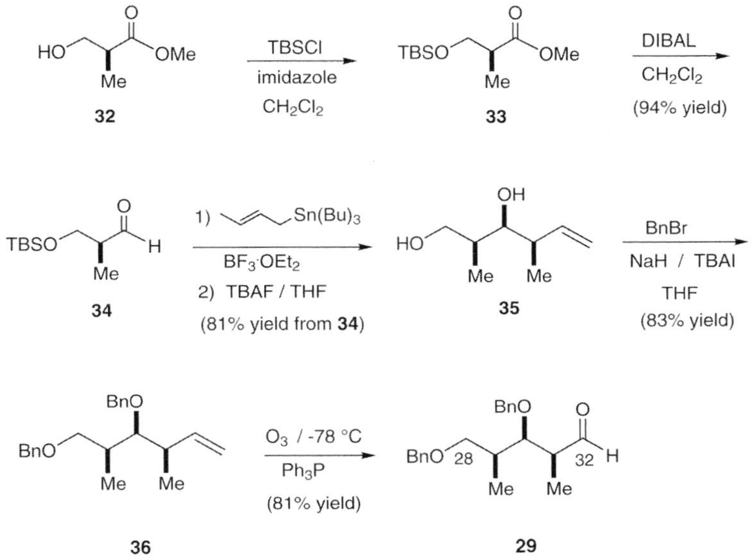

Scheme 7. Synthesis of the C28-C32 aldehyde intermediate for okadaic acid

spanning C29–C31 was established in a direct fashion using Keck's crotylstannane methodology (Scheme 7). Silylation of commercially available methyl (*S*)-3-hydroxy-2-methylpropionate (**32**), followed by reduction of the resulting ester **33** with DIBAL, gave aldehyde **34**. Keck had shown that the use of a silyl-protecting group in aldehydes similar to **34** was required for high syn-syn diastereoselectivity in $BF_3{\cdot}OEt_2$-mediated crotylstannane additions (Keck and Abbott 1984). However, uniform protection of the hydroxyl groups as benzyl ethers was desired here to facilitate a subsequent reduction-spiroketalization sequence. Thus, the crude product of crotyl addition to **34** was desilylated with TBAF to afford diol **35** in 81% overall yield (ca. 18:1 syn-syn:syn-anti) from **34**. After **35** had been converted into dibenzyl ether **36**, ozonolytic cleavage of the alkene and reductive workup gave aldehyde **29**.

Coupling of **29** and **30** under modified Horner-Wadsworth-Emmons conditions provided *trans*-enone **28** without detectable epimerization of the α-aldehyde stereogenic center of **29** (Scheme 8). A two-step conver-

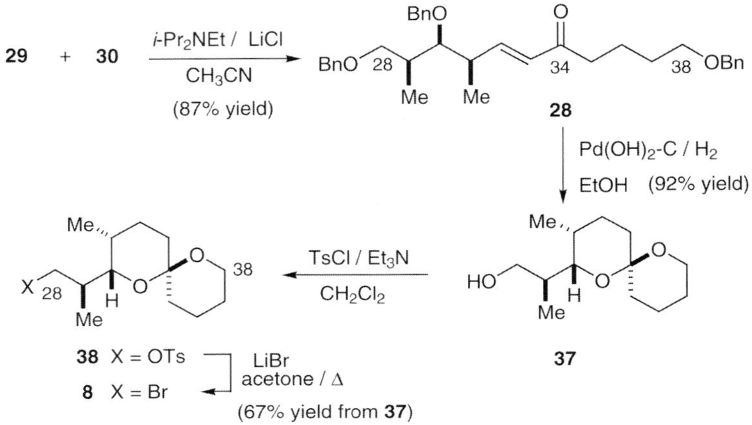

Scheme 8. Completion of the C28-C38 lipophilic domain intermediate for okadaic acid

sion of the tribenzyloxy enone **28** into the spiroketal **37** was planned. This was to involve reductive cleavage of the benzyl ethers and concomitant saturation of the alkene under Pd-catalyzed hydrogenation conditions, followed by a discrete acid-induced dehydration step. However, vigorous stirring of enone **28** with 20% palladium hydroxide on carbon in absolute ethanol under 1 atm of H_2 not only reduced the alkene and cleaved the three benzyl ethers but also induced spiroketalization to give **37**. This in situ reduction-spiroketalization sequence gave the expected product of thermodynamic spiroketalization (**37**) in greater than 90% yield. It is likely that the commercially obtained $Pd(OH)_2$ on carbon used in this sequence provided a source of acid that promoted in situ thermodynamic spiroketalization.

Having successfully constructed the fully functionalized C28-C38 domain of OA in eight steps from methyl (*S*)-3-hydroxy-2-methylpropionate, we used two additional steps to convert the primary alcohol **37** into the corresponding bromide **8**. With multi-gram quantities of **8** thus available, direct C27-C28 bond formation could be thoroughly examined.

3.5.2.3 C27-C28 Bond Formation and β-Ketophosphonate Installation

The direct addition of an organometallic species derived from bromide **8** to aldehyde **7** would provide the fully functionalized C16-C38 portion of OA. Intramolecular metal chelation involving the C27 aldehyde and the α-pyranyl oxygen would be expected to preferentially reveal the pro-*S* face of the aldehyde to nucleophilic species (Fig. 3). Hence, the (27*S*) configuration of the natural product might result from a chelation-controlled addition of a C28 nucleophile to the C27 aldehyde.

However, for any nucleophilic species to be useful for C27-C28 bond formation, enolization of the C27 aldehyde would have to be avoided. Although simple alkyl Grignard and higher-order cuprate nucleophiles could be added successfully to **7**, attempts to convert **8** into such species were unsuccessful. The organolithium derived from **8** by metal-halogen exchange with *tert*-butyllithium was itself short-lived and too basic to undergo useful 1,2-addition to **7**. Instead, an α,β-unsaturated aldehyde and numerous other by-products were formed. However, the organocerium species (Corey and Ha 1988), derived from lithium-halogen exchange of **8** followed by addition of a THF suspension of $CeCl_3$ at

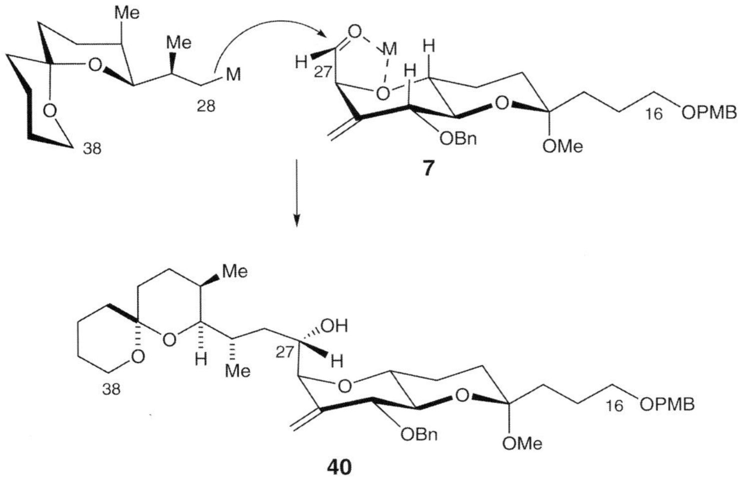

Fig. 3. Anticipated chelation-controlled formation of the C27-C28 bond of okadaic acid

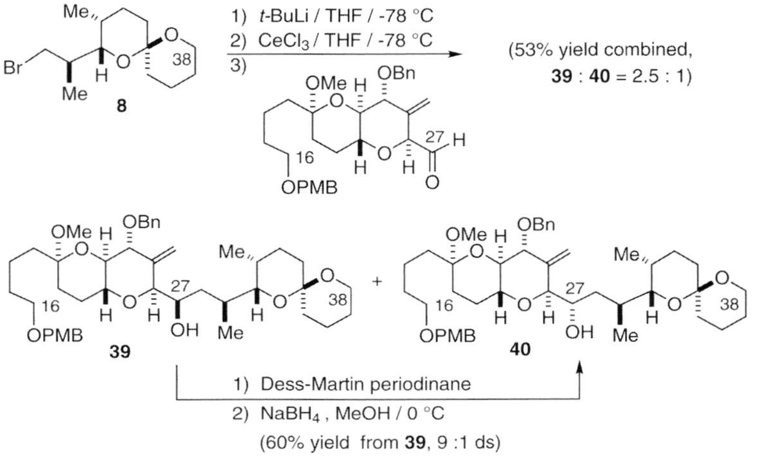

1) t-BuLi / THF / -78 °C
2) $CeCl_3$ / THF / -78 °C
3) [aldehyde 7]

(53% yield combined, **39** : **40** = 2.5 : 1)

1) Dess-Martin periodinane
2) $NaBH_4$, MeOH / 0 °C
(60% yield from **39**, 9 :1 ds)

Scheme 9. Convergent formation of the C27-C28 bond for okadaic acid

–78°C, added smoothly to aldehyde **7** to give the C27 carbinols **39** and **40** (2.5:1 diastereomeric ratio, respectively) in moderate yield (Scheme 9). Inversion of the C27 carbinol configuration of **39** via oxidation-reduction not only supported the stereochemical assignment but also provided a facile preparative route to overcome the unfavorable coupling diastereoselectivity. Although modest in both yield and diastereoselectivity, this coupling-inversion protocol represents a reliable and useful method for the direct joining of OA intermediates **7** and **8**. Unfortunately, the use of an organocerium species for the formation of the C27-C28 bond did not provide the desired chelation-control outcome to set the C27 stereogenic center.

With the central and C38 terminal fragments of OA joined, preparation for the attachment of the C1-C14 domain required only elaboration of the C16 terminus into a β-ketophosphonate. This was preceded by conversion of the C27 hydroxyl of **40** into the corresponding benzyl ether **41** (Scheme 10). Oxidative removal of the *p*-methoxy benzyl group using DDQ (Nakajimi et al. 1988) liberated the C16 hydroxyl to give **42**. Unless the DDQ reaction mixture and the alcohol product were protected against low pH, spontaneous formation of the C16-C23 spi-

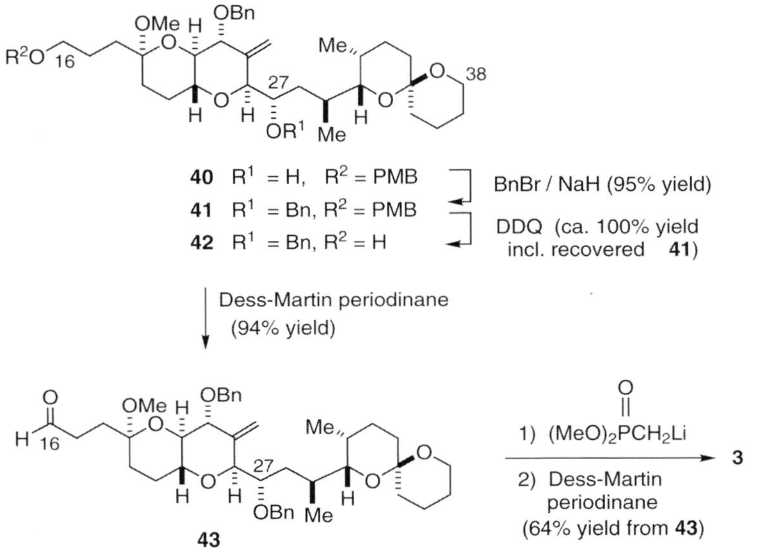

40 R^1 = H, R^2 = PMB
41 R^1 = Bn, R^2 = PMB
42 R^1 = Bn, R^2 = H

BnBr / NaH (95% yield)
DDQ (ca. 100% yield incl. recovered **41**)

Dess-Martin periodinane (94% yield)

43

1) $(MeO)_2P(O)CH_2Li$
2) Dess-Martin periodinane
(64% yield from **43**)
→ **3**

Scheme 10. Formation of the C15-C38 ketophosphonate intermediate (**3**) for okadaic acid

roketal occurred. Oxidation of **42** with Dess-Martin periodinane buffered with $NaHCO_3$ gave the aldehyde **43** uneventfully. Thereafter, addition of lithiomethylphosphonate, followed by a second oxidation with Dess-Martin periodinane, delivered the β-ketophosphonate **3**. This route provides the C15-C38 OA intermediate **3** in approximately 19 linear steps from methyl 3-*O*-benzyl-α-D-altropyranoside **9**. Only the complementary C1-C14 building block and five additional steps were required to advance intermediate **3** into OA.

3.5.3 Design of the C1-C14 Domain

The strategy for the preparation of **4** is outlined in Scheme 11. The intact C1-C2 α-hydroxyl, α-methyl carboxylate moiety would be installed by diastereoselective alkylation of Seebach's lactate pivalidene **44** with a C3-C14 spiroketal-bearing fragment (**45**). The electrophilic C3 carbon

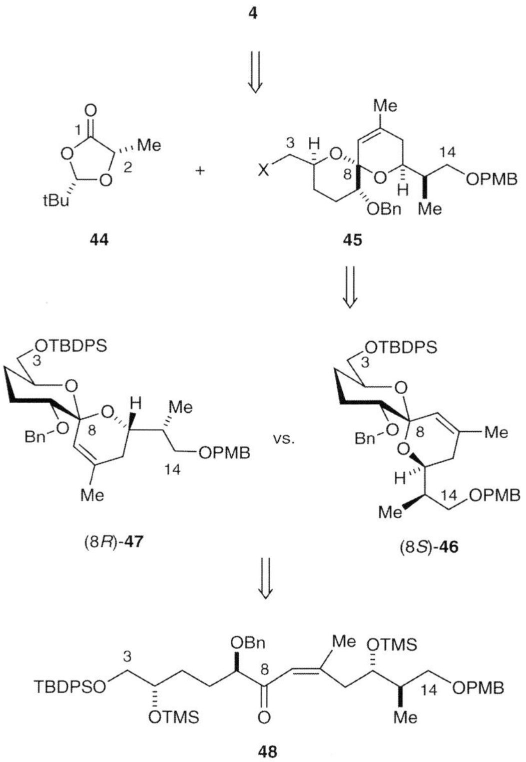

Scheme 11. Retrosynthesis of the C1-C14 intermediate (**4**) of okadaic acid

of **45** would bear an α-stereogenic center and be equatorially disposed on the C4-C8 oxane ring of the 1,7-dioxaspiro[5.5]undec-4-ene system. Hence, potential complications arising from steric hindrance and double diastereoselective matching (Masamune et al. 1985) could be anticipated. However, the expediency of incorporating the fully functionalized, stereochemically correct and appropriately masked C1-C2 moiety in such a direct fashion would avoid reliance upon alternative multi-step sequences (Isobe et al. 1986, 1987) to incorporate the biologically essential (Nishiwaki et al. 1990) α-hydroxyl carboxylate moiety. Once installed, the protected α-hydroxyl carboxylate moiety would be subjected to only a few mild transformations en route to OA. Whereas the pivalidene group would be expected to resist premature cleavage during late-stage C16-C23 spiroketalization catalyzed by mildly acidic conditions, it could be removed readily by saponification at the penultimate step in the total synthesis.

A hydroxyl group at C3 could act as a versatile precursor to a variety of potential leaving groups in **45** for lactate addition. Differential protection of the latent hydroxyl at C14 from those at C2, C3, and C7 would facilitate final installation of the C14 aldehyde. Hence, the (8*S*)-spiroketal **46** was targeted as a key intermediate in the preparation of **4**. Thermodynamic equilibration would be relied upon to generate preferentially the (8*S*) configuration of the natural product in **46** upon acid catalyzed dehydration of δ,δ'-dihydroxy enone **48**, or its equivalent (Deslongchamps et al. 1981). In the context of the original synthesis of OA, Isobe had reported that the yield of the desired and anticipated major diastereomer obtained via acid catalyzed spiroketalization to form a similar fragment was limited to 30%–40% (Isobe et al. 1985a). This indicated a need to explore potential improvements for the formation of the targeted OA intermediate **46**.

A reliable route to the (*Z*)-enone **48** would involve methyl cuprate addition upon the corresponding ynone (**49**) (Isobe et al. 1985a; Corey and Katzenellenbogen 1969), which, in turn, could be obtained from the monoaddition of alkyne **50** to lactone **13** (Scheme 12). Subsequent acid catalyzed spiroketalization would complete the synthesis of **46**. The projected synthesis of the C1-C14 domain of OA was thus focused on three stages: the formation of (*Z*)-enone **48** from lactone and alkyne precursors, thermodynamic spiroketalization to provide **46**, and installation of the α-hydroxyl, α-methyl carboxylate moiety using lactate **44**.

3.5.4 Synthesis of the C1-C14 Aldehyde (4)

Lactone **13**, representing carbons 3–8 of OA, could be obtained from the known (Isobe et al. 1985b) isopropyl glycoside used in Isobe's total synthesis (Sabes et al. 1998), or more directly from glycidol epoxide (Dounay and Forsyth 1999). Similarly, carbons 9–14 of OA (**50**) could be derived from the known (Guindon et al. 1994) pentylidene-protected

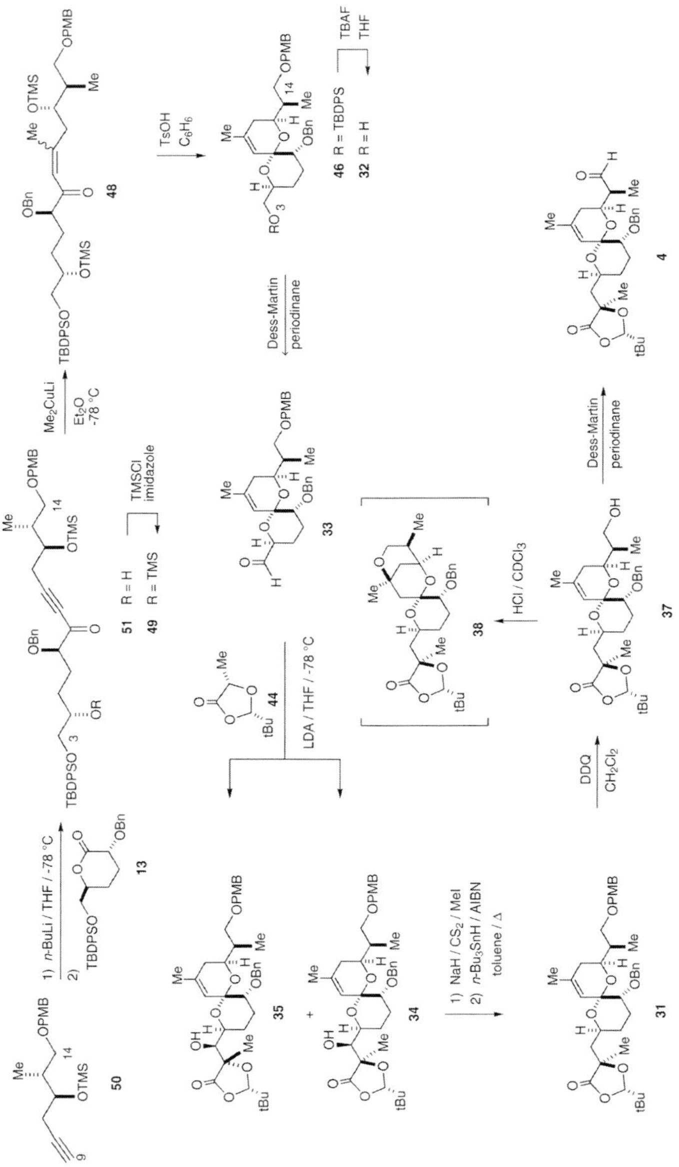

Scheme 12. Formation of the C1-C14 intermediate for okadaic acid

triol (Sabes et al. 1998), or in fewer steps from *trans*-hex-2-en-5-yn-1-ol (Dounay et al. 1999). Addition of the C9-C14 acetylide to lactone **13** followed by conjugate addition of dimethylcuprate to the resultant ynone was originally expected to give the corresponding (*Z*)-enone stereoselectively (Isobe et al. 1985a; Corey and Katzenellenbogen 1969). In fact, addition of the lithium acetylide derived from **50** to lactone **13** gave ynone **51**. Without purification of **51**, the newly formed hydroxyl at C4 was silylated to provide bis-TMS ether. Conjugate addition of dimethylcuprate then gave the β-methylenone **48** in nearly quantitative yield. Spiroketalization was induced by treatment of the enone with TsOH·H_2O in benzene at room temperature to give only modest yields of the spiroketal **46**.

It was not originally clear which among the alternative potential reaction pathways were responsible for the limited yields of **46**. Subsequent study, however, indicated that it was the presence of the C7 substituent that was hindering the key spiroketalization step. Hence, spiroketalization of a 7-deoxy intermediate proceeded in 76% yield, versus the 30%–40% obtained with the 7-benzyloxy intermediate (Dounay and Forsyth 1999).

Completion of the synthesis of the C1-C14 intermediate **4** required elaboration of the 1,7-dioxaspiro[5.5]undec-4-ene **46** to incorporate the C1-C2 α-hydroxyl, α-methyl carboxylate, and C14 aldehyde moieties. Cleavage of the C3-silyl ether of **46** with TBAF yielded the primary alcohol, which served as a precursor to a variety of potential leaving groups. Numerous attempts to displace halide and sulfonate groups at C3 with enolates arising from lactate pivalidene **44** were unsuccessful. Instead of the desired alkylation product, unreacted species were generally recovered. This may well reflect the crowded steric environment about the C3 carbon.

In contrast, the aldehyde (**33**) obtained by treatment of **32** with Dess-Martin periodinane (Dess and Martin 1991) participated readily in carbon-carbon bond formation. Addition of **33** to an excess of the pre-formed lithium enolate derived from (*S*)-lactate pivalidene **44** gave alkylation products **34** and **35** in good combined yield. However, the facial selectivity of lactate alkylation was much lower (2.2:1.0) than originally anticipated. That **34** and **35** were diastereomeric at C2 was confirmed by removal of the C3 stereogenic center via Barton deoxygenation.

With the installation of C1-C2 successfully completed, only conversion of C14 into an aldehyde remained to complete the synthesis of **4**. This seemingly straightforward task was accompanied by an unexpected acid-catalyzed isomerization. Oxidative cleavage of the *p*-methoxybenzyl ether of **31** using DDQ predictably gave primary alcohol **37**, which could be oxidized efficiently to aldehyde **4** with $NaHCO_3$-buffered Dess-Martin periodinane. However, exposure of **37** to acidic chloroform revealed the propensity of the δ-alkenyl alcohol to undergo intramolecular addition of the alcohol across the alkene, irretrievably generating tricyclic ether **38**. This presumably involves activation of the latent Michael acceptor by protonation of a spiroketal oxygen followed by 1,4-addition of the proximal C14 hydroxyl group to C10 of the allylically stabilized oxonium species. Formation of the trioxatricycle **38** was prevented by simply avoiding exposure of **31** to acidic conditions. Hence, the C1-C14 intermediate **4** could be prepared in 11 steps and approximately 20% overall yield from lactone **13**.

3.5.5 Completion of the Total Synthesis of Okadaic Acid

The complete carbon skeleton of OA was assembled by joining aldehyde **4** (Sabes et al. 1998) with ketophosphonate **3** (Urbanek et al. 1998). The use of a mildly basic Horner-Wadsworth-Emmons reaction allowed the fully functionalized coupling partners to be joined reliably to give (*E*)-enone **5**. Only assembly of the central C16-C23 spiroketal and deprotection remained as the final synthetic tasks. Diastereoselective reduction of the C16 carbonyl was a necessary prelude to spiroketal formation. Corey's CBS reagent system (Corey et al. 1987) seemed ideal for the diastereoselective reduction of enone **5**, given that the ketone was flanked by unbranched aliphatic carbons on one side and an (*E*)-alkene on the other. Thus, regio- and stereoselective reduction of **5** was accomplished using the (*S*)-CBS / BH_3 combination to give predominately the (16*R*)-allylic alcohol **2**. The reduction product mixture was subjected directly to acid catalyzed spiroketalization without rigorous prior purification. This provided spiroketal **6** in 81% yield from **5**. A minor spiroketal stereoisomer was generated along with **6** in this two-step sequence, which arose from the ketone reduction step.

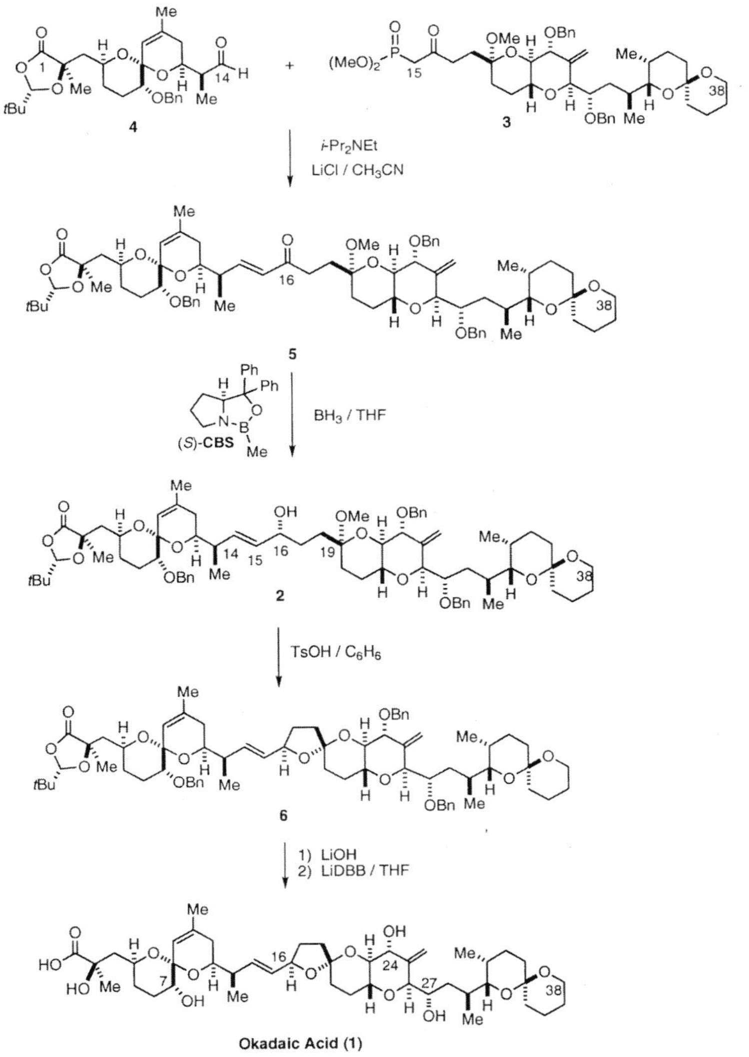

4 + 3

i-Pr_2NEt
LiCl / CH_3CN

5

(*S*)-**CBS**
BH_3 / THF

2

TsOH / C_6H_6

6

1) LiOH
2) LiDBB / THF

Okadaic Acid (1)

Scheme 13. Completion of the total synthesis of okadaic acid

Removal of the carboxylate and hydroxyl protecting groups was required to complete the synthesis of OA. The C1-C2 pivalidene protecting group was easily detached from **6** by mild aqueous base-induced saponification to free both the C2 hydroxyl and C1 carboxylate moieties. Saponification prior to dissolving metal cleavage of the benzyl ethers allowed the carboxylate to be protected from reduction as its carboxylate salt. Benzyl ethers were selected early in the synthetic planning as useful protecting groups for the C7, C24, and C27 hydroxyls. However, reductive cleavage of the benzyl ethers in the final step of the synthesis of OA has been complicated by overreduction using lithium in liquid ammonia (Isobe et al. 1986; Forsyth et al. 1997a; Ley et al. 1998; Ichikawa et al. 1988). In contrast, the use of lithium di-*tert*-butylbiphenylide (Freeman and Hutchinson 1980) (LiDBB) solutions in THF for controlled debenzylation (Ireland and Smith 1988) was far less problematic. Thus, the three benzyl ethers were reductively cleaved without substantial complications from the other potentially sensitive functional groups to deliver OA reproducibly in approximately 70% yield after isolation and purification. Aside from the initial difficulties associated with liberating the three secondary hydroxyl groups of OA, the end-game strategy for a rapid completion of the total synthesis was executed as planned.

3.6 Conclusion: Biotherapeutic Potential of Okadaic Acid and Synthetic Analogs

OA itself has been a tremendously useful tool for implicating the roles of serine/threonine phosphatases in many essential cellular processes. Its greatest biotherapeutic potential, however, may reside in its role as a well-defined scaffold for elucidating the structural requirements for potent *and selective* phosphatase inhibition. Small molecule inhibitors that would be specific for each subclass of phosphatases, as well as their various isoforms, would be invaluable tools for the selective intervention in the cellular and disease processes that the specific enzymes mediate. Towards this end we have developed a relatively flexible and concise entry into the OA architecture. Extension of this work towards developing an empirical understanding of the structural basis of enzyme inhibition has begun by addressing problems that emerged from our original synthetic effort. These included the moderate yields of C8 spiroketalization and C27-C28 bond formation and, to a lesser extent, the potential of over-reduction in the final scission of the benzyl ether protecting groups. These issues have been largely overcome by targeting 7-deoxy-OA and developing an alternative construct of the central C15-C27 intermediate.

The natural product 7-deoxy-okadaic acid has an affinity towards PP1 and PP2A that closely mimics that of OA (Takai et al. 1992, 1995). In addition to being biologically dispensable, the absence of the C7 functionality facilitates both the C8 spiroketalization process (Dounay and Forsyth 1999) and the final debenzylation (Dounay et al. 1999). These observations support the use of OA analogs lacking the C7-hydroxyl moiety in further studies involving OA-sensitive phosphatases.

Although the active sites of PP1, PP2A, and their isoforms are expected to be quite similar to one another in terms of amino acid composition and topology, significant structural differences exist outside of the highly conserved catalytic site. For example, two cystine residues, Cys 127 and Cys 202, line the hydrophobic groove proximal to the active site in PP1, but they are absent in PP2A. Most docking models of OA and PP1 simultaneously place the carboxylate moiety of OA near the bimetallic active site and the C28-C38 lipophilic domain in the proximal lipophilic surface groove of PP1 (Gauss et al. 1997; Frydrychowski et al. 1998; Forsyth et al. 1997a; Bagu et al. 1997; Lindvall et al. 1997). Hence, hydrophobic-hydrophobic interactions may contribute not only to binding affinity but also to specificity. To take advantage of this possibility, we have revised the order of fragment couplings so that variable lipophilic domains may be attached at a late stage to the intact C1-C27 portion (**53**) of 7-deoxy-OA (Scheme 13). This simply required modification of the C24 and C27 hydroxyl protecting groups in the central core intermediate (**56**), so that after coupling with the C1-C14 fragment **55** and C19 spiroketalization, as before, C27 may be selectively converted into the aldehyde **53**. In order to preserve the carboxylate functionality, organochromic-mediated couplings (Kishi 1992) of various unsaturated lipophilic side chains (**54**) are envisioned. This strategy will provide access to chimeric inhibitors (**52**) bearing constant active site binding elements but variable, and perhaps phosphatase subtype-specific, lipophilic domains (Scheme 14). The generation and assay of such non-natural analogs of OA varying in structure of the

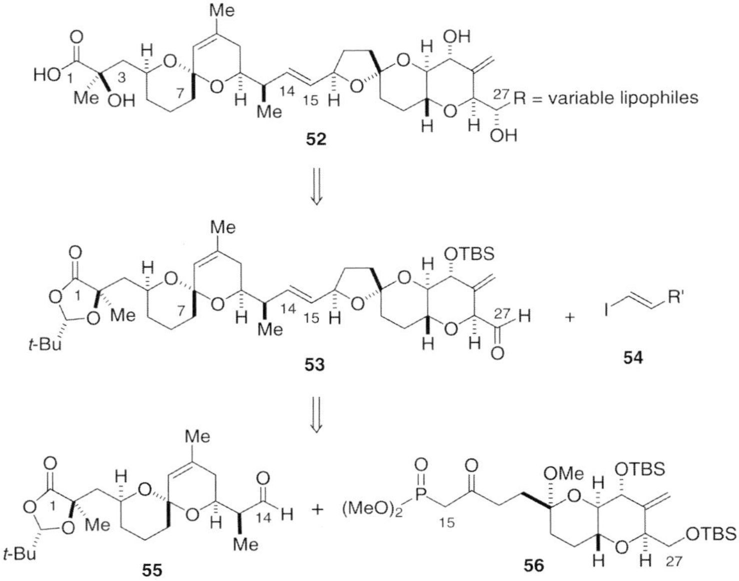

Scheme 14. Strategy for the synthesis of variable lipophilic analogs of okadaic acid

lipophilic domain, as well as in the core functionality, is currently being pursued in our laboratories.

Since it was first isolated from the crude extracts of the black sponge *H. okadai*, okadaic acid has had a profound impact on the fields of natural product, cellular biological, and biomedical research. It has also inspired the development and application of innovative synthetic chemistry, which may ultimately be required to realize the full biotherapeutic potential of okadaic acid.

Acknowledgements. Financial support from the US National Institutes of Health (CA62195), the University of Minnesota, AstraZeneca, and Bristol-Myers Squibb is gratefully acknowledged. We also thank the American Chemical Society, Division of Medicinal Chemistry, for Graduate Fellowship support to ABD sponsored by Parke-Davis.

References

Adler HT, Nallaseth FS, Walter G, Tkachuk DC (1997) HRX leukemic fusion proteins form a heterocomplex with the leukemia-associated protein SET and protein phosphatase 2A. J Biol Chem 272:28407–28414

Aicher TD, Buszek KR, Fang FG, Forsyth CJ, Jung SH, Kishi Y, Scola PM, Spero DM, Yoon SK (1992) Total synthesis of halichondrin B and norhalichondrin B. J Am Chem Soc 114:3162–3164

Ainbinder E, Bergelson S, Pinkus R, Daniel V (1997) Regulatory mechanisms involved in activator-protein-1 (AP-1)-mediated activation of glutathione-S-transferase gene expression by chemical agents. Eur J Biochem 243:49–57

Andjelkovic N, Zolnierowicz S, Van Hoof C, Goris J, Hemmings BA (1996) The catalytic subunit of protein phosphatase 2A associates with the translation termination factor. EMBO J 15:7156–7167

Armstrong CG, Browne GJ, Cohen P, Cohen PTW (1997) PPP1R6, a novel member of the family of glycogen-targetting subunits of protein phosphatase 1. FEBS Lett 418:210–214

Bagu JR, Sykes BD, Craig MM, Holmes CFB (1997) A molecular basis for different interactions of marine toxins with protein phosphatase-1. Molecular models for bound motuporin, microcystins, okadaic acid, and calyculin A. J Biol Chem 272:5087–5097

Bennek JA, Gray GR (1987) An efficient synthesis of anhydroalditols and allyl *C*-glycosides. J Org Chem 52:892–897

Bialojan C, Takai A (1988) Inhibitory effect of a marine-sponge toxin, okadaic acid, on protein phosphatases. Specificity and kinetics. Biochem J 256:283–290

Blumberg PM (1988) Protein kinase C as the receptor for the phorbol ester tumor promoters: sixth Rhodes memorial award lecture. Cancer Res 48:1–8

Borgne A, Meijer L (1996) Sequential dephosphorylation of p34cdc2 on Thr-14 and Tyr-15 at the prophase/metaphase transition. J Biol Chem 271:27847–27854

Brady MJ, Printen JA, Mastick CC, Saltiel AR (1997) Role of protein targeting to glycogen (PTG) in the regulation of protein phosphatase-1 activity. J Biol Chem 272:20198–20204

Cohen P (1989) The structure and regulation of protein phosphatases. Annu Rev Biochem 58:453–508

Cohen P, Cohen PTW (1989) Protein phosphatases come of age. J Biol Chem 264:21435–21438

Cohen P, Holmes CFB, Tsukitani Y (1990) Okadaic acid: a new probe for the study of cellular regulation. Trends Biochem Sci 15:98–102

Corey EJ, Ha D-C (1988) Total synthesis of venustatriol. Tetrahedron Lett 29:3171–3174

Corey EJ, Katzenellenbogen JA (1969) A new stereospecific synthesis of trisubstituted and tetrasubstituted olefins. The conjugate addition of dialkyl-copper-lithium reagents to α,β-acetylenic esters. J Am Chem Soc 91:1851–1852

Corey EJ, Bakshi RK, Shibata S, Chen C-P, Singh VK (1987) A stable and easily prepared catalyst for the enantioselective reduction of ketones. Applications to multistep synthesis. J Am Chem Soc 109:7925–7926

Csortos C, Zolnierowicz S, Bako E, Durbin SD, DePaoli-Roach AA (1996) High complexity in the expression of the B' subunit of protein phosphatase 2A0. Evidence for the existence of at least seven novel isoforms. J Biol Chem 271:2578–2578

Depaoli-Roach A, Park I-K, Cerovsky V, Csortos C, Durbin SD, Kuntz MJ, Sitikov A, Tang PM, Verin A, Zolnierowicz S (1994) Serine/threonine protein phosphatases in the control of cell function. Adv Enzyme Regul 34:199–224

Deslongchamps P, Rowan DR, Pothier N, Sauve T, Saunders JK (1981) 1,7-Dioxaspiro[5.5]undecanes. Can J Chem 59:1105

Dess DB, Martin JC (1991) A useful 12-I-5 triacetoxyperiodinane (the Dess-Martin periodinane) for the selective oxidation of primary or secondary alcohols and a variety of related 12-I-5 species. J Am Chem Soc 113:7277–7287

Dickey RW, Bobzin SC, Faulkner DJ, Bencsath FA, Andrzejewski D (1990) Identification of okadaic acid from a Caribbean dinoflagellate, Prorocentrum concavum. Toxicon 28:371–377

Dohadwala M, da Cruz e silva, EF, Hall FL, Williams RT, Carbonaro-Hall DA, Nairn AC, Greengard P, Berndt N (1994) Phosphorylation and inactivation of protein phosphatase 1 by cyclin-dependent kinases. Proc Natl Acad Sci U S A 91:6048–6412

Dounay AB, Forsyth CJ (1999) Abbreviated synthesis of the C3-C14 (substituted 1,7-dioxaspiro-[5.5]undec-3-ene) system of okadaic acid. Org Lett 1:451–453

Dounay AB, Urbanek RA, Sabes SF, Forsyth CJ (1999) Total synthesis of the natural product 7-deoxy-okadaic acid: a potent inhibitor of protein serine/threonine phosphatases. Angew Chem Int Ed Engl 38:2258–2262

Durfee T, Becherer K, Chen P-L, Yeh S-H, Yang Y, Kilburn AE, Lee W-H, Elledge SJ (1993) The retinoblastoma protein associates with the protein phosphatase type 1 catalytic subunit. Genes Dev 7:555–569

Egloff M-P, Cohen PTW, Reinemer P, Barford D (1995) Crystal structure of the catalytic subunit of human protein phosphatase 1 and its complex with tungstate. J Mol Biol 254:942–959

Egloff M-P, Johnson DF, Moorhead G, Cohen PTW, Cohen P, Barford D (1997) Structural basis for the recognition of regulatory subunits by the catalytic subunit of protein phosphatase 1. EMBO J 16:1876–1887

Floer M, Stock J (1994) Carboxyl methylation of protein phosphatase 2A from Xenopus eggs is stimulated by cAMP and inhibited by okadaic acid. Biochem Biophys Res Commun 198:372–379

Forsyth CJ, Frydrychowski V, Sabes SF, Urbanek RA (1997a) Novel protein phosphatase inhibitors based upon okadaic acid. Abstracts of Papers, 212th American Chemical Society National Meeting, San Francisco, CA, American Chemical Society, Washington, DC; ORGN 108

Forsyth CJ, Sabes SF, Urbanek RA (1997b) An efficient total synthesis of okadaic acid. J Am Chem Soc 119:8381–8382

Freeman PK, Hutchinson LL (1980) Alkyllithium reagents from alkyl halides and lithium radical anions. J Org Chem 45:1924–1930

Frydrychowski VA, Dounay AB, Plummer K, Urbanek RA, Sabes SF, Forsyth CJ (1998) Design, synthesis, and evaluation of protein phosphatase inhibitors related to okadaic acid. Abstracts of Papers, 215th American Chemical Society National Meeting, Boston, MA, American Chemical Society, Washington, DC

Fujiki H (1992) Is the inhibition of protein phosphatase 1 and 2 A activities a general mechanism of tumor promotion in human cancer development? Mol Carcinog 5:91–94

Fujiki H, Suganuma M, Suguri H, Yoshizawa S, Takagi K, Uda N, Wakamatsu K, Yamada K, Murata M (1988) Diarrhetic shellfish toxin, dinophysistoxin-1, is a potent tumor promoter on mouse skin. Jpn J Cancer Res 79:1089–1093

Fujiki H, Suganuma M, Suguri H, Yoshizawa S, Takagi K, Sassa T, Uda N, Wakamatsu K, Yamada K (1989) New tumor promoters from marine sources: the okadaic acid class. Bioact Mol 10:453–460

Gauss C-M, Sheppeck JE Jr, Nairn AC, Chamberlin R (1997) A molecular modeling analysis of the binding interactions between the okadaic acid class of natural product inhibitors and the Ser-Thr phosphatases, PP1 and PP2A. Bioorg Med Chem 5:1751–1773

Goldberg J, Huang H, Kwon Y, Greengard P, Nairn AC, Kuriyan J (1995) Three-dimensional structure of the catalytic subunit of protein serine/threonine phosphatase-1. Nature 376:745–753

Guindon Y, Yoakim C, Gorys V, Ogilvie WW, Delorme D, Renaud J, Robinson G, Lavallee J-F, Slassi A, Jung G, Rancourt J, Durkin K, Liotta D (1994) Stereoselective hydrogen transfer reactions involving acyclic radicals. Tandem substituted tetrahydrofuran formation and stereoselective reduction: synthesis of the C17-C22 subunit of ionomycin. J Org Chem 59:1166–1178

Hecker E (1978) Carcinogenesis – a comprehensive survey, vol 2. Raven, New York, pp 11–48
Helps N, Barker H, Elledge SJ, Cohen PTW (1995) Protein phosphatase 1 interacts with p53BP2, a protein which binds to the tumor suppressor p53. FEBS Lett 377:295–300
Hemmings HCJ, Greengard P, Tung HYL, Cohen P (1984) DARP-32, a dopamine-regulated neuronal phosphoprotein, is a potent inhibitor of protein phosphatase-1. Nature 310:503–505
Hemmings HC jr, Nairn AC, Elliot JI, Greengard P (1990) Synthetic peptide analogs of DARP-32 (M_r32,000 dopamine- and cAMP-regulated phosphoprotein), an inhibitor of protein phosphatase-1. Phorphorylation, dephosphorylation, and inhibitory activity. J Biol Chem 265:20369–20376
Heriche J-K, Lebrin F, Rabillous T, Leroy D, Chambaz EM, Goldberg Y (1997) Regulation of protein phosphatase 2A by direct interaction with casein kinase 2a. Science 276:952–955
Herschman HR, Lim RW, Brankow DW, Fujiki H (1989) The tumor promoters 12-*O*-tetradecanoylphorbol 13-acetate and okadaic acid differ in toxicity, mitogenic activity, and induction of gene expression. Carcinogenesis 10:1495–1498
Hicks DR, Fraser-Reid B (1974) Selective sulphonylation with N-tosylimidazole. A one-step preparation of methyl 2,3-anhydro-4,6-*O*-benzylidene-α-d-mannopyranoside. Synthesis 203
Holmes CFB, Luu HA, Carrier F, Schmitz FJ (1990) Inhibition of protein phosphatases-1 and -2A with acanthifolicin. Comparison with diarrhetic shellfish toxins and identification of a region on okadaic acid important for phosphatase inhibition. FEBS Lett 270:216–218
Hu T, Doyle J, Jackson D, Marr J, Nixon E, Pleasance S, Quilliam MA, Walter JA, Wright JLC (1992) Isolation of a new diarrhetic shellfish poison from Irish mussels. J Chem Soc Chem Commun 39–41
Ichikawa Y, Isobe M, Goto T (1984) Synthetic studies toward marine toxic polyethers [2] synthesis of the B-segment of okadaic acid and coupling with the C-segment. Tetrahedron Lett 25:5049–5052
Ichikawa Y, Isobe M, Goto T (1987a) Synthesis of a marine polyether toxin, okadaic acid [2] – synthesis of segment B. Tetrahedron 43:4749–4758
Ichikawa Y, Isobe M, Masaki H, Kawai T, Goto T (1987b) Synthesis of a marine polyether toxin, okadaic acid (3) – synthesis of segment C. Tetrahedron 43:4759–4766
Ichikawa Y, Isobe M, Goto T (1988) Transformation of a marine toxic polyether, okadaic acid. Agric Biol Chem 52:975–981
Ireland RE, Smith MG (1988) 3-Acyltetramic acid antibiotics. 2. Synthesis of (+)-streptolic acid. J Am Chem Soc 110:854–860

Ishihara H, Martin BL, Brautigan DL, Karaki H, Ozaki H, Kato Y, Fusetani N, Watabe S, Hashimoto K (1989) Calyculin A and okadaic acid: inhibitors of protein phosphatase activity. Biochem Biophys Res Commun 159:871–877

Isobe M, Ichikawa Y, Bai D-L, Goto T (1985a) Synthetic studies toward marine toxic polyethers 2 synthesis of segment-A of okadaic acid via anti-selectivity by heteroconjugate addition. Tetrahedron Lett 26:5203–5206

Isobe M, Ichikawa Y, Goto T (1985b) Synthetic studies toward marine toxic polyethers (3) stereocontrol for segment-A_1 of okadaic acid by means of oxymercuration and epoxidation. Tetrahedron Lett 26:5199–5202

Isobe M, Ichikawa Y, Goto T (1986) Synthetic studies toward marine toxic polyethers (5) the total synthesis of okadaic acid. Tetrahedron Lett 27:963–966

Isobe M, Ichikawa Y, Bai D-L, Masaki H, Goto T (1987) Synthesis of a marine polyether toxin, okadaic acid [4] – total synthesis. Tetrahedron 43:4767–4776

Kamibayashi C, Estes R, Slaughter C, Mumby MC (1991) Subunit interactions control protein phosphatase 2A. Effects of limited proteolysis, N-ethylmaleimide, and heparin on the interaction of the B subunit. J Biol Chem 266:13251–13260

Kawabe T, Muslin AJ, Korsmeyer SJ (1997) HOX11 interacts with protein phosphatases PP2A and PP1 and disrupts a G2/M cell cycle checkpoint. Nature 385:454–458

Keck GE, Abbott DE (1984) Stereochemical consequences for the Lewis acid mediated additions of allyl and crotyl tri-n-butyl stannane to chiral β-hydroxy aldehyde derivatives. Tetrahedron Lett 25:1883–1886

Kishi Y (1992) Applications of Ni(II)/Cr(II)-mediated coupling reactions to natural products syntheses. Pure Appl Chem 64:343–350

Kleinberger T, Shenk T (1993) Adenovirus E4orf4 protein binds to protein phosphatase 2A and the complex down regulates E1A-enhanced *junB* transcription. J Virol 67:7556–7560

Kunz H, Weissmueller J (1984) Synthese des methylethers der herzgiftmethylreduktinsäure aus d-glucose. Liebigs Ann Chem: 66–77

Lee J, Chen Y, Tolstykh T, Stock J (1996) A specific protein carboxyl methyltransferase that demethylates phosphoprotein phosphatase 2A in bovine brain. Proc Natl Acad Sci U S A 93:6043–6047

Ley SV, Humphries AC, Eick H, Downham R, Ross AR, Boyce RJ, Pavey JBJ, Pietruszka J (1998) Synthesis of okadaic acid. J Chem Soc Perkin Trans 1:3907–3911

Li M, Makkinje A, Damuni Z (1996) Molecular identification of I1PP2A, a novel potent heat-stable inhibitor protein of protein phosphatase 2A. Biochemistry 35:6998–7002

Li Y-M, Casida JE (1992) Cantharidin-binding protein: Identification as protein phosphatase 2A. Proc Natl Acad Sci U S A 89:11867–11870

Lindvall MK, Pihko PM, Koskinen AMP (1997) The binding mode of calyculin A to protein phosphatase-1. A novel spiroketal vector model. J Biol Chem 272:23312–23316

MacKintosh C, Klumpp S (1990) Tautomycin from the bacterium Streptomyces verticillatus. Another potent and specific inhibitor of protein phosphatases 1 and 2A. FEBS Lett 277:137–140

MacKintosh C, Beattie KA, Klumpp S, Cohen P, Codd GA (1990) Cyanobacterial microcystin-LR is a potent and specific inhibitor of protein phosphatases 1 and 2A from both mammals and higher plants. FEBS Lett 264:187–192

Masamune S, Choy W, Petersen JS, Sita LR (1985) Double asymmetric synthesis and a new strategy for stereocontrol in organic synthesis. Angew Chem Int Ed Engl 24:1–30

Matsumori N, Murata M, Tachibana K (1995) Conformational analysis of natural products using long-range carbon-proton coupling constants: three dimensional structure of okadaic acid in solution. Tetrahedron 51:12229–12238

Matsushima R, Yoshizawa S, Watanabe MF, Harada K, Furusawa M, Carmichael WW, Fujiki H (1990) In vitro and in vivo effects of protein phosphatase inhibitors, microcystins and nodularin, on mouse skin and fibroblasts. Biochem Biophys Res Commun 171:867–874

Matsuzawa S, Suzuki T, Suziki M, Matsuda A, Kawamura T, Mizuno Y, Kikuchi K (1994) Thyrsiferyl 23-acetate is a novel specific inhibitor of protein phosphatase PP2A. FEBS Lett 356:272–274

Mayer-Jaekel RE, Hemmings BA (1994) Protein phosphatase 2A-α 'menage a trois'. Trends Cell Biol 4:287–291

McCright B, Rivers AM, Audlin S, Virshup DM (1996) The B56 family of protein phosphatase 2A (PP2A) regulatory subunits encodes differentiation-induced phosphoproteins that target PP2A to both nucleus and cytoplasm. J Biol Chem 271:22081–22089

Merrick SE, Demoise DC, Lee VMY (1996) Site-specific dephosphorylation of tau protein at Ser202/Thr205 in response to microtubule depolymerization in cultured human neurons involves protein phosphatase 2A. J Biol Chem 271:5589–5594

Mihara K, Cao XR, Yen A (1989) Cell cycle-dependent regulation of phosphorylation of the human retinoblastoma gene product. Science 246:1300–1303

Miljkovic M, Glisin D (1975) Synthesis of macrolide antibiotics. II. Stereoselective synthesis of methyl 4,6-*O*-benzylidene-2-deoxy-2-*C*,3-*O*-dimethyl-α-D-glucopyranoside. J Org Chem 40:3357–3359

Moens U, Seternes OM, Johansen B, Rekvig OP (1997) Mechanisms of transcriptional regulation of cellular genes by SV40 large T- and small T-antigens. Virus Genes 15:135–154

Moorhead G, Douglas P, Morrice N, Scarabel M, Aitken A, MacKintosh C (1996) Phosphorylated nitrate reductase from spinach leaves is inhibited by 14–3-3 proteins and activated by fusicoccin. Curr Biol 6:1104–1113

Mumby MC, Walter G (1993) Protein serine/threonine phosphatases: structure, regulation, and functions in cell growth. Physiol Rev 73:673–699

Murakami Y, Oshima Y, Yasumoto T (1982) Identification of okadaic acid as a toxic component of a marine dinoflagellate *Prorocentrum lima*. Jpn Soc Sci Fish 48:69–72

Nakajimi N, Abe, R, Yonemitsu O (1988) 3-Methoxybenzyl (3-MPM) and 3,5-dimethoxybenzyl (3,5-DMPM) protecting groups for the hydroxy function less readily removable than 4-methoxybenzyl (MPM and 3,4-dimethoxybenzyl (DMPM) protecting groups by DDQ oxidation. Chem Pharm Bull 36:4244

Nishiwaki S, Fujiki H, Suganuma M, Furuya-Suguri H, Matsushima R, Iida Y, Ojika M, Yamada K, Uemura D, Yasumoto T, Schmitz FJ, Sugimura T (1990) Structure-activity relationship within a series of okadaic acid derivatives. Carcinogenesis 11:1837–1841

Nishizuka Y (1984) The role of protein kinase C in cell surface signal transduction and tumor promotion. Nature 308:693–698

Norte M, Gonzalez R, Fernandez JJ, Rico M (1991) Okadaic acid: a proton and carbon NMR study. Tetrahedron 47:7437–7446

Ogris E, Gibson DM, Pallas DC (1997) Protein phosphatase 2A subunit assembly: the catalytic subunit carboxy terminus is important for binding cellular B subunit but not polyomarvirus middle tumor antigen. Oncogene 15:911–917

Ohuchi K, Tamura T, Ohashi M, Watanabe M, Hirasawa N, Tsurufuji S, Fujiki H (1989) Okadaic acid and dinophysistoxin-1, non-TPA-type tumor. Biochim Biophys Acta 1013:86–91

Okamoto K, Kamibayashi C, Serrano M, Prives C, Mumby MC, Beach D (1996) p53-Dependent association between cyclin G and the B' subunit of protein phosphatase 2A. Mol Cell Biol 16:6593–6602

Orgad S, Brewis ND, Alphey L, Axton JM, Dudai Y, Cohen P (1990) The structure of protein phosphatase 2A is as highly conserved as that of protein phosphatase 1. FEBS Lett 275:44–48

Pallas DC, Shahrik LK, Martin BL, Jaspers S, Miller TB, Brautigan DL, Roberts TM (1990) Polyoma small and middle T antigens and SV 40 small T antigen form stable complexes with protein phosphatase 2A. Cell 60:167–176

Park K, Chung M, S.-J. K (1992) Inhibition of myogenesis by okadaic acid, an inhibitor of protein phosphatases, 1 and 2A, correlates with the induction of AP1. J Biol Chem 267:10810–10815

Parsons R (1998) Phosphatases and tumorigenesis. Curr Opin Oncol 10:88–91

Peng J, Bowden GT, Domann FE (1997) Activation of AP-1 by okadaic acid in mouse keratinocytes associated with hyperphosphorylation of c-jun. Mol Carcin 18:37–43

Printen JA, Brady MJ, Saltiel AR (1997) PTG, a protein phosphatase 1-binding protein with a role in glycogen metabolism. Science 275:1475–1478

Roberge M, Tudan C, Hung SMF, Harder KW, Jirik FR, Anderson H (1994) Antitumor drug fostriecin inhibits the mitotic entry checkpoint and protein phosphatases 1 and 2A. Cancer Res 54:6115–6121

Rosenberger SF, Bowden GT (1996) Okadaic acid stimulated TRE binding activity in a papilloma producing mouse keratinocyte cell line involves increased AP-1 expression. Oncogene 12:2301–2308

Sabes SF, Urbanek RA, Forsyth CJ (1998) Efficient synthesis of okadaic acid. 2. Synthesis of the C1-C14 domain and completion of the total synthesis. J Am Chem Soc 120:2534–2542

Sakai R, Rinehart K (1995) A new polyether acid from a cold water marine sponge, a phakellia species. J Nat Prod 58:773–777

Sassa T, Richter WW, Uda N, Suganuma M, Suguri H, Yoshizawa S, Hirota M, Fujiki H (1989) Apparent "activation" of protein kinases by okadaic acid class tumor promoters. Biochem Biophys Res Commun 159:939–944

Scheuer P (1995) Marine natural products research: a look into the dive bag. J Nat Prod 58:335–342

Schmitz FJ, Prasad RS, Gopichand Y, Hossain MB, van der Helm D, Schmidt P (1981) Acanthifolicin, a polyether from the sponge Pandoras acanthifolium. J Am Chem Soc 103:2467

Schonthal A (1992) Okadaic acid-a valuable new tool for the study of signal transduction and cell cycle regulation? New Biologist 4:16–21

Schonthal A, Tsukitani Y, Feramisco JR (1991) Transcriptional and post-transcriptional regulation of c-fos expression by the tumor promoter okadaic acid. Oncogene 6:423–430

Sheppeck JE, II, Gauss C-M, Chamberlin AR (1997) Inhibition of the Ser-Thr phosphatases PP1 and PP2A by naturally occurring toxins. Bioorg Med Chem 5:1739–1750

Shima H, Tohda H, Aonuma S, Nakayasu M, DePaoli-Roach A, Sugimura T, Nagao M (1994) Characterization of the PP2Aa gene mutation in okadaic acid-resistant variants of CHO-K1 cells. Proc Natl Acad Sci U S A 91:9267–9271

Sontag E, Federov S, Kamibayashi C, Robbins D, Cobb M, Mumby M (1993) The interaction of SV40 small tumor antigen with protein phosphatase 2A stimulates the MAP kinase pathway and induces cell proliferation. Cell 75:887–897

Sontag E, Nunbhakdi-Craig V, Lee G, Bloom GS, Mumby MC (1996) Regulation of the phosphorylation state and microtubule-binding activity of tau by protein phosphatase 2A. Neuron 17:1201–1207

Sontag E, Sontag J-M, Garcia A (1997) Protein phosphatase 2A. EMBO J 16:5662–5671

Suganuma M, Fujiki H, Suguri H, Yoshizawa S, Hirota M, Nakayasu M, Ojika M, Wakamatsu K, Yamada K, Sugimura T (1988) Okadaic acid: an additional non-phorbol-12-tetradecanoate. Proc Natl Acad Sci U S A 85:1768–1771

Tachibana K, Scheuer PJ, Tsukitani Y, Kikuchi H, Engen DV, Clardy J, Gopichand Y, Schmitz FJ (1981) Okadaic acid, a cytotoxic polyether from two marine sponges of the genus Halichondria. J Am Chem Soc 103:2469

Takai A, Murata M, Torigoe K, Isobe M, Mieskes G, Yasumoto T (1992) Inhibitory effect of okadaic acid derivatives on protein phosphatases. A study on structure-affinity relationship. Biochem J 284:539–544

Takai A, Sasaki K, Nagai H, Mieskes G, Isobe M, Isono K, Yasumoto T (1995) Inhibition of specific binding of okadaic acid to protein phosphatase 2A by microcystin-LR, calyculin-A and tautomycin: method of analysis of interactions of tight-binding ligands with target proteins. Biochem J 308:1039

Thevenin C, Kim S-J, Kehrl JH (1991) Inhibition of protein phosphatases by okadaic acid induces AP1 in human T cells. J Biol Chem 266:9363–9366

Tsuboi K, Ichikawa Y, Isobe M (1997) Synthesis of okadaic acid-tautomycin hybrid. Synlett 713–715

Turowski P, Favre B, Campbell KS, Lamb NJC, Hemmings BA (1997) Modulation of the enzymic properties of protein phosphatase 2A catalytic subunit by the recombinant 65-kDa regulatory subunit PR65.alpha. Eur J Biochem 248:200–208

Uemura D, Hirata Y (1988) Antitumor polyethers from marine animals. In: Rahman AU (ed) Studies in natural products chemistry, vol 5. Elsevier, Amsterdam, pp 377–401

Uemura D, Takahashi K, Yamamoto T, Katayama C, Tanaka J, Okumura Y, Hirata Y (1985) Norhalichondrin A: an antitumor polyether macrolide from a marine sponge. J Am Chem Soc 107:4796–4798

Urbanek RA, Sabes SF, Forsyth CJ (1998) Efficient synthesis of okadaic acid. 1. Convergent assembly of the C15-C38 domain. J Am Chem Soc 120:2523–2533

Virshup DM, Kelly TJ (1989) Purification of replication protein C, a cellular protein involved in the initial stages of simian virus 40 DNA replication in vitro. Proc Natl Acad Sci U S A 86:3584–3588

Virshup DM, Kauffman MG, Kelly TJ (1989) Activation of SV40 DNA replication in vitro by cellular protein phosphatase 2A. Eur Mol Biol J 8:3891–3898

Walter G, Ruediger R, Slaughter C, Mumby M (1990) Association of protein phosphatase 2 A with polyoma virus medium tumor antigen. Proc Natl Acad Sci U S A 87:2521–2525

Windust A, Quilliam MA, Wright JLC, McLachlan JL (1997) Comparative toxicity of the diarrhetic shellfish poisons, okadaic acid, okadaic acid diolester and dinophysistoxin-4, to the diatom Thalassiosira weissflogii. Toxicon 35:1591–1603

Yang S-I, Lickteig RL, Estes RC, Rundell K, Walter G, Mumby MC (1991) Control of protein phosphatase 2A by Simian virus 40 Small-t antigen. Mol Cell Biol 11:1988–1995

Yasumoto T, Murata M, Oshima Y, Sano M, Matsumoto GK, Clardy J (1985) Diarrhetic shellfish toxins. Tetrahedron 41:1019

Zhang L, Lee EYC (1997) Mutational analysis of substrate recognition by protein phosphatase 1. Biochemistry 36:8209–8214

Zhang Z, Zhao S, Long F, Zhang L, Bai G, Shima HN, M., Lee EYC (1994) A mutant of protein phosphatase-1 that exhibits altered toxin sensitivity. J Biol Chem 269:16997–11700

Zhang L, Zhang Z, Long F, Lee EYC (1996) Tyrosine-272 is involved in the inhibition of protein phosphatase-1 by multiple toxins. Biochemistry 35:1606–1611

Zolnierowicz S, Van Hoof C, Andjelkovic N, Cron P, Stevens I, Merlevede W, Goris J, Hemmings BA (1996) The variable subunit associated with protein phosphatase 2A0 defines a novel multimember family of regulatory subunits. Biochem J 317:187–194

4 Total Synthesis of Marine Natural Products Driven by Novel Structure, Potent Biological Activity, and/or Synthetic Methodology*

D. Romo, R.M. Rzasa, W.D. Schmitz, J. Yang, S.T. Cohn,
I.P. Buchler, H.A. Shea, K. Park, J.M. Langenhan,
N.B. Messerschmidt, M.M. Cox

4.1 Bioactive Marine Natural Products: Relatively Untapped Biodiversity and an Important Forum for the Development of New Synthetic Strategies and Methods

A number of structurally complex and unique natural products have been extracted from a variety of marine organisms (Kerr and Kerr 1999; Alam and Thomson 1997; Scheuer 1978, 1983, 1987, 1989; Okaichi et al. 1989; Faulkner and Fenical 1977; Chemical Reviews, vol. 93). While the utility of these compounds as health products has not yet been proven, except in only a few cases*, many marine natural products or derivatives are currently in clinical trials (e.g., bryostatin 1, dolostatin 10, AE941, ecteinsascidin-743). Moreover, besides their use as drugs, many marine natural products have proven to be valuable probes of biological processes (e.g., didemnins, discodermolide, elutherobin, manolide; Hung et al. 1996a; Caterall and Gainer 1985; Debonnel et al. 1989; Soriente et al. 1999; Berlinck et al. 1998). Nevertheless, many question whether there are any truly new and unique structures yet to be isolated from the sea. While some may argue that the rate of discovery of truly novel molecular architectures from marine sources has dwindled in recent years, a recent article in *The Scientist* gives a resounding "yes" to the question of whether new molecular structures remain to be found in the oceans (Rayl 1999). On the other hand, with the surge in combinatorial synthesis, it appears that drug companies are relying more on this chemical tool at the risk of missing potential important lead compounds to be found in the slower interest-bearing marine environment. Despite concerns about supply issues regarding marine-based pharmaceuticals, some researchers go so far as to say that the ocean will likely provide the greatest yield of new medicines in the new millennium (Rayl 1999). This statement seems justified when one considers that approximately 70% of the earth's surface is ocean, that many marine organisms are sessile and thus require potent chemicals for all facets of survival including reproduction, defense, and communication,

---

* One marine natural product, pseudopterosin C, from the sea whip *Pseudopterogorgia elisaethae*, has made it to the marketplace as an anti-inflammatory agent used in Estee Lauder's Resilience; see: Rouhi (1995)

and that only 1%–2% of cells found in seawater have been analyzed for the compounds they produce (Rayl 1999). Thus, it would appear that the oceans have not yet divulged all their secrets and that many new and exciting marine natural products will be found for many years to come. Directly linked to these natural product discoveries is the impetus to develop synthetic strategies and methods for the synthesis of these compounds in order to fully understand and utilize their often potent biological activity.

My research group has been engaged in the total synthesis of bioactive marine natural products, and the reasons for a particular target selection originate from the structural novelty, the desire to study and understand their biological activity, and/or the desire to demonstrate the utility of developed methodologies or strategies. Three total synthesis projects fueled by these motivations that my group has been engaged in during my first 6 years at Texas A&M University are presented in this chapter.

4.2 Total Synthesis of the Thiazole-Containing Macrodiolide, (-)-Pateamine A: a Potent Immunosuppressive Agent from the Sponge *Mycale* sp

4.2.1 Introduction

Natural products have proven to be useful probes of various biological processes (Hung et al. 1996b; Hinterding et al. 1998; Schreiber 1992). In particular, agents that exhibit specific cellular effects have served as powerful biochemical tools for dissecting molecular mechanisms of signal transduction pathways involved in various cellular functions. Immunosuppressive agents are examples of such natural products which have made it possible to discover a number of key signaling molecules involved in the orchestration of the immune response (Thomson and Starzl 1994). Prominent among them are FK506, cyclosporin A (CsA), and rapamycin, which have proven extremely useful in shedding new light on intracellular signaling pathways involved in T-cell activation (Fig. 1; Schreiber 1991; Belshaw et al. 1994). These immunosuppressants have an unusual mode of action; their immunosuppressive activi-

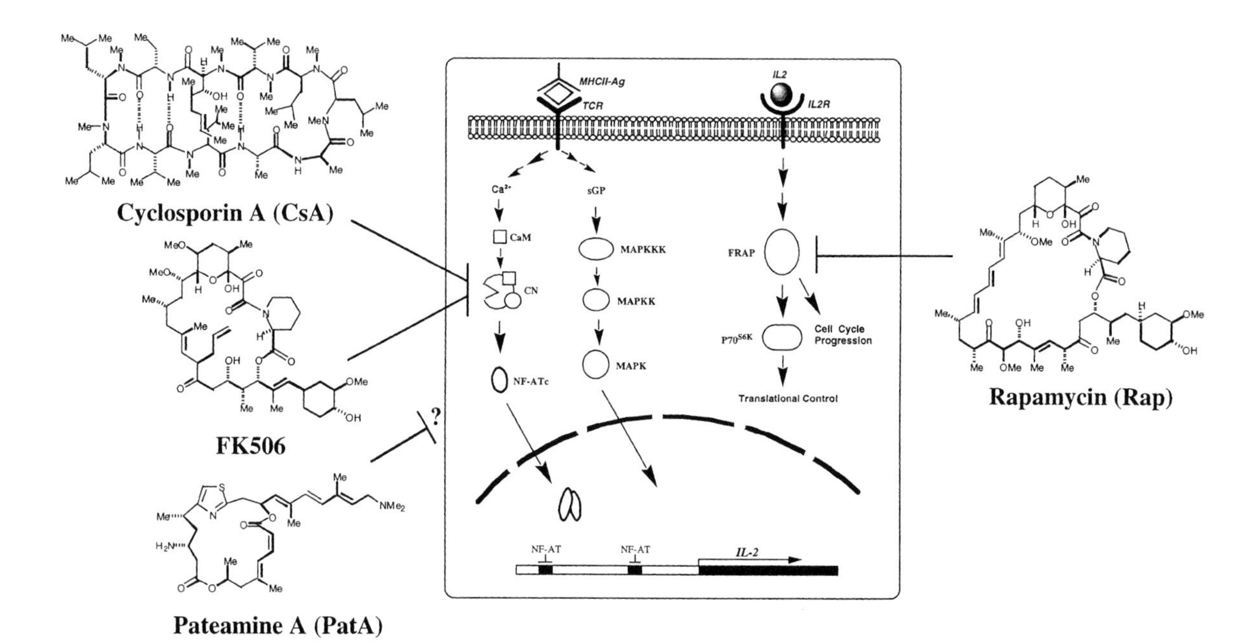

Fig. 1. Signal transduction pathways involved in T-lymphocyte activation and the site of action of the immunosuppressants cyclosporin A, FK506, rapamycin, and pateamine A. *MHCII-Ag* Major histocompatibility complex II-antigen complex, *TCR* T-cell receptor, *IL2* interleukin 2, *IL2R* IL2 receptor, *CaM* calmodulin, *CN* calcineurin, *NF-AT* nuclear factor of activated T cells, *NF-ATc* the cytoplasmic subunit of NF-AT, *FRAP* FKBP-rapamycin-associated protein, *sGP* small GTP binding proteins, *MAPK* generic MAP kinases, *MAPKK* MAPK kinase, *MAPKKK* MAPKK kinase

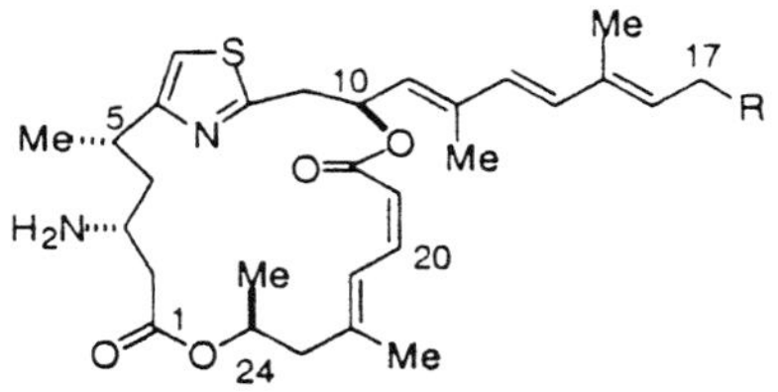

Pateamine A (**1**) R = NMe_2
B (**2**) R = NHMe
C (**3**) R = $N(O)Me_2$

Fig. 2. Structure of pateamines

ties are mediated by immunophilins. CsA binds cyclophilins while FK506 and rapamycin bind to FKBPs. The cyclophilin-CsA and FKBP-FK506 complexes bind to and inhibit the phosphatase activity of calcineurin, a critical enzyme involved in intracellular signal transduction emanating from the T-cell receptor and leading to the production of cytokines, including interleukin 2. The FKBP-rapamycin complex inhibits a kinase called FRAP (also known as RAFT) that is involved in interleukin-2 receptor-mediated T-cell proliferation with no effect on T-cell receptor-mediated signaling.

A novel immunosuppressant that likewise promises to be quite useful as a biochemical probe is the marine natural product, pateamine A (**1**, Fig. 2; Northcote et al. 1991). Pateamine A displays potent immunosuppressive properties [mixed lymphocyte reaction (MLR) IC_{50}=2.6 n*M*] with low cytotoxicity [lymphocyte viability assay (LCV)]/MLR ratio >1000; G.T. Faircloth 1994, private communication). However, which signaling pathway(s) is or are affected by pateamine remains unknown. Furthermore, the amounts of pateamine A from the natural source are limited, making it difficult to carry out structural modifications of the compound to prepare biochemical probes.

We completed the first total synthesis of (-)-pateamine A in 1998 which confirmed its relative and absolute stereochemistry. (For other synthetic studies of pateamine A, see Critcher and Pattenden 1996. Our synthesis was previously communicated in Rzasa et al. 1998.) More recently, it was found that pateamine A specifically inhibits an intracellular step of the T-cell receptor signal transduction pathway leading to

IL-2 transcription, making pateamine A an attractive molecular probe to potentially uncover a new signaling molecule(s) in this pathway (Fig. 1; Romo et al. 1998). As a first step toward identifying the pateamine A target, we also found that the primary C3-amino group in pateamine A can tolerate further chemical modification without significant loss of immunosuppressive activity. With this information in hand, a pateamine A-dexamethasone hybrid to be utilized in the yeast three-hybrid system (Licitra and Liu 1996) and a 2,4-dinitroaniline-pateamine A hybrid have been synthesized. Both of these derivatives are useful for putative receptor isolation and receptor-ligand interaction studies.

Background. Pateamine A was isolated off the shores of New Zealand by Munro and co-workers from the marine sponge *Mycale* sp, and in their initial report in 1991 only its two-dimensional structure was described (Northcote et al. 1991). This unique natural product bears a thiazole and an *E,Z*-dienoate within a 19-membered macrocycle and a trienylamine side chain. Two minor constituents, pateamines B and C, were also isolated, and their structures differ from pateamine A only in the nature of the terminal group of the trienylamine side chain (Fig. 2; M.H.G. Munro, P.T. Northcote, J.W. Blunt, 1993, personal communication). Other natural products isolated from this same species of sponge include the mycalamides, which are structurally related to the pederins. (For a lead reference, see Abell et al. 1997.) Pateamine A exhibits antifungal and selective cytotoxic activity (Northcote et al. 1991); however, our interest in a synthesis of this natural product was sparked by its unique structure, its potent immunosuppressive effects, and – importantly – its low toxicity. On comparison to cyclosporin A in the mouse skin graft rejection assay, pateamine A resulted in a 15-day survival period, whereas cyclosporin A led to only a 10-day survival of skin grafts (G.T. Faircloth 1994, private communication). For these reasons, we embarked on a total synthesis of this natural product as a means of determining the relative and absolute stereochemistry, of providing further quantities of the natural product, and of providing access to structural derivatives for further biological studies, including isolation of a putative cellular receptor. (For our initial communication on stereochemical determination of the C24 stereocenter of pateamine A in collaboration with the New Zealand group, see Rzasa et al. 1995.)

4.2.2 Total Synthesis Employing a β-Lactam-Based Macrocyclization

Retrosynthetic Analysis. Several considerations guided our retrosynthetic plan, including the reported lability of the C3 amino group (Northcote et al. 1991), the potential for isomerization of the *E,Z*-dienoate, the desire to introduce and liberate the polar amino groups at a late stage of the synthesis, and the unknown stereochemistry of the natural product. With these considerations in mind, our retrosynthesis called for the synthesis and subsequent union of three principal fragments, namely dienylstannane **4**, β-lactam **7**, and enyne acid **8** (Fig. 3). These fragments thus became our initial synthetic targets. An enyne was used to allow a late-stage introduction of the *E*, *Z*-dienoate in order to minimize the potential for isomerization. A late-stage Stille coupling would append the trienylamine side chain and would allow a variety of side chains to be introduced onto the macrocyclic core structure. A key strategy to be employed to construct the 19-membered dilactone macrolide of pateamine A was a β-lactam-based macrocyclization that was suggested by the pateamine A structure. Thus, a β-lactam would be used to install the C3-amino group and would then serve as an activated acyl group for macrocyclization. To address the issue of stereochemical uncertainty, we elected to use reagent-based control to set the four stereogenic centers contained in pateamine A to provide the most stereoflexible approach. In addition, we realized some inherent flexibility was present at C10, based on our synthetic plan, since the hydroxyl could be acylated or alternatively inverted by a Mitsunobu process to provide both epimers at this position. Two stereogenic centers at C3 and C10 would be introduced using the chiral auxiliary-based acetate aldol methodology of Nagao (Nagao et al. 1986), and the C5 stereocenter would be introduced by an asymmetric conjugate addition by the method of Hruby (Nicolas et al. 1993; Li et al. 1993). The C24 stereochemistry would originate from yeast reduction (Seebach et al. 1993) or Noyori asymmetric hydrogenation (Noyori et al. 1987) of ethyl acetoacetate.

As mentioned above, Munro and co-workers had described only the two-dimensional structure of pateamine A, and attempts to produce crystalline derivatives were unsuccessful (Northcote et al. 1991). In order to embark on a total synthesis of pateamine A, a tentative stereo-

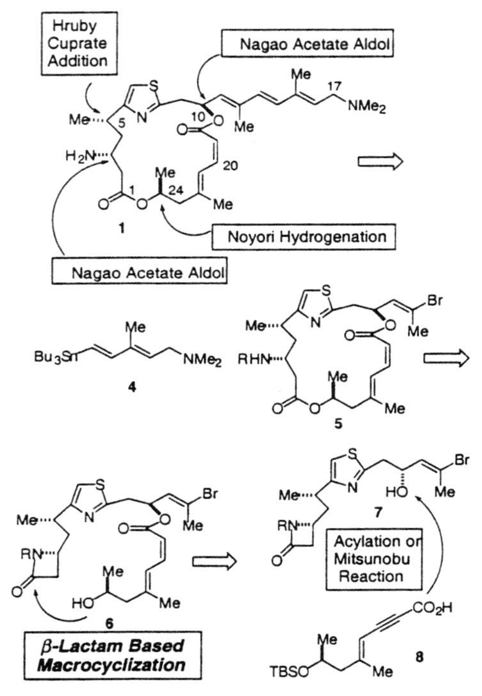

Fig. 3. Retrosynthesis of (–) –pateamine A showing key disconnections including the β-lactam-based macrocyclization strategy and stereoflexible methods for introduction of stereogenic centers. R=Boc (*t*-butoxycarbonyl); R=TCBoc (trichlor-*t*-butoxycarbonyl)

chemical assignment was required to define an initial synthetic target. In collaboration with the New Zealand groups, we set out to determine the absolute stereochemistry at C24. To accomplish this task, we envisioned that a hydroxy dienoate corresponding to the C18–C25 fragment of pateamine A could be secured by methanolysis of pateamine A, and a synthesis of this fragment was devised in order to make a direct comparison of natural and synthetic materials. We were also mindful of the possibility of using the same intermediate for our projected total synthesis.

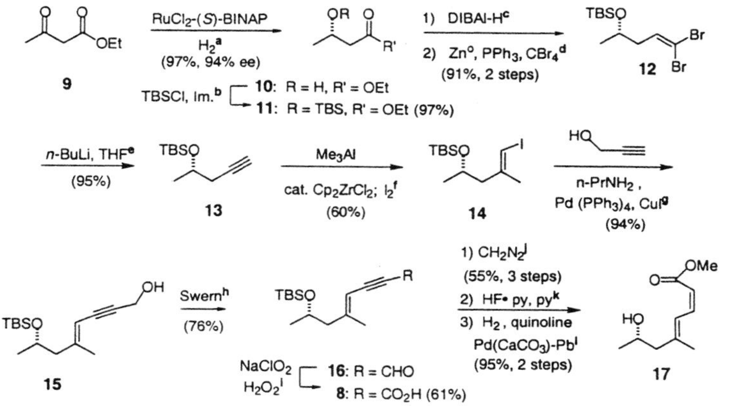

9 → $RuCl_2$-(*S*)-BINAP, H_2[a] (97%, 94% ee) → 10: R = H, R' = OEt; TBSCl, Im.[b] → 11: R = TBS, R' = OEt (97%)

1) DIBAl-H[c] 2) Zn°, PPh_3, CBr_4[d] (91%, 2 steps) → 12

n-BuLi, THF[e] (95%) → 13

Me_3Al, cat. Cp_2ZrCl_2; I_2[f] (60%) → 14

HO–propargyl, n-$PrNH_2$, Pd $(PPh_3)_4$, CuI[g] (94%) → 15

Swern[h] (76%) → 16: R = CHO; $NaClO_2$, H_2O_2[i] → 8: R = CO_2H (61%)

1) CH_2N_2[j] (55%, 3 steps) 2) HF• py, py[k] 3) H_2, quinoline Pd($CaCO_3$)-Pb[l] (95%, 2 steps) → 17

[a] (a) EtOH, Dowex-50 resin, 200 psi H_2, 130 °C; (b) DMF, 25 °C; (c) CH_2Cl_2, -90 °C; (d) CH_2Cl_2, 25 °C; (e) -78 → 25 °C; (f) CH_2Cl_2, 0 → 25 °C then I_2, Et_2O, 0 °C ; (g) C_6H_6, 0 → 25 °C; (h) $(COCl)_2$, DMSO, CH_2Cl_2, -78 → 0 °C then Et_3N, -78 °C; (i) CH_3CN, NaH_2PO_4 buffer, *t*-butanol, 0 °C; (j) Et_2O, 25 °C (yield includes steps (h) and (i)); (k) THF, 25 °C; (l) MeOH, 1 atm H_2, 25 °C.

Scheme 1. Synthesis of the enyne acid fragment **8** and hydroxydienoate **17**

Synthesis of Enyne Acid 8 and Hydroxy Dienoate 20 (C18–C25): Defining a Stereochemical Target. Our approach to the *E,Z*-hydroxy dienoate corresponding to the C18–C25 fragment of pateamine A began with the known (*S*)-2-hydroxybutyrate **10**. This was prepared by yeast reduction (Seebach et al. 1993) or by Noyori hydrogenation of ethyl acetoacetate using the modified conditions of Taber (Scheme 1; Taber and Silverber 1991; Taber et al. 1992). The modified Noyori method was readily performed on a large scale and the enantiomeric purity and product yield were superior to those with the yeast reduction method (94% ee vs. 84%–90% ee, chiral GC), and it was therefore employed for large-scale preparations. After silylation, the known silyloxy ester **11** (Ireland and Wardle 1987) was reduced to the aldehyde and converted to the acetylene **13** by the Corey-Fuchs procedure (Corey and Fuchs 1972). A carboalumination/iodination sequence (Nicolas et al. 1993; Li et al. 1993) delivered the vinyl iodide **14**, which was then subjected to a Sonogashira coupling (Sonogashira et al. 1975) with propargyl alcohol to provide the enyne alcohol **15**. A two-step oxidation gave the pivotal intermediate, enyne acid **8**. This intermediate proved useful for the total synthesis of pateamine A as well as for the synthesis of the *E,Z*-hydroxydienoate for stereochemical determination of C24. Towards the latter goal, methylation with diazomethane, deprotection of the silyl group, and Lindlar reduction of enyne acid **8** gave the required *E*, *Z*-dienoate **17** for comparison with the same compound derived from methanolysis of pateamine A. Comparative HPLC analysis on a chiral stationary phase of the derived Mosher esters of both natural and synthetic hydroxydienoate **17** confirmed the (*S*)-configuration at C24 (Rzasa et al. 1995). This result, in conjunction with molecular modeling studies and extensive two-dimensional NMR experiments by the New Zealand group, provided a tentative absolute stereostructure of pateamine A as 3*R*, 5*S*, 10*R*, 24*S*. Thus, this stereoisomer became our initial synthetic target (Blincoe 1994; a full account of this work will be forthcoming: M.H.G. Munro 1998, private communication).

Synthesis of the C3–C12 Fragment. The synthesis of the C3–C12 fragment began with a Nagao acetate aldol reaction (Nagao et al. 1986) with known aldehyde **19** (Fischer et al. 1990) and the *N*-acetylthiazolidinethione **18** (Calo et al. 1994; Scheme 2). This proceeded smoothly to give an 84% yield of the aldol adduct **21** in a highly diastereoselective fashion (>19:1 dr) through the presumed transition state arrangement **20**. Silylation and transamidation with gaseous ammonia gave amide **23** in high overall yield. Treatment of amide **23** with Belleau's reagent (Lajoie et al. 1983) gave the desired thioamide **24** in 72% yield, accompanied by ~15% of the corresponding nitrile. Conversion to the thiazole **25** was effected using the modified Hantzsch conditions reported by Meyers (Aguilar and Meyers 1994).

After studying several strategies for introduction of the C5 stereocenter, we found that the method of Hruby (Nicolas et al. 1993; Li et al. 1993) involving an asymmetric conjugate addition to an Evans oxazolidinone-derived eneimide provided an efficient means of introducing this stereogenic center. Half-reduction following Horner-Emmons olefination (Z:E, >19:1) of thiazole ester **25** simultaneously added the required two carbons and introduced the chiral auxiliary required for the asymmetric conjugate addition (Scheme 3). (For the same homologation strategy applied to the (*S*)-phenylalanine-derived oxazolidinone

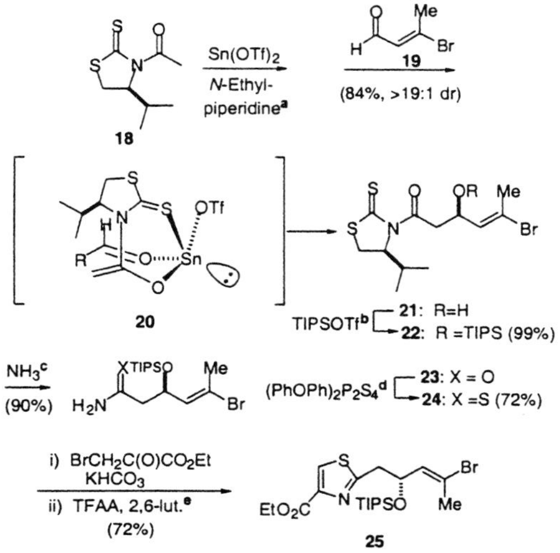

[a](a) CH_2Cl_2, -40°C then **22**; (b) CH_2Cl_2, 0 °C; (c) CH_2Cl_2, 0 °C; (d) THF, 25 °C; (e) DME, -20 °C → 0 °C.

Scheme 2. Synthesis of the thiazole ester fragment **25**

auxiliary, see Broka and Ehrler 1991.) Treatment of eneimide **28** with an excess of MeMgBr and an equivalent amount of CuBr•DMS, according to the conditions of Hruby, led to the methylated product **31** with good diastereoselectivity (6.4:1, dr). The major diastereomer could be isolated in 77% yield. The stereochemical outcome, as evidenced by a subsequent X-ray analysis of a crystalline intermediate, is consistent with initial formation of a Cu(I) π-olefin complex-Mg(II) chelate on the face of the *s*-trans eneimide opposite the phenyl group of the auxiliary (cf. **30**), as proposed by Hruby based on NMR studies (Nicolas et al. 1993). A fortuitous crystalline by-product (not shown) was obtained during transamidation to the Weinreb amide **32** (Basha et al. 1977), which results from aluminum amide attack at the endocyclic carbonyl. This by-product allowed verification of the stereochemistry at C5 and C10 (pateamine numbering) as 5*S*, 10*R* by X-ray crystallography. Half-

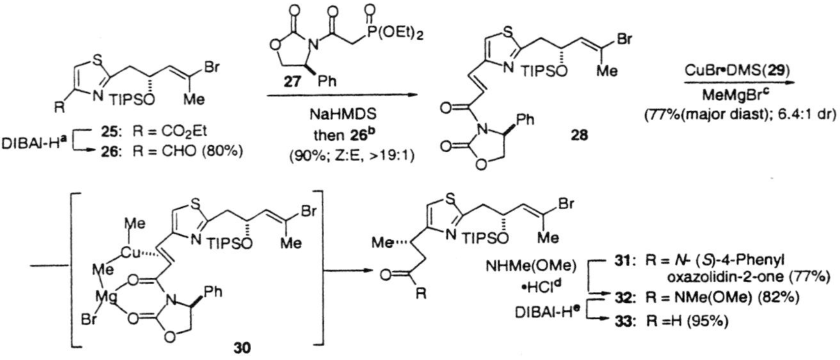

[a] (a) CH_2Cl_2, -90 °C; (b) THF, 25 °C; (c) THF/DMS (3:2), -20 °C; (d) Me_3Al, CH_2Cl_2, 0 °C; (e) CH_2Cl_2, -90 °C.

Scheme 3. Elaboration of thiazole ester **25** to the aldehyde **33**

reduction of amide **32** to aldehyde **33** provided the substrate required for introduction of the C3 stereocenter.

Synthesis of the Dienylstannane 4. The synthesis of the dienyl amino stannane **4** was accomplished in two steps from the known enyne alcohol **34** (Scheme 4). [(*E*)-3-methyl-2-penten-4-yn-1-ol (1-pentol) was generously provided by Dr. E. Gutknecht and Dr. P. Weber, F. Hoffman-La Roche, Switzerland.) A one-pot tosylation and displacement with dimethylamine gave the allylic amine **35**. Stannylcupration and quenching at low temperature according to the method of Oehlschlager (Kasela and Oehlschlager 1991) provided the desired dienylstannane **4** as a mixture of regioisomers (9:1).

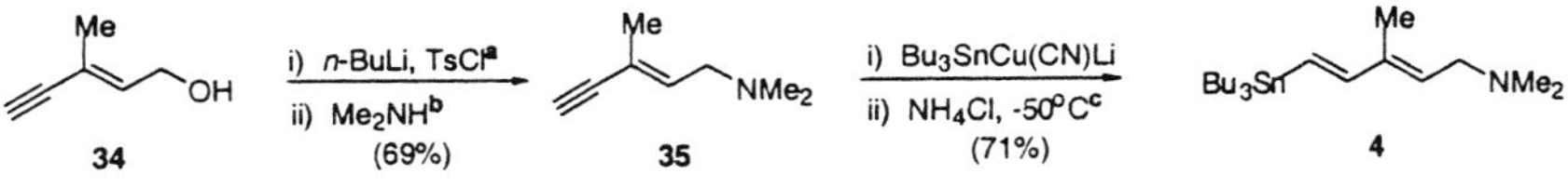

[a](a) THF, -78 °C; (b) THF, -78 → 0 °C; (c) THF, -60 → -40 °C.

Scheme 4. Synthesis of the dienyl stannane **4**

One concern with a late-stage Stille coupling was the potential for competing π-allyl formation, since the substrate would be an allylic acetate (cf. **5**) and the product would be a triallylic acetate (cf. pateamine A). (A very recent report describes Stille reactions in the presence of allylic acetates; see Nicolaou et al. 1998.) In order to address these concerns, we carried out model studies of the proposed Stille reaction with vinyl bromide **25** bearing an acetate protecting group at C10 (pateamine numbering) rather than TIPS group an stannane **4**. Using the conditions of Farina and Krishnan (1991), we were pleased to find that Stille coupling proceeded competitively with π-allyl formation under these conditions to provide 46%–53% yield of the expected triene (plus recovered starting material). With these results in hand, we proceeded with our plan to perform a late-stage Stille reaction to append the trienylamine side chain.

Completion of the β-Lactam Fragment. What remained to complete the synthesis of the final fragment for pateamine A synthesis was introduction of the C3 stereocenter (pateamine numbering) and construction of the β-lactam ring. We envisioned the use of a second Nagao acetate aldol to introduce the C3 stereocenter followed by the Miller intramolecular Mitsunobu reaction to construct the β-lactam. Utilizing the same auxiliary previously employed to introduce the C10 stereocenter (i.e., **18**), the Nagao acetate aldol reaction with aldehyde **33** provided a high degree of stereocontrol (>19:1, dr) in excellent yield (Scheme 5). Transamidation of aldol adduct **36** to the N-benzyloxy amide **37**, followed by an intramolecular Mitsunobu, gave the desired N-benzyloxy β-lactam **38** in excellent overall yield. We found that diisopropylazodicarboxylate (DIAD) gave superior yields compared with the use of carbon tetrachloride (Guzzo and Miller 1994) to activate triphenylphoshine in this Mitsunobu reaction. Although model studies had indicated that an *N*-benzyloxy substituent was sufficient for activation of a β-lactam towards intermolecular nucleophilic attack by sodium isopropoxide, we elected to convert this group to a *t*-butylcarbamate in order to utilize deprotection conditions that were more likely to be compatible with late intermediates. In model studies, we had found that SmI_2 in the presence of H_2O cleanly effects N-O cleavage of an *N*-benzyloxy β-lactam using the conditions reported by Keck et al. (1995) and Marco-Contelles (Chiara et al. 1996)for N-O cleavage of O-alkyl hydroxylamines

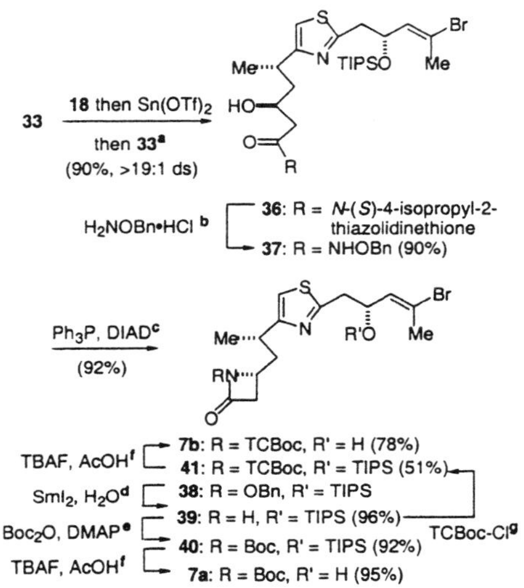

33 → 18 then $Sn(OTf)_2$ then 33[a] (90%, >19:1 ds)

$H_2NOBn{\bullet}HCl$ [b]

36: R = N-(S)-4-isopropyl-2-thiazolidinethione

37: R = NHOBn (90%)

Ph_3P, DIAD[c] (92%)

TBAF, AcOH[f] → 7b: R = TCBoc, R' = H (78%)

41: R = TCBoc, R' = TIPS (51%)

SmI_2, H_2O[d]: 38: R = OBn, R' = TIPS

39: R = H, R' = TIPS (96%) → TCBoc-Cl[g]

Boc_2O, DMAP[e] → 40: R = Boc, R' = TIPS (92%)

TBAF, AcOH[f] → 7a: R = Boc, R' = H (95%)

[a] (a) N-ethyl piperidine, CH_2Cl_2, -40 °C; (b) Me_3Al, CH_2Cl_2, 0 °C; (c) THF, 25 °C; (d) THF, 0 °C; (e) CH_2Cl_2, 25 °C; (f) THF, 0 °C; (g) Et_3N, DMAP, CH_2Cl_2, 25°C.

Scheme 5. Completion of the Boc- and TCBoc-β-lactam fragments **7a** and **7b**

to amines. To the best of our knowledge, this was the first application of these conditions to the deprotection of an *N*-benzyloxy-β-lactam. (Several metholds have been described for reductive cleavage of the N-O bond of N-benzylosy-β-lactams, but they are in general quite harsh. For a lead reference, see Georg 1993.) This procedure, when applied to the benzyloxy lactam **38**, led smoothly to the unprotected β-lactam **39** in excellent yield. Protection of the β-lactam nitrogen as the *t*-butylcarbamate followed by desilylation of the C10 silylether with TBAF provided the final fragment, alcohol **7** (R = Boc), required for the total synthesis of pateamine A.

Coupling of β-Lactam 7 (R = Boc) and Enyne Acid 8 (Acylation vs Mitsunobu Reaction): Revision of the C10 Stereochemistry. Further degradation and derivatization work, including mandelic amide analysis

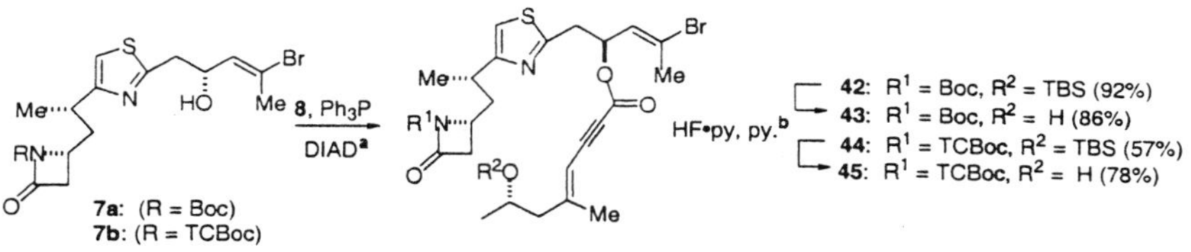

7a: (R = Boc)
7b: (R = TCBoc)

8, Ph_3P, DIAD[a]

HF•py, py.[b]

42: R^1 = Boc, R^2 = TBS (92%)
43: R^1 = Boc, R^2 = H (86%)
44: R^1 = TCBoc, R^2 = TBS (57%)
45: R^1 = TCBoc, R^2 = H (78%)

[a] (a) THF, -20 °C; (b) THF, 25 °C.

Scheme 6. Fragment coupling and synthesis of the macrocyclization substrate **45**

at C3 and Mosher ester analysis at C10 of the product of methanolysis, led to a stereochemical revision at C10 (M.H.G. Munro 1996, personal communication). Since a Nagao acetate aldol had introduced the (*R*)-stereochemistry at this center, an inversion was required to obtain the correct stereochemistry. This was readily accomplished by a Mitsunobu reaction, which simultaneously introduced the enyne acid fragment **8** and inverted the C10 stereochemistry to the required (*S*)-configured ester **42** in 92% yield (Scheme 6). A model system verified that the Mitsunobu coupling proceeded with inversion of configuration and minimal loss of stereochemical integrity. A subsequent desilylation of the TBS ether provided the substrate, alcohol **43**, for the key β-lactam-based macrocyclization.

A β-Lactam-Based Macrocyclization Strategy Applied to the Pateamine A Macrocycle. β-Lactams have been utilized as acylating agents in both inter- and intramolecular reactions with oxygen (Palomo et al. 1995; Ojima et al. 1992, 1993), nitrogen (Bhupathy et al. 1995; Ojima et al. 1994), and carbon (Ojima et al. 1995; Palomo et al. 1994) nucleophiles. (For reviews of the utility of β-lactams as synthetic intermediates, see Wasserman 1987; Hesse 1991; Manhas et al. 1988.) Wasserman has described elegant applications of intramolecular transamidation reactions of β-lactams to access a number of alkaloid natural products (Wasserman 1987). A well-known example of an intermolecular β-lactam-based acylation is the side-chain attachment of baccatin III and derivatives to give taxol or taxotere. (For a lead reference, see Ojima 1995 and Holton 1990.) In addition, Kahn reported an intramolecular transamidation approach to **7** and 10-membered β-turn mimetics employing acyl hydrazides as the nucleophilic partner (Gardner et al. 1993). However, the use of a β-lactam to acylate a secondary alcohol in a macrocyclization strategy has not been reported to our knowledge.

Our initial studies of the β-lactam-based intermolecular acylation focused on the conditions of Ojima that had been developed for the attachment of the side chain of taxol (Ojima et al. 1995). In order to determine possible nitrogen substituents that would activate the β-lactam towards acylation and serve as a suitable protecting group for late-stage deprotection, we studied several simple model β-lactams as acylating agents. (For related studies of intermolecular acylations with α-substituted β-lactams, see Ojima et al. 1992.) We employed isopropanol as the nucleophile to mimic the secondary alcohol at C24 (pateamine numbering) that would serve as the nucleophilic partner in the proposed pateamine A macrocycle synthesis (Fig. 3). These model studies indicated that both the *N*-benzyloxy and *N*-*t*-Boc-β-lactams provided good yields (66%–80%) of the corresponding isopropyl ester using the conditions of Ojima (Ojima et al. 1995). We also explored the conditions of Palomo that were utilized for the intermolecular alcoholysis of β-lactams (Palomo et al. 1995). While NaN_3 gave no reaction, even on heating in DMF, KCN in DMF gave good yields of the corresponding isopropyl esters from the *N*-*t*-Boc-β-lactams (67%–84%).

Turning to the real system, we initially studied the conditions of Ojima that had proven successful in model studies (Ojima et al. 1995). Using the same conditions but under high dilution (~0.002 *M*), we obtained a 37% yield of the desired macrocycle **46** and 17% of an isomeric macrocycle (not shown) which was not characterized completely but appears to be due to isomerization about the C21, C22 olefinic bond (Scheme 7). This may arise from intramolecular deprotonation at the C22 methyl group by the alkoxide formed at C24. Other strong bases led to no improvements, and the yields obtained in these attempts were unsatisfactory to complete the synthesis, so we studied other conditions that were ideally neutral pH. Thus, we applied Palomo's conditions to alcohol **43** and were pleased to find that syringe pump addition of this alcohol to a 0.5–0.6 *M* KCN in DMF solution (~0.002 *M*, final substrate concentration) led cleanly to the desired macrocycle in 52%–72% yield (<5 mg scale). An observed polar by-product in this reaction that was not isolated is presumably the amino acid derived from hydrolysis of the acyl cyanide intermediate.

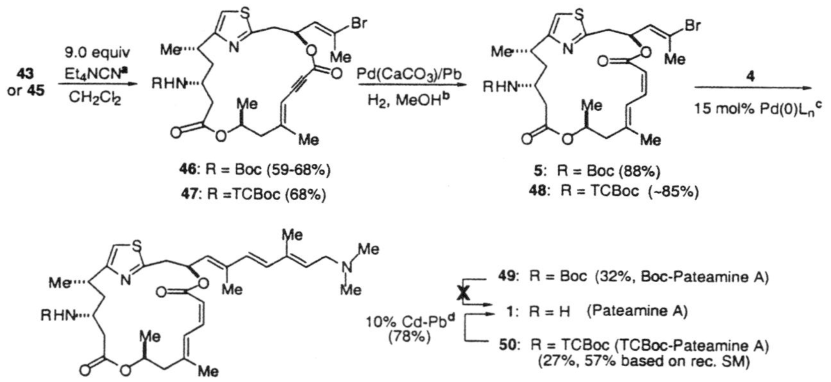

43 or 45 —(9.0 equiv Et_4NCN^a, CH_2Cl_2)→ 46: R = Boc (59-68%); 47: R = TCBoc (68%) —($Pd(CaCO_3)/Pb$, H_2, $MeOH^b$)→ 5: R = Boc (88%); 48: R = TCBoc (~85%) —(4, 15 mol% $Pd(0)L_n^c$)→

49: R = Boc (32%, Boc-Pateamine A) —(10% $Cd\text{-}Pb^d$ (78%)) ✗→ 1: R = H (Pateamine A)

50: R = TCBoc (TCBoc-Pateamine A) (27%, 57% based on rec. SM) —(10% $Cd\text{-}Pb^d$ (78%))→ 1: R = H (Pateamine A)

[a] (a) Final concentration 0.0002 M, 4Å MS, 25°C; (b) 1 atm H_2, 25 °C; (c) (1:4, Pd :ligand), TFP or Ph_3As, THF, 25 °C; (d) 1M NH_4OAc, THF, 25 °C.

Scheme 7. Synthesis and failed deprotection of Boc-pateamine A and completion of the synthesis via TCBoc-pateamine A

Despite this advance, we were somewhat concerned about product isolation from the large volumes of DMF that would be required, as the reaction was performed on larger scale. We therefore explored the use of Et_4NCN, a CH_2Cl_2 soluble source of cyanide ion. To our delight, use of 9.0 equivalents of Et_4NCN in CH_2Cl_2 (0.002 *M*, final substrate concentration) led to practical rates of formation of the desired macrocycle **46** in 59%–68% yield. Stirring both the alcohol **43** and Et_4NCN in CH_2Cl_2 over freshly activated, powdered molecular sieves prior to reaction to remove traces of H_2O led to slight improvements in yields. The use of CH_2Cl_2 as the solvent in this macrocyclization greatly simplified product isolation.

We have recently obtained (Romo et al. 1998) ^{13}CNMR evidence for an acyl cyanide intermediate in a model acylation reaction (**51**→**53**) (unpublished results of R.M. Rzasa). As shown in Fig. 4, addition of Et_4NCN to a CD_2Cl_2 solution of β-lactam **51** leads to appearance of a new set of carbonyl signals after 1 h (Fig. 4b) corresponding to the presumed acyl cyanide intermediate **52**. The chemical shift for this acyl cyanide, carbonyl carbon (δ 175.9) correlates well with that for acetyl cyanide (δ 173.8). Some quantity of the carboxylic acid (δ 173.1)

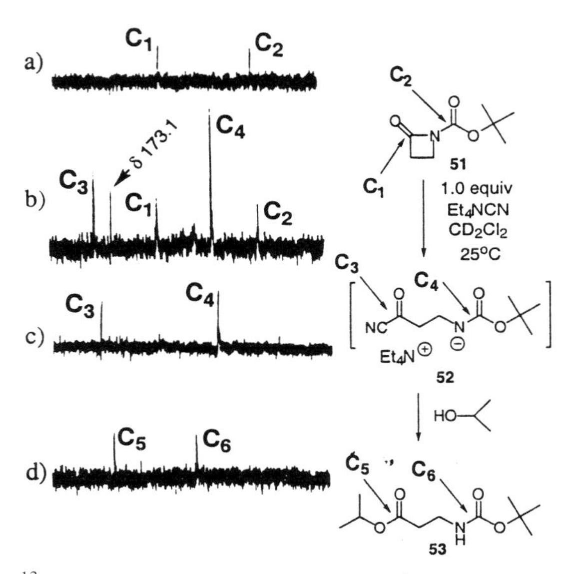

Fig. 4a–d. ^{13}C NMR spectra (carbonyl region) of β-lactam **51**, presumed acyl cyanide **52**, and isopropyl ester **53**. (**a**) β-Lactam **51** (C_1 δ 165.0; C_2 δ 143.3). (**b**) One hour after addition of 1.0 equiv Et_4NCN (the signal at δ,173.1 corresponds to the carbonyl carbon of the carboxylic acid formed by adventitious water); (**c**) 5 h after addition of 1.0 equiv Et_4NCN in a separate experiment (C_3 δ 175.9; C_4 δ 155.7); (**d**) after addition of isopropanol (C_5 δ 172.0; C_6 δ 155.9). (Chemical shifts are relative to TMS and spectra are at slightly different spectral widths)

derived from hydrolysis of the acyl cyanide by adventitious water can also be observed in this spectrum (Fig. 4b). In a different experiment after 5 h (Fig. 4c), the β-lactam **51** was consumed and signals corresponding to the starting β-lactam were replaced with those corresponding to the presumed acyl cyanide intermediate **52**. Upon addition of isopropanol, the acyl cyanide signals disappeared and a new set of signals emerged, corresponding to the independently prepared isopropyl ester **53**. These experiments provide direct evidence for the intermediacy of an acyl cyanide intermediate in Palomo's cyanide-promoted acylations (Palomo et al. 1995) and the present β-lactam-based macrocyclization.

Synthesis of Boc-Pateamine A. With the macrocycle **46** in hand, hydrogenation with Lindlar's catalyst led to slow reduction of the alkyne to the *E*, Z-dienoate (Scheme 7). Due to the presence of the thiazole, it was necessary to determine the optimal Lindlar catalyst to substrate ratio (~600 mg Lindlar catalyst/ mmol substrate) for a satisfactory hydrogenation rate. Stille coupling, using methods established previously in the model studies, appended the trienylamine side chain to provide Boc-pateamine A **49** in 32% yield, which correlated well with the same compound derived from the natural product as determined by comparison of 300 MHz ^{1}H NMR. (We thank Profs. Munro and Blunt for providing a 300 MHz ^{1}H NMR spectrum of Boc-pateamine A derived from natural material.) No further characterization data were available for Boc-pateamine A derived from natural material, but this gave us our first indication that we were pursuing the correct stereoisomer. Unfortunately, attempts to deprotect Boc-pateamine A or a simpler model system led only to decomposition or no reaction. Importantly, however, Boc-pateamine A provided an interesting insight into the structural requirements for immunosuppressive activity (see below) and also led us to explore novel methods for simultaneous deprotection and purification of Boc-amines (Liu et al. 1998).

Completion of the Synthesis of (-)-Pateamine A. In light of the inability to deprotect Boc-pateamine A, we searched for alternative amine-protecting groups that could ideally be removed under essentially neutral conditions. In this regard, we were attracted to a recent report by Ciufolini on deprotection of the trichloroethoxy carbamate (TROC), involving the use of Cd-Pb couple and a buffered solvent system of NH_4OAc and THF. (Dong et al. 1995; we thank Prof. Ciufolini for helpful discussions concerning the suitability of these deprotection conditions for the problem at hand.) In addition, it had been demonstrated that these conditions were tolerated by reducible functionality such as a α-bromoenone; this gave us the confidence to apply it to the projected substrate, which would contain a triene and a dienoate. We settled on the use of the trichloro-*t*-butoxy carbamate (TCBoc) rather than the TROC group to ensure a regioselective acylation during the β-lactam-based macrocyclization.

Towards this end, nitrogen protection of β-lactam **39** with TCBoc-Cl led to incomplete acylation, and this could not be forced to completion by excess reagents or gentle heating (Scheme 5). After one recycle of starting material, we obtained a 51% (unoptimized) yield of the *N*-TCBoc-β-lactam **41**. Desilylation, followed by the Mitsunobu reaction to append the enyne acid **8** and a second desilylation, delivered alcohol **45** (Scheme 6).

Macrocyclization of alcohol **45** using the previously developed conditions gave the TCBoc macrocycle **47**. Lindlar reduction proceeded, as before, to deliver the diene **48** in good yield. Stille coupling, using conditions established previously, delivered TCBoc-pateamine A (**50**) in 27% yield (57% based on recovered bromide **48**). It proved beneficial to stop the Stille reaction prior to completion and recycle recovered bromide **5** (R = TCBoc), since allowing the reaction to proceed to completion yielded only 25%–32% of TCBoc pateamine A (**50**) and no recovered starting bromide **5**. This result indicates that π-allyl formation may indeed be competitive with the Stille reaction in this system. Final deprotection of TCBoc pateamine A (**50**) by the method of Ciufolini cleanly provided (-)-pateamine A (**1**) after purification by reverse-phase column chromatography and passage through an amino cartridge to free-base the final product. Synthetic pateamine A correlated well with the natural product in all respects, including ^{1}H, ^{13}C NMR, IR, UV, $[\alpha]_D$, CD, HRMS, TLC, and IC_{50} in the IL-2 reporter gene assay (see below). [We thank Prof. Peter Northcote (Victoria University of Wellington, New Zealand), Mr. Lyndon West, and Dr. Chris Attershill (New Zealand National Institute of Water and Atmospheric Research, Ltd.) for kindly providing a sample of natural pateamine A.]

4.2.3 Biomechanistic Studies and Synthesis of Derivatives for Putative Cellular Receptor Isolation

Activity in the IL-2 Reporter Gene Assay of (-)-Pateamine A and Related Compounds. Pateamine A has previously been shown to inhibit both mouse and human mixed lymphocyte reactions, indicating that it is immunosuppressive (G.T. Faircloth 1994, private communication). It is not clear, however, which signaling pathway involved in T-cell activation is affected by pateamine A. Pateamine A inhibits T-cell receptor-mediated IL-2 production but does not appear to affect IL-2 receptor-mediated cell proliferation (J.O. Liu, L. Sun, private communication). Furthermore, pateamine A potently blocks PMA/ionomycin-stimulated IL-2 promoter activation (Su et al. 1994). The IC_{50} values for pateamine A in the IL-2 reporter gene assay were determined to be 0.33 and 0.46 n*M* for the synthetic and natural pateamine A, respectively. These results suggest that pateamine A inhibits an intracellular signaling step in the TCR signaling pathway leading to IL-2 production (Fig. 1). The mode of action of pateamine A is therefore similar to that of FK506 and CsA but distinct from that of rapamycin (Fig. 1). Furthermore, the fact that Boc-pateamine A retains a significant amount of immunosuppressive activity, as measured by the IL-2 reporter gene assay (IC_{50}=2.1 n*M*), suggested a site of attachment for the preparation of biochemical probes for detection and isolation of the molecular target of pateamine A via modification of the C3 amino group in pateamine A. This latter result has guided our synthesis of various derivatives outlined below.

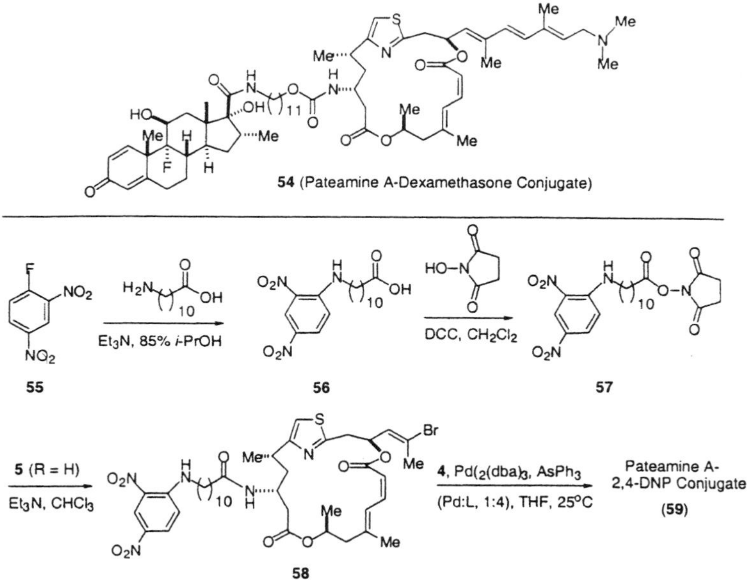

54 (Pateamine A-Dexamethasone Conjugate)

55 → ($H_2N(CH_2)_{10}COOH$; Et_3N, 85% *i*-PrOH) → **56** → (N-hydroxysuccinimide; DCC, CH_2Cl_2) → **57** → (**5** (R = H); Et_3N, $CHCl_3$) → **58** → (**4**, $Pd_2(dba)_3$, $AsPh_3$; (Pd:L, 1:4), THF, 25°C) → Pateamine A-2,4-DNP Conjugate (**59**)

Scheme 8. Structure of the pateamine-dexamethasone conjugate and synthesis of the pateamine A-2,4-DNP conjugate

Synthesis of Pateamine A Hybrid Molecules. One approach to the isolation of a putative cellular receptor involves the use of the recently reported yeast three-hybrid assay developed in the lab of our collaborator, Professor Jun Liu (Licitra and Liu 1996). To this end, we developed a synthesis of a pateamine A-dexamethasone hybrid **9** (Scheme 8, synthesis not shown) to be utilized in this assay either for target identification or for future studies of the interaction between pateamine A and its target. The synthesis involved simple acylation of our synthetic pateamine A with an appropriate dexamethasone coupling agent (Romo et al. 1998).

More interestingly, we were able to make use of our synthetic intermediates to prepare targeted derivatives (Scheme 8). Attachment of the trienylamine side chain of pateamine A greatly increases the polarity and thus the difficulty in handling these tertiary amine-containing intermediates. We therefore elected to prepare a required dinitrophenyl (2,4-DNP) derivative by first acylating the amino group of macrocycle **5** (R=H) and then, in a final step, attaching the polar side chain via Stille coupling. This allowed the synthesis of the 2,4-DNP conjugate **59** required for ongoing affinity chromatography experiments (Scheme 8). (These experiments are being performed in the laboratory of our collaborator, Prof. Jun Liu, Center for Cancer Research and Departments of Biology and Chemistry, Massachusetts Institute of Technology.)

4.3 Total Synthesis of (8*S*, 21*S*, 22*S*, 23*R*) and (8*R*, 21*S*, 22*S*, 23*R*)-Okinonellin B via a β-Lactone Intermediate

Although β-lactones undergo a number of unique and stereospecific reactions (Pommier and Pons 1993), they have had limited use as intermediates in the context of natural product total synthesis. (We are aware of only a few examples of the use of β-lactones as intermediates in asymmetric, natural product total synthesis: huperizine A: Kozikowski et al. 1996. bourgeanic acid: White and Johnson 1994. lactacystin: Corey et al. 1993; Taunton et al. 1996; Fukuyama and Xu 1993.) This may in part be due to the lack of direct and general methods for their synthesis in optically pure form. (For some recent asymmetric methods of β-lactone synthesis, see Bates et al. 1991; Case-Green et al.

1991; Capozzi et al. 1993; Tamai et al. 1994a,b; Dirat et al. 1995; Zemribo and Romo 1995; Arrastia et al. 1996; Dymock et al. 1996; Yang and Romo 1998.) As part of a program aimed at the application of β-lactones as intermediates in natural product synthesis, we utilized a β-lactone-based strategy for the total syntheses of (8*S*, 21*S*, 22*S*, 23*R*)-okinonellin B (**60**) and (8*R*, 21*S*, 22*S*, 23*R*)-okinonellin B (**61**). Isolated by Fusetani and co-workers from the marine sponge *Spongionella* sp (Kato et al. 1986), the cytotoxin okinonellin B is a member of a family of marine furanosesterterpenes that display a variety of biological activities including antibacterial, cytotoxic, and antispasmodic activity (Crews and Naylor 1985; Rochfort et al. 1996). The reduced butenolide of okinonellin B makes it unique among other sesterterpenoids in this class of marine natural products. Fusetani described the relative stereochemistry of the butyrolactone, but the relative stereochemistry between the butyrolactone and the C8 stereocenter and the absolute stereochemistry were not determined. The present synthesis demonstrates the utility of β-lactones as intermediates in the synthesis of natural products and, specifically, in the concise synthesis of all syn-trisubstituted butyrolactones. The synthesis utilized two methods developed in our laboratories. The first is a tandem Mukaiyama aldol-lactonization reaction which provides a highly diastereoselective and stereocomplementary route to *trans* ($ZnCl_2$ as Lewis acid promoter; Hirai et al. 1994; Yang and Romo 1997a,b; Rouhi 1995)and *cis* ($SnCl_4$ as Lewis acid promoter; Wang et al. 1999) β-lactones from aldehydes and thiopyridyl ketene acetals. The second reaction allows a one-pot conversion of benzyloxy-substituted β-lactones to γ- and δ-lactones with concomitant debenzylation (Arrastia et al. 1996; Zemribo et al. 1996).

4.3.1 Retrosynthetic Analysis of Okinonellin B

It was envisioned that a tandem Mukaiyama aldol-lactonization (TMAL) reaction (**64**→**65**) (Hirai et al. 1994; Yang and Romo 1997d; Yang et al. 1997; Rouhi 1995) and a tandem transacylation-debenzylation of a benzyloxy-substituted β-lactone (**65**→**63**) (Arrastia et al. 1996; Zemribo et al. 1996) would deliver the butyrolactone **63** in a highly stereocontrolled manner (Fig. 5). A Negishi coupling of the (*R*) and (*S*)-vinyl iodide **62** and the butyrolactone fragment **63** would then com-

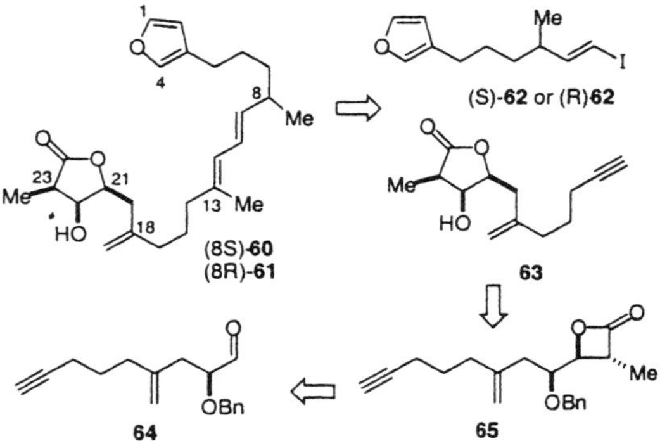

Fig. 5. Retrosynthesis of okinonellin B showing the two-step β-lactone-based strategy for the synthesis of butyrolactone **63** from aldehyde **64**

plete the synthesis in a concise fashion (Negishi et al. 1978). Both enantiomers of vinyl iodide **62** would be synthesized and coupled, in order to assign the relative configuration at C8 as well as the absolute configuration of the natural product.

The synthesis of the enantiomeric vinyl iodides **73** began with two sequential alkylations of 1,3-dithiane using the readily available iodides (*S*) and (*R*)-**68** (the enantiomeric iodides **67**, **68** are availabe in three steps from (*S*) and (*R*)-methyl-3-hydroxy-2-methylpropanoate, see

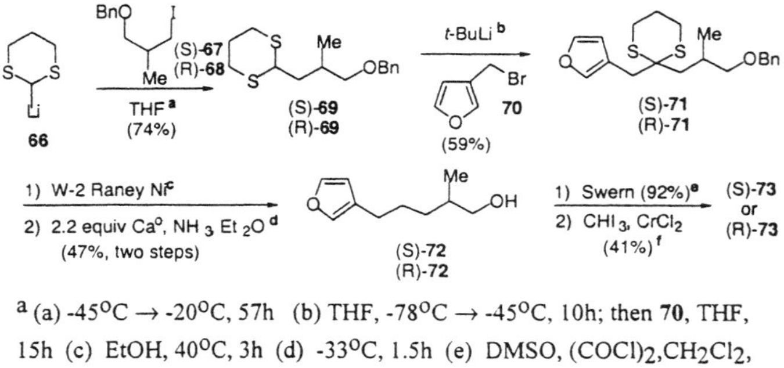

[a] (a) -45°C → -20°C, 57h (b) THF, -78°C → -45°C, 10h; then **70**, THF, 15h (c) EtOH, 40°C, 3h (d) -33°C, 1.5h (e) DMSO, $(COCl)_2$, CH_2Cl_2, -78°C; Et_3N, -78°C → 0°C, 25 min (f) THF, 0°C, 4h.

Scheme 9. Synthesis of the enantiomeric furans **73**

White and Kawasaki 1992) and 3-bromomethyl furan **70** (the unstable furanyl bromide **70** was prepared immediately prior to use from 3-furan-methanol using PPh_3/Br_2: Bernasconi et al. 1986) to give the dialkylated dithianes **71** (Scheme 9). A two-stage reduction involving Raney nickel and dissolving metal reduction with calcium cleaved the dithiane and benzylether and provided the alcohols **72**. Swern oxidation (Mancuso and Swern 1981), followed by Takai reaction (Takai et al. 1986), provided the required vinyl iodides **73** for Negishi coupling to the butyrolactone **63**.

4.3.2 Synthesis of the Butyrolactone Fragment via a β-Lactone Intermediate

The synthesis of the butyrolactone **63** began with conversion of the known lactone **74** (Lactone **74** is available in 4 steps from -(-)-malic acid, see Bernardi et al. 1990) to the Weinreb amide; this was followed by silylation to give amide **75** (Scheme 10). A carefully controlled addition (best results were obtained after titration of the generated Grignard reagent; see Bergbreiter and Pendergrass 1981) of the Grignard reagent derived from bromide **76** [bromide **76** was obtained by a four-step sequence from commercially available 4-pentyn-1-ol: (a) DHP, TsOH (b) *n*-BuMgBr, thexyldimethylsilylchloride (c) $BF_3{\bullet}OEt_2$, EtSH, (d) PPh_3, Br_2] to amide **75** gave the desired ketone **77** in 87% yield. Tebbe methylenation (Tebbe et al. 1978), simultaneous desilylation of the silylether and silylacetylene, and Swern oxidation delivered the aldehyde **64**, required for the TMAL reaction. Treatment of aldehyde **64** with $ZnCl_2$ and the ketene acetal **80** at ambient temperature for 14 h gave the β-lactone **65** as a single diastereomer (no minor diastereomers were detected in either $CDCl_3$ orC_6D_6 (300 MHz ^{1}H-NMR) in 73% yield and with <49% epimerization. (The enantiomeric purity was determined by chiral HPLC (Chiralcel OD) on comparison with the racemic β-lactone. The latter was obtained by TMAL reaction of racemic aldehyde **64** via silyl enol ether formation and desilylation of the enantiopure aldehyde **64**.) The stereochemical outcome is consistent with a chelation-controlled initial aldol reaction as determined by conversion to the *all*-syn-butyrolactone **63**. (Our current mechanistic understanding of the Zn(II)-mediated TMAL reaction has been described in a

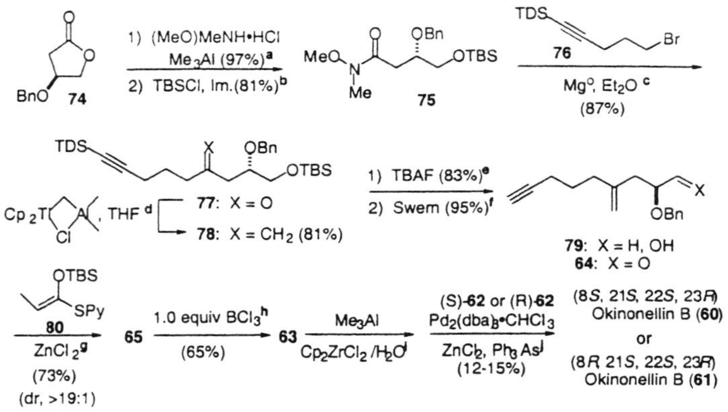

[a] (a) C_6H_6, 0°C, 1.5 h (b) DMF, 24°C, 2.5 h (c) reflux then added to **75**, Et_2O, -78°C → 0°C, 39 h (d) -40°C, 3h (e) THF, 0°C → 24°C, 3h (f) DMSO, $(COCl)_2$, CH_2Cl_2, -78°C; Et_3N, -78°C, 1h (g) CH_2Cl_2, 25°C, 14h (h) CH_2Cl_2, -78°C, 15 min (i) CH_2Cl_2, -23°C, 2 h (j) (1:4, Pd (8 mol%):ligand), NMP, 25°C. 7.5 h.

Scheme 10. Synthesis of the key β-lactone intermediate **65** and its conversion to okinonellin

separate report: Zhao and Romo 2000.) The tandem transacylation-debenzylation of β-lactone **65** to the *all*-syn-butyrolactone was effected using BCl_3,which gave improved results over the originally reported $FeCl_3$ (29%) (Zemribo et al. 1996; Arrastia et al. 1996). Thus, in two steps aldehyde **64** is transformed into the highly functionalized, *all*-syn-butyrolactone **63** in a highly stereocontrolled fashion.

4.3.3 Completion of the Synthesis of (8*S*, 21*S*, 22*S*, 23*R*) and (8*R*, 21*S*, 22*S*,23*R*)-Okinonellin B

For the fragment coupling to provide okinonellin B, we relied on the one-pot procedure of Negishi involving alkyne carboalumination, followed by transmetallation to Zn(II) and a Pd(0)-mediated coupling to a vinyl halide (Negishi et al. 1978). Carboalumination of alkyne **63**, using the water-accelerated conditions developed by Wipf et al. (1993), followed by addition of $ZnCl_2$, the Pd(0)/Ph_3As catalyst system reported by Farina (Farina and Krishnan 1991), and vinyl iodides (*S*)-**73** or (*R*)-**73**, gave (8*S*)-okinonellin B (**60**) and (8*R*)-okinonellin B (**61**), respectively (Scheme 10). Significant amounts (~40%) of quenched vinyl metallic species derived from **63** and methylated products derived from vinyl iodides **62** account in part for the low yields obtained in this carboalumination/coupling sequence. [An authentic sample of okinonellin B was no longer available, as it had degraded. In addition, recent re-extraction of the sponge did not afford any detectable okinonellin B (private communication from N. Fusetani and S. Matsunaga, University of Tokyo). We have also noted the instablility of okinonellin B even when frozen in benzene.]

Not unexpectedly, the diastereomers of okinonellin B (**60** and **61**) did not exhibit any differences by either ^{1}H or ^{13}C NMR. A significant difference in the CD spectrum of the two diastereomers was observed. However, neither a CD spectrum of the natural product nor the natural product itself was available for comparison. At this time, we can only speculate, based on the optical rotation data, that natural okinonellin B possesses the 8*R*, 21*R*, 22*R*, 23*S* stereochemistry that is enantiomeric to our synthetic (8*S*, 21*S*, 22*S*, 23*R*)-okinonellin B (**60**).

[Natural okinonellin B: $[\alpha]^{D}_{20}$ = 17.9 (*c* 0.15, EtOH).

(8*S*, 21*S*, 22*S*, 23*R*)-Okinonellin B (**1**): $[\alpha]^{D}_{24}$ = –7.6 (*c* 0.19, EtOH).

(8*R*, 21*S*, 22*S*, 23*R*)-Okinonellin B (**2**): $[\alpha]^{D}_{24}$ = –98.0 (*c* 0.15, EtOH)].

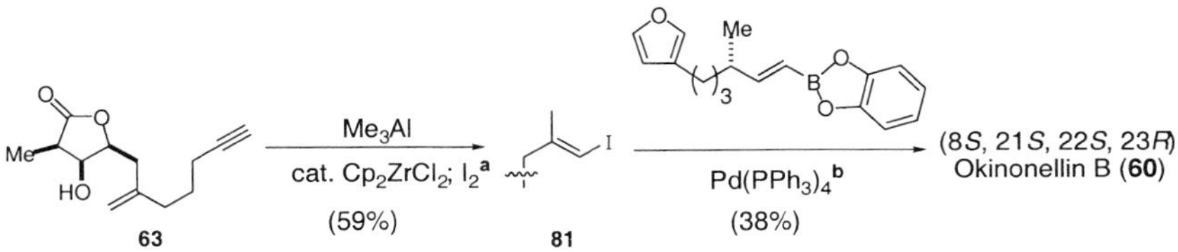

[a](a) H_2O_2. CH_2Cl_2, 25°C (b) TlOEt, hexane-THF, 25°C

Scheme 11. Improved end-game strategy to okinonellin using a Suzuki coupling

More recently, we have applied the Suzuki coupling (Suzuki 1998) to the final coupling of the furan and butyrolactone fragments. Following conversion of the acetylene **63** to the vinyl iodide **81**, Suzuki coupling with the vinyl boronate **82** obtained by Swern oxidation, Corey-Fuchs (Corey and Fuchs 1972) reaction, and hydroboration of alcohol (*S*)-**72** gave improved yields of okinonellin B (Scheme 11).

4.4 (-)-Gymnodimine: a Potent Marine Toxin Possessing a Unique 2-Azine Spiro[5.5]undecane Moiety from the Dinoflagellate, *Gymnodinium* cf. *Mikimotoi*

4.4.1 Introduction

Marine toxins have proven to be useful biochemical probes for the study of a variety of biological systems. Examples include the use of tetrodotoxin and brevetoxin in studies of sodium channels (Catterall 1992), okadaic acid in studies of protein phosphatases (Bialojan and Takai 1988; Haystead et al. 1989), and domoic acid in studies of neuronal receptors (Debonnel et al. 1989; Wenthold and Hapson 1988). Gymnodimine (**83**) is a member of a class of recently isolated marine toxins that have in common spirocyclic imines as part of a macrocycle that also contains a polyether subunit (Seki et al. 1995). Other members of this family include the spirolides (Hu et al. 1995) and pinnatoxins (Uemura et al. 1995), and these natural products have attracted much synthetic interest, culminating recently in an elegant synthesis of pinnatoxin A by Kishi employing an intramolecular Diels-Alder reaction. (For other synthetic studies toward gymnodimine, see Ishihara et al. 1997; Rainier and Wu 1999. For synthetic studies of pinnatoxin, see Ishihara et al. 1998, 1999; Nitta et al. 1999; Sugimoto et al. 1999. Total synthesis of pinnatoxin A: McCauley et al. 1998.) Recently, Munro and Blunt disclosed the relative and absolute stereochemistry of gymnodimine that was determined by X-ray analysis of the *p*-bromobenzamide **85** of gymnodimine (**84**), the reduction product of gymnodimine (Stewart et al. 1997). A new analog of gymnodimine, named gymnodimine B, has recently been isolated (Miles et al. 1999).

As a means to explore concise approaches to the interesting spirocyclic imines found in these toxins and to elucidate possible mechanisms

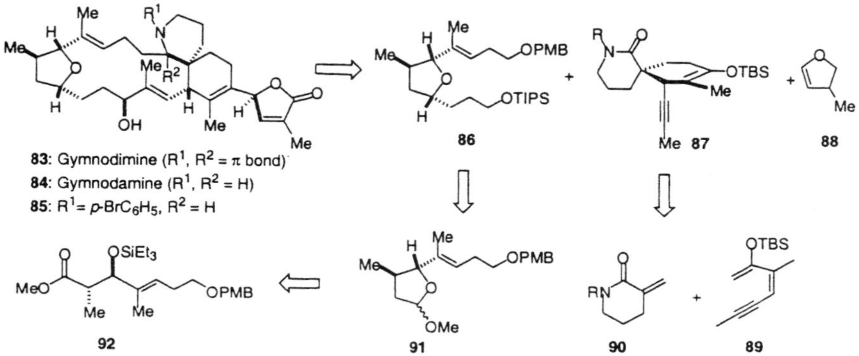

Fig. 6. Retrosynthetic analysis of gymnodimine

of action, we initiated a total synthesis of (-)-gymnodimine. Towards these goals, we have developed a concise Diels-Alder strategy to the spirocyclic imine moiety of gymnodimine that may be applicable to other members of this family of toxins. In addition, we have completed an asymmetric synthesis of the tetrahydrofuran fragment of gymnodimine, employing a Heathcock anti-aldol reaction and an allylation of a furanose.

4.4.2 Retrosynthetic Analysis

In our retrosynthetic analysis of gymnodimine (Fig. 6), we envisioned the coupling of three fragments to provide the most convergency and the most flexibility in preparing derivatives for biomechanistic studies. Coupling of the tetrahydrofuran fragment, the spirocyclic lactam, and the dihydrofuran would deliver the target molecule. Assuming an endo transition state, the spirocyclic lactam **87** with the required regio- and stereochemistry would be prepared by the Diels-Alder reaction of a (*Z*)-diene **89** and the α-methylene lactam **90**. The recent compilation of acyclic (*Z*)-dienes that participate in Diels-Alder reactions by Roush and Barda (1997) suggested that the proposed (*Z*)-diene would participate in the projected Diels-Alder reaction. A Heathcock anti-aldol (Walker and Heathcock 1991; Raimundo and Heathcock 1995; Heathcock et al.

1990) reaction and a stereoselective allylation of the furanose **91** would be utilized for the preparation of the tetrahydrofuran fragment **86**.

4.4.3 A Diels-Alder Approach to Spirocyclic Imines

The α-methylene lactams required for the Diels-Alder strategy were prepared by acylation (TCBoc-Cl) or tosylation (TsCl) of the known δ-lactam **96** (Klutchko and Hoefle 1981) (Scheme 12). This lactam is ultimately derived from diester **93**, readily available in >100 g quantities by Michael addition of diethylmalonate to acrylonitrile using a known procedure (Koelsch 1943; Hesse and Bücking 1949; Albertson and Fillman 1949). Although lactam **94** is commercially available, it is somewhat expensive and can be readily prepared in large quantities by Raney nickel reduction of the nitrile **93** (Koelsch 1943; Hesse and Bücking 1949; Albertson and Fillman 1949). Subsequent reduction and dehydration gave the known α-methylene lactam **96** (Klutchko and Hoefle 1981). The lactam **96** is a pivotal intermediate, as several nitrogen protecting groups can be added and studied in the Diels-Alder reaction. Initially, the trichloro-*t*-butoxycarbamate-protected dienophile **97** and the tosylated dienophile **98** were prepared. Using this sequence, the latter dienophile was readily prepared on a large scale (>30 g).

The required diene for the Diels-Alder strategy was prepared in a concise fashion from propyne using tellurium chemistry as outlined in Scheme 13. Hydrotelluration (Dabdoub et al. 1992; Comasseto et al.

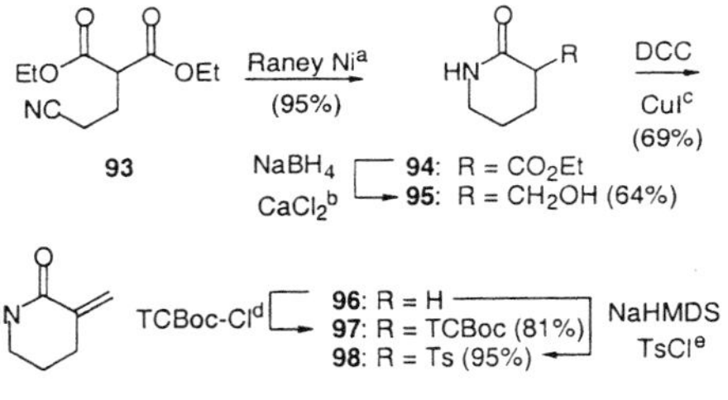

EtO OEt NC **93** Raney Ni[a] (95%) HN R DCC CuI[c] (69%)
NaBH4 CaCl2[b] **94**: R = CO_2Et **95**: R = CH_2OH (64%)
RN TCBoc-Cl[d] **96**: R = H **97**: R = TCBoc (81%) **98**: R = Ts (95%) NaHMDS TsCl[e]

[a](a) EtOH, 80^0C, H_2 at 1000 psi, 8 h; (b) MeOH, $0° \rightarrow 23°$C, 12 h; (c) PhMe, reflux, 25 min; (d) THF, -78°C, 1 h; (e) THF, -78°C, 1 h

Scheme 12. Large-scale synthesis of the dienophile

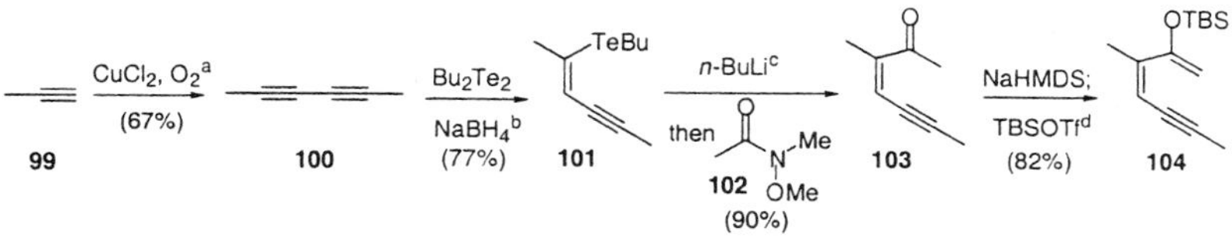

[a] (a) py.,DMF,DBU, -42° → 35°C, 3 h; (b) abs. EtOH, reflux, 3 h; (c) THF, -78°C, 1 h then **102**, -78°C, 0.75 h; (d) THF, -78°C, 0.5 h.

Scheme 13. Large-scale synthesis of the diene

1996) of hexa-2,4-diyne (Brandsma et al. 1991) followed by metallation and quenching of the vinyl anion with Weinreb amide (Weinreb and Nahm 1981; Harris and Oster 1983) gave ketone **103**. Silyl enol ether formation proceeded without isomerization; however, silica gel chromatography of this silyl enol ether led to varying degrees of isomerization. In this way, diene **104** could be readily prepared on a large scale (up to 3 g). However, due to the instability of this enol silyl ether, material is typically advanced and stored at the stage of the ketone **103**. Using this procedure, ~20 g of ketone **103** has been prepared.

Attempted Diels-Alder reactions with the TCBoc-protected lactam **97** failed under a variety of conditions. We reasoned that further activation of the dienophile was required, and tosylation of lactam dienophiles has previously been employed for increased reactivity in Diels-Alder reactions (Torisawa et al. 1996; Casamitjana et al. 1997; Torisawa et al. 1994; Oppolzer et al. 1984). Gratifyingly, the Diels-Alder reaction of tosylated dienophile **98** and diene **104** (Scheme 14) was found to proceed efficiently with Et_2AlCl at –30°C to give a single diastereomer of a crystalline cycloadduct **105** (67%). This spirocyclic lactam was found to possess the correct regio- and relative stereochemistry required for the synthesis of gymnodimine as determined by single-crystal X-ray analysis. Interestingly, we found that the same diastereomer is obtained regardless of the geometry of the diene, indicating the possibility that this is in fact a formal Diels-Alder reaction proceeding through a stepwise process. Deprotection of the lactam was accomplished using sodium napthalenide in 83% yield (Closson et al. 1967; Alonso and Andersson 1998).

This overall sequence provides a concise entry into the spirocyclic array of gymnodimine and is functionalized appropriately for elabora-

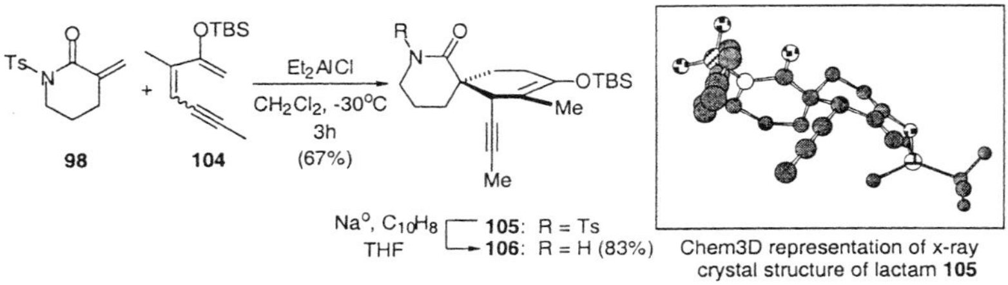

Ts–N, 98 + 104 —Et$_2$AlCl, CH_2Cl_2, -30°C, 3h (67%)→ R, N, O, OTBS, Me, Me

Na°, $C_{10}H_8$ THF: 105: R = Ts → 106: R = H (83%)

Chem3D representation of x-ray crystal structure of lactam **105**

Scheme 14. The formal Diels-Alder reaction leading to the spirocycle of gymnodimine

tion to the complete spirocyclic fragment. Chiral Lewis acids are currently being screened for their ability to promote this presumed formal Diels-Alder reaction. Importantly, this approach would appear to be applicable to the synthesis of the related spirolides.

4.4.4 Asymmetric Synthesis of the Tetrahydrofuran Fragment

The synthesis of the tetrahydrofuran fragment commenced with a Heathcock anti-aldol reaction (Walker and Heathcock 1991; Raimundo and Heathcock 1995; Heathcock et al. 1990) employing the ephedrine-derived auxiliary **107** (Evans and Mathre 1985; Scheme 15) and the known aldehyde **108** (Ito et al. 1988). This provided a mixture of three diastereomers (dr, 5:1:0.5) from which the major diastereomer **109** was isolated in 68% yield. The stereochemistry of this diastereomer was determined to be anti by ^1H-^1H coupling constant analysis which showed a coupling constant $J_{Ha,Hb}$ = 8.7–9.3 Hz (typical values: syn, J=4.8–6.4 Hz; anti, J=8.4–9.6 Hz) Walker and Heathcock 1991; Raimundo and Heathcock 1995; Heathcock et al. 1990). This aldol reaction set the stereochemistry corresponding to the C15 and C16 stereocenters of gymnodimine, and what was now required was homologation to the furanose. After some experimentation, it was determined that removal of the chiral auxiliary could best be achieved by methanolysis, which gave the ester **110** in 80% yield. Silylation, followed by a reduction-oxidation sequence, provided the aldehyde **112** in good overall yield. Wittig olefination proceeded smoothly to give the

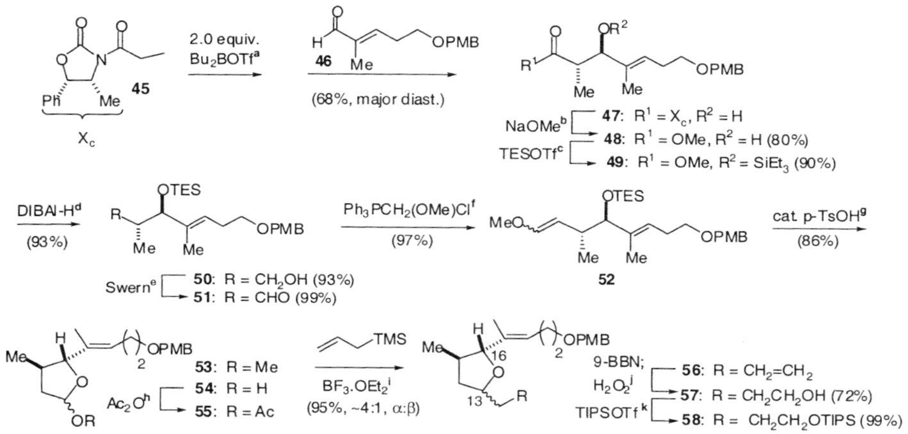

[a] (a) *i*-Pr_2NEt, Et_2O, 0°C, 45 min, -78°C, then **46**, Et_2O, -78°C, 6h; (b) MeOH, -78°C → -10°C, 2 h; (c) 2,6-lut., CH_2Cl_2, -78°C, 10 min; (d) CH_2Cl_2, -78°C 30 min; (e) DMSO, $(COCl)_2$, CH_2Cl_2, -78°C, 10 min, then **50**, Et_3N, -78°C, 1 h; (f) K-O*t*Bu, THF, 0°C → 22°C, 2 h; (g) MeOH, 25°C (for **53**), or THF, H_2O, 25°C (for **54**); (h) Et_3N, CH_2Cl_2, cat. DMAP, 22°C, 2h; (i) Et_2O, 0°C (for **53**) or Et_2O, -78°C (for **55**); (j) THF, 0°C, 24 h; NaOH; (k) 2,6-lut. CH_2Cl_2, -78°C, 40 min.

Scheme 15. Synthesis of the tetrahydrofuran fragment of gymnodimine

methoxy olefins **113** as a 3:1 mixture (300 MHz ^1H NMR) of geometric isomers in 97% yield. (For a detailed procedure for this Wittig reaction, see Pyne et al. 1982.) Acid-promoted desilylation and cyclization in MeOH provided the epimeric furanoses **91** (For a related deprotection/cyclization, see Haltiwanger et al. 1988.). Allylation with $BF_3{\bullet}OEt_2$ and allylsilane did not proceed at $-78^{\circ}C$ but required warming to $0^{\circ}C$ and delivered a 4:1 mixture of diastereomers (Kozikowski et al. 1983). The stereochemistry of the major diastereomer was determined to be α, based on a ROESY experiment which showed a cross-peak for H13 and H16 for the major diastereomer. With the hope of performing the allylation at lower temperature, the acetoxyfuranoses **116** were prepared. While allylation of these substrates did proceed at $-78^{\circ}C$, the ratio of epimers did not improve. This is surprising, in light of an allylation by Ishihara et al. (1997) on a related substrate which differs only by the nature of the substituent at C5. In their substrate, an

sp^3 hybridized carbon is present at C16, and allylation led to very high diastereoselectivity. One rationalization for this disparity invokes $A^{1,2}$ strain, which effectively makes the sp^2 hybridized side chain at C5 sterically larger than the sp^3 hybridized carbon in the Ishihara substrate. The diastereomeric tetrahydrofurans **115** were inseparable at this stage, so they were processed to the silyl ethers **117** by a hydroboration/oxidation sequence, followed by silylation. This provided a protected version, silyl ether **86**, of the tetrahydrofuran fragment of gymnodimine.

4.5 Conclusion

The synthetic opportunities provided by the novel molecular structures isolated from the marine world will undoubtedly continue to attract the interest of synthetic chemists who are interested in developing new synthetic methods and strategies and/or who wish to understand how these compounds exert their potent biological effects. It is hoped that government agencies and pharmaceutical companies will continue to spend the extra time required (relative to combinatorial chemistry) to isolate and characterize the myriad of unique molecular entities to be discovered in the oceans. Certainly, it can be argued that the success rate for discovering truly novel molecular scaffolds continues to be greater in the ocean than in the laboratory. Thus nature continues to “set the goal” for many synthetic endeavors, and the prospects for new discoveries ultimately leading to drug discovery are very good indeed.

Several marine natural products have attracted the interest of my research group over the past 6 years. This interest stems from our desire to develop methods or strategies that would be applicable to the compounds of interest and from the need to develop efficient syntheses for biomechanistic studies of these agents. Alternatively, the natural products provided a useful proving ground for methods that were developed in our group.

We developed a convergent, stereoflexible strategy for the total synthesis of (-)-pateamine A and derivatives which led to structural verification of this marine sponge isolate as the 3*R*, 5*S*, 10*S*, 24*S* isomer. This synthesis provides an avenue to explore the potent immunosuppressive properties of pateamine A and derivatives. During the course of this synthesis, we developed a β-lactam based macrocyclization that employs CH_2Cl_2-soluble Et_4NCN as promoter, and we provided evidence for the intermediacy of an acyl cyanide intermediate in this type of acylation reaction. We also determined that a Stille coupling is competitive with potential π-allyl formation from an allylic and even a triallylic acetate. (For another report of a Stille reaction with an allylic acetate substrate that appeared after our initial report, see: Nicolaou et al. 1998). A mild SmI_2 reduction was applied to the reductive cleavage of the N-O bond of *N*-benzyloxy-β-lactams (Yang and Romo 1999).

Pateamine A inhibits T-cell receptor-mediated IL-2 transcription, indicating that its mode of action is similar to that of FK506 and CSA but distinct from that of rapamycin. The fact that Boc-pateamine A retains significant activity in the IL-2 reporter gene assay suggested a site of attachment (the C3 amino group) for hybrid molecules useful for isolation of a putative cellular receptor. Two such hybrid molecules, a dexamethasone-pateamine A hybrid and a 2,4-dinitrophenyl hybrid to be utilized either for target identification or for further studies of the interaction between pateamine A and its target, were synthesized based on this finding. In addition preliminary studies of pateamine A derivatives suggest that both the macrocycle and trienylamine side chain are necessary for a competent immunosuppressive agent (Romo et al. 1998). The synthesis of designed derivatives for further understanding of the structural features required for immunosuppressive activity and for isolation of a putative cellular receptor characterize our ongoing studies of the pateamines.

Application of the tandem Mukaiyama-aldol lactonization and a tandem transacylation/debenzylation sequence led to the first total syntheses of (8*S*, 21*S*, 22*S*, 23*R*) and (8*R*, 21*S*, 22*S*, 23*R*)-okinonellin B. The synthesis demonstrated the utility of β-lactones as intermediates in natural product synthesis and specifically their use for the synthesis of highly substituted and functionalized butyrolactones. A two-step procedure efficiently converted a chiral aldehyde to an *all*-syn-butyrolactone in a highly diastereoselective manner.

In an effort to understand the potent toxic effects of a new family of marine agents that uniquely contain a 2-azine spiro[5.5]undecane or 2-azine spiro[5.6]dodecane moiety, we have initiated synthetic studies involving the simplest member of this class, (-)-gymnodimine. An apparent "formal" Diels-Alder reaction provides a rapid entry into the spirocyclic moiety of gymnodimine and promises to be applicable to other members of this family of marine toxins. An asymmetric synthesis of the tetrahydrofuran fragment of gymnodimine was accomplished but suffered from moderate diastereoselectivity (4:1) in a furanose allylation step. These preliminary studies lay the foundation for an eventual total synthesis of gymnodimine and derivatives. Importantly, our synthetic studies already provide routes to compounds that may prove useful for understanding the reactivity of the presumed cyclic imine pharmacophore and to develop immunoassays to detect these toxins prior to outbreak.

Acknowledgements. I would like to express my deepest appreciation for the dedication shown by the students with whom I have had the pleasure to work during my early years at Texas A&M and whose work is recounted in this manuscript. Special thanks to Ms. Ingrid Buchler for her help in the preparation of this manuscript. These studies were made possible by generous support from the NIH (GM 52964–01), the Robert A. Welch Foundation (A-1280), the Alfred P. Sloan Foundation (BR-3787), Zeneca Pharmaceuticals, and the Camille and Henry Dreyfus Foundation. We are grateful to our collaborators Prof. Jun Liu (MIT), Profs. Murray M.H.G. Munro, and John W. Blunt (Univ. of Canterbury, NZ) for their contributions to the studies described herein. We would like to thank the many skilled undergraduate students who provided technical assistance for these projects: Kristofor A. Voss, Erin Chavez, W. Chad Shear, Dora Rios, Noel Crenshaw, and Chris Youngkin. Mr. Joe Langenhan and Mr. Kristofor A. Voss were supported by the NSF through the REU program (CHE-9322109) at Texas A&M. We thank Dr. Joe Reibenspies (Texas A&M) for performing X-ray analyses using instruments obtained with funds from the NSF (CHE-9807975) and Dr. Lloyd Sumner and Dr. Barbara Wolf of the Texas A&M Center for characterization for mass spectral analyses obtained on instruments acquired by funding from the NSF (CHE-8705697) and the TAMU Board of Regents Research Program.

References

Abell AD, Blunt JW, Foulds GJ, Munro MHG (1997) Chemistry of the mycalamides – antiviral and antitumor compounds from a New Zealand marine sponge. 6. The synthesis and testing of analogues of the C(7)–C(10) fragment. J Chem Soc Perkin Trans 1:1647–1654

Aguilar E, Meyers AI (1994) Reinvestigation of a modified Hantzsch thiazole synthesis.Tetrahedron Lett 35:2473–2476

Aksela R, Oehlschlager AC (1991) Stannylmetallation of conjugated enynes. Tetrahedron 47:1163–1167

Alam M, Thomson RH (1997) Handbook of natural products from marine invertebrates, part I: phylum mollusca. Harwood Academic, Amsterdam

Albertson NF, Fillman JL (1949) A synthesis of DL-proline. J Am Chem Soc 71:2818–2820

Alonso DA, Andersson PG (1998) Deprotection of sulfonyl aziridines. J Org Chem 63:9455–9461

Arrastia I, Lecea B, Cossio FP (1996) Highly stereocontrolled synthesis of substituted propiolactones and butyrolactones from achiral lithium enolates and homochiral aldehydes. Tetrahedron Lett 37:245–248

Basha A, Lipton M, Weinreb SM (1977) A mild, general method for the conversion of esters to amides. Tetrahedron Lett: 4171–4174

Bates RW, Fernandez-Moro R, Ley SV (1991) Synthesis of the β-lactone esterase inhibitor valilactone using π-allyltricarbonyliron lactone complexes. Tetrahedron Lett 32:2651

Belshaw PJ, Meyer SD, Johnson DD, Romo D, Ikeda Y, Andrus M, Alberg DG, Schultz LW, Clardy J, Schreiber SL (1994) Synthesis, structure and mechanism in immunophilin research. Synlett 381–392

Bergbreiter DE, Pendergrass E (1981) Analysis of organomagnesium and organlithium reagents using N-phenyl-l-naphthlamine. J Org Chem 46:219–220

Berlinck RGS, Britton R, Piers E, Lim L, Roberge M, da Rocha RM, Andersen RJ (1998) Granulatimide and isogranulatimide, aromatic alkaloids with G2 checkpoint inhibition activity isolated from the Brazilian ascidian Didemnum granulatum: structure elucidation and synthesis. J Org Chem 63:9850–9856

Bernardi A, Cardani S, Scolastico C, Villa R (1990) Asymmetric synthesis of malic acid-type synthons via chiral norephedrine-derived oxazolidines. Tetrahedron 46:1987–1998

Bernasconi S, Colombo M, Jommi G, Sisti M (1986) A convenient synthesis of perillenal. Gazz Chem Ital 116:69–71

Bhupathy M, Bergan JJ, McNamara JM, Volante RP, Reider PJ (1995) A convergent synthesis of a novel non-peptidyl growth hormone secretagogue, L-692,429. Tetrahedron Lett 36:9445–9448

Bialojan C, Takai A (1988) Inhibitory effect of a marine sponge toxin, okadaic acid, on protein phosphatases – specificity and kinetics. Biochem J 256:283–290

Blincoe SN (1994) MSc thesis, University of Canterbury, New Zealand

Brandsma L, Verkruijsse HD, Walda B (1991) A procedure for the oxidative "dimerization" of aliphatic 1-alkynes. Syn Commun 21:137–139

Broka CA, Ehrler J (1991) Enantioselective total syntheses of begnamides B and E. Tetrahedron Lett 32:5907–5910

Calo V, Fiandanese V, Nacci A, Scilimati A (1994) Chelation-controlled chemo-, regio-, and enantioselective synthesis of homoallylic alcohols. Tetrahedron 50:7283–7292

Capozzi G, Roelens S, Talami S (1993) A protocol for the efficient synthesis of enantiopure β-substituted β-lactones. J Org Chem 58:7932–7936

Casamitjana N, Jorge A, Pérez CG, Bosch J, Espinosa E, Molins E (1997) Diels-Alder reactions of 5,6-dihydro-2(^{1}H)-pyridones. Preparation of partially reduced *cis*-isoquinolones and *cis*-3,4-disubstituted piperidines. Tetrahedron Lett 38:2295–2298

Case-Green SC, Davies SG, Hedgecock CJR (1991) Asymmetric synthesis of homochiral β-lactones via the iron chiral auxiliary [(η^5-C_5H_5) Fe (CO) (PPh_3)]. Synlett 779–780

Catterall WA (1992) Cellular and molecular biology of voltage-gated sodium channels. Physiol Rev 72:S15–S48

Caterall WA, Gainer M (1985) Interaction of brevetoxin A with a new receptor site on the sodium channel. Toxicon 23:497–504

Chiara JL, Destabel C, Gallego P, Marco-Contelles J (1996) Cleavage of N-O bonds promoted by samarium diodide: reduction of free or N-acylated O-alkylhydroxylamines. J Org Chem 61:359–360

Closson WD, Bank S, Battisti A, Waring A, Gortler LB, Ji S (1967) Cleavage of sulfonamides with sodium naphthalene. J Am Chem Soc 89:5311–5312

Comasseto JV, Chieffi A, Tucci FC (1996) Preparation of Z-vinylic cuprates from Z-vinylic tellurides and their reaction with enones and epoxides. J Org Chem 61:4975–4989

Corey EJ, Fuchs P (1972) A synthetic method for formyl→ethynyl conversion (RCHO→RC≡CH or RC≡CR'). Tetrahedron Lett 3769–3772

Corey EJ, Reichard GA, Kania R (1993) Studies on the total synthesis of lactacystin. An improved aldol coupling reaction and a β-lactone intermediate in thiol ester formation. Tetrahedron Lett 34:6977–6980

Crews P, Naylor S (1985) Fortschr Chem Org Naturst 48:203–268

Critcher DJ, Pattenden G (1996) Synthetic studies towards pateamine, a novel thiazole-based 19-membered bis-lactone from Mycale sp. Tetrahedron Lett 37:9107–9110

Dabdoub JJ, Dabdoub VB, Comasseto JV (1992) Synthesis of (E)-1,4-bis(organyl) but-1-en-3-ynes by lithium-tellurium exchange reaction on (Z)-1-butyltelluro-1,4-biy(organyl)but-1-en-3-ynesl. Tetrahedron Lett 33:2261–2264

Debonnel G, Beauchesne L, de Motigny C (1989) Domoic acid, the alleged "mussel toxin", might produce its neurotoxic effect through kainate receptor activation: an electrophysiological study in the rat dorsal hippocampus. Can J Physiol Pharmacol 67:29–33

Dirat O, Berranger T, Langlois Y (1995) [2+3] Cycloadditions of enantiomerically pure oxazoline-N-oxides: stereoselective synthesis of functionalized β-lactones. Synlett 935–937

Dong Q, Anderson CE, Ciufolini MA (1995) Reductive cleavage of TROC groups under neutral conditions with cadmium-lead couple. Tetrahedron Lett 36:5681–5682

Dymock BW, Kocienski PJ, Pons J-M (1996) 3-(Trimethylsilyl)oxetan-2-ones via enantioselective [2+2] cycloaddition of (trimethylsilyl)ketene to aldehydes catalysed by methylaluminoimidazolines. J Chem Soc Chem Commun 1053–1054

Evans DA, Mathre DJ (1985) Asymmetric synthesis of the enkephalinase inhibitor thiorphan. J Org Chem 50:1830–1835

Farina V, Krishnan B (1991) Large rate accelerations in the Stille reaction with tri-2-furylphophine and triphenylarsine as palladium ligands: mechanistic and synthetic implications. J Am Chem Soc 113:9585–9595

Faulkner DJ, Fenical WH (eds) (1977) Marine natural products chemistry. Plenum, New York

Fischer H, Klippe M, Lerche H, Severin T, Wanninger G (1990) Electrophilic β-bromination and nucleophilic α-methoxylation of α,β-unsaturated carbonyl compounds. Chem Ber 123:399–404

Fukuyama T, Xu L (1993) Total synthesis of (-)-tantazole. J Am Chem Soc 115:8449–8450

Gardner B, Nakanishi H, Kahn M (1993) Conformationally constrained nonpeptide β-turn mimetics of enkephalin. Tetrahedron 49:3433–3448

Georg GI (1993) The organic chemistry of β-lactams. VCH, New York

Guzzo PR, Miller MJ (1994) Catalytic asymmetric synthesis of the carbacephem framework. J Org Chem 59:4862–4867

Haltiwanger RC, Thurmes WN, Walba DM (1988) A highly stereocontrolled route to the monensin spiroketal ring system. J Org Chem 53:1046–1056

Harris TM, Oster TA (1983) Acetylations of strongly basic and nucleophilic enolate anions with N-methoxy-N-methylacetamide. Tetrahedron Lett 24:1851–1854

Haystead TAJ, Sim ATR, Carling D, Honnor RC, Tsukitani Y, Cohen P, Hardie DG (1989) Effects of the tumour promoter okadaic acid on intracellular protein phosphorylation and metabolism. Nature 337:78–81

Heathcock CH, Hansen MM, Danda H (1990) Reversal of stereochemistry in the aldol reactions of a chiral boron enolate. J Org Chem 55:173–181

Hesse G, Bücking E (1949) Justus Liebigs Ann Chem 563:31–37

Hesse M (1991) Ring enlargements in organic chemistry. VCH, New York

Hinterding K, Alonso-Diaz D, Waldmann H (1998) Organic synthesis and biological signal transduction. Angew Chem Int Ed 37:688–749

Hirai K, Homma H, Mikoshiba I (1994) Reactivity of the pyridylthiosilyl enol ether route to β-lactone and β-lactam. Heterocycles 38:281–282

Holton RA (1990) Eur Pat App EP 400,971, Chem Abs (1990) 114, 164568q

Hu T, Curtis JM, Oshima Y, Quilliam MA, Walter JA, Wright WM, Wright JLC (1995) Spirolides B and D, two novel macrocycles isolated from the digestive glands of shellfish. J Chem Soc Chem Comm 2159–2161

Hung DT, Chem J, Schreiber SL (1996a) (+)-Discodermolide binds to microtubules in stoichiometric ratio to tubulin dimers, blocks taxol binding and results in mitotic arrest. Chem Biol 3:287–293

Hung DT, Jamison TF, Schreiber SL (1996b) Understanding and controlling the cell cycle with natural products. Chem Biol 3:623–639

Huperizine A, Kozikowski A, Campiani G, Nacci V, Sega A, Saxena A, Doctor B (1996) An approach to modified heterocyclic analogues of huperizine A and isohuperizine A. Synthesis of the primidone and pyrazole analogues, and their anticholinesterase activity. J Chem Soc Perkin Trans 1:1287

Ireland RE, Wardle RB (1987) 3-Acyltetramic acid antibiotics. 3. An approach to the synthesis of BU-2313. J Org Chem 52:1780–1789

Ishihara J, Miyakawa J, Tsujimoto T, Murai A (1997) Synthetic study on gymnodimine: highly stereoselective construction of substituted tetrahydrofuran and cyclohexene moieties. Synlett 1417–1419

Ishihara J, Sugimoto T, Murai A (1998) Studies on the stability of 1,7,9-trioxadispiro[5.1.5.2]pentadecane system: the common tricyclic acetal moiety in pinnatoxins. Synlett 603–606

Ishiwata A, Sakamoto S, Noda T, Hirama M (1999) Synthetic study of pinnatoxin A: intramolecular Diels-Alder approach to the AG-ring. Synlett 692–694

Itô S, Takeishi S, Noda T, Hirama M (1988) Reinvestigation of the diastereoselectivity of kinetic aldol condensation of cyclohexanone lithium enolate with benzaldehyde. Bull Chem Soc Jpn 61:2645–2646

Kato Y, Fusetani N, Matsugaga S, Hashimoto K (1986) Okinonellins A and B, two novel furanosesterterpenes, which inhibit cell division of fertilized starfish eggs, from the marine sponge Spongionella sp. Experientia 42:1299–1300

Keck GE, McHardy SF, Wager TT (1995) Reductive cleavage of N-O bonds in hydroxylamine and hydroxamic acid derivatives using Sm_2I2/THF. Tetrahedron Lett 41:7419–7422

Kerr RG, Kerr SS (1999) Marine natural products as therapeutic agents. Expert Opin Ther Patents 9:1207–1222

Klutchko S, Hoefle ML (1981) Synthesis and angiotensin-converting enzyme inhibitory activity of 3-(mercaptomethyl)-2-oxo-1-pyrrolidineacetic acids and 3-(mercaptomethyl-2-oxo-1-piperidineacetic acids. J Med Chem 24:104–109

Koelsch CF (1943) A synthesis of 3-alkylpiperidones. J Am Chem Soc 65:2458–2459

Kozikowski AP, Sorgi KL, Wang BC, Xu Z (1983) An improved method for the synthesis of anomerically allylated C-glycopyranosides and C-glycofuranosides. Tetrahedron Lett 24:1563–1566

Lajoie G, Lepine F, Maziak L, Belleau B (1983) Facile regioselective formation of thiopeptide linkages from oligopeptides with new thionation reagents. Tetrahedron Lett 24:3815–3818

Li G, Patel D, Hruby VJ (1993) Asymmetric synthesis of (2R, 3S) and (2S, 3R) precursors of β-methylhistidine, β-phenylalanine and β-tyrosine. Asymmetry. Tetrahedron 4:2315–2318

Licitra EJ, Liu JO (1996) A three-hybrid system for detecting small ligand-protein receptor interactions. Proc Natl Acad Sci U S A 93:12817–12821

Liu Y-S, Zhao C, Bergbreiter DE, Romo D (1998) Simultaneous deprotection and purification of BOC-amines based on ionic resin capture. J Org Chem 63:3471–3473

Mancuso AJ, Swern D (1981) Activated dimethyl sulfoxide: useful reagent for synthesis. Tetrahedron Lett 35:2473–2476

Manhas MS, Wagle DR, Chiang J, Bose AK (1988) Conversion of β-lactams to versatile synthons via molecular rearrangement and lactam cleavage. Heterocycles 27:1755–1802

McCauley JA, Nagasawa K, Lander PA, Mischke SG, Semones MA, Kishi Y (1998) Total synthesis of pinnatoxin A. J Am Chem Soc 120:7647–7648

Miles CO, Wilkins AL, Stirling DJ, MacKenzie AL (1999) Proceedings of the marine biotoxin science, workshop no 11. Ministry of Agriculture and Forestry, Wellington, New Zealand, pp 103–106

Nagao Y, Hagiwara Y, Kumagai T, Ochiai M, Inoue T, Hashimoto K, Fujita E (1986) New C4-chiral 1,2-thiazolidine-2-thiones: excellent chiral auxiliaries for highly diastereocontrolled aldol-type reactions of acetic acid and α,β-unsaturated aldehydes. J Org Chem 51:2391–2393

Negishi E, Okukado N, King AO, Van Horn DE, Speigel BI (1978) Double metal catalysis in the cross-coupling reaction and its application to the stereo and regioselective synthesis of trisubstituted olefins.J Am Chem Soc 100:2254–2256

Nicolaou KC, He Y, Roschagar G, King NP, Vourloumis D, Li T (1998) Total synthesis of epothilone E and analogues with modified side chains through the Stille coupling reaction. Angew Chem Int Ed Engl 37:84–87

Nicolas E, Russell KC, Hruby VJ (1993) Asymmetric 1,4-addition of organocuprates to chiral α,β-unsaturated N-acyl-4-phenyl-2-oxazolidinones: a new approach to the synthesis of chiral β-branched carboxylic acids. J Org Chem 58:766–770

Nitta A, Ishiwata A, Noda T, Hirama M (1999) Intramolecular alkylation approach to the G-ring. Synlett 695–696

Northcote PT, Blunt JW, Munro MHG (1991) Pateamine: a potent cytotoxin from the New Zealand marine sponge. Tetrahedron Lett 32:6411–6414

Noyori R, Ohkuma T, Kitamura M, Takaya H, Sayo N, Kumobayashi H, Akutagawa S (1987) Asymmetric hydrogenation of β-keto carboxylic esters. A practical, purely chemical access to β-hydroxy esters in high enantiomeric purity. J Am Chem Soc 109:5856–5858

Ojima I (1995) Recent advances in the β-lactam synthon method. Acc Chem Res 28.383–389

Ojima I, Habus I, Zhao M, Zucco M, Park YH, Sun CM, Brigaud T (1992) New and efficient approaches to the semisynthesis of taxol and its C-13 side chain analogs by means of β-lactam synthon method. Tetrahedron 48:6985–7012

Ojima I, Sun CH, Zucco M, Park YH, Duclos O, Kuduk S (1993) A highly efficient route to taxotere by the β-lactam synthon method. Tetrahedron Lett 34:4149–4152

Ojima I, Sun CM, Park YH (1994) New and efficient coupling method for the synthesis of peptides bearing the norstatine residue and their analogs. J Org Chem 59:1249–1250

Ojima I, Ng EW, Sun CM (1995) Novel route to hydroxy(keto)ethylene dipeptide isosteres through the reaction of N-tBoc-β-lactams with enolates. Tetrahedron Lett 36:4547–4550

Okaichi T, Anderson DM, Nemoto T (eds) (1989) International symposium on red tides. Elsevier, New York

Oppolzer W, Chapuis C, Bernardinelli G (1984) Camphor-derived N-acryloyl and N-crotonoyl sultams: practical activated dienophiles in asymmetric Diels-Alder reactions. Helv Chem Acta 67:1397–1401

Palomo C, Aizpurua JM, Garcia JM, Iturburu M, Odriozola JM (1994) Concise and general synthesis of α-disubstituted-β-amino ketones from β-lactams. J Org Chem 59:5184–5188

Palomo C, Aizpurua HM, Cuevas C, Mielgo A, Galarza R (1995) A mild method for the alcoholoysis of β-lactams. Tetrahedron Lett 36:9027–9030

Pommier A, Pons J-M (1993) Recent advances in β-lactone chemistry. Synthesis 441–449

Pyne SG, Hensel MJ, Fuchs PL (1982) Chiral and stereochemical control via intramolecular Diels-Alder reaction of Z dienes. J Am Chem Soc 104:5719–5728

Raimundo B, Heathcock CH (1995) Further studies on the anti-selective aldol reaction of chiral imides. Synlett 1213–1214

Rainier JD, Xu Q (1999) A novel anionic condensation, fragmentation, and elimination reaction of bicyclo[2.2.1]heptanone ring systems. Org Lett 1:27–29

Rayl AJS (1999) Oceans: medicine chests of the future? Scientist 13(19):1

Rochfort SJ, Atkin D, Hobs L, Capon RJ (1996) Hippospongins A–F: new furanoterpenes from a southern Australian marine sponge Hippospongia sp. J Nat Prod 59:1024–1028

Romo D, Rzasa RM, Shea HA, Park K, Langenhan JM, Sun L, Akhiezer A, Liu JO (1998) Total synthesis and immunosuppressive activity of (-)-pateamine A and related compounds: implementation of a β-lactam-based macrocyclization. J Am Chem Soc 120:12237–12254

Rouhi AM (1995) Supply issues complicate trek of chemicals from sea to market. Chem Eng News 42–44

Roush WR, Barda DA (1997) Highly selective Lewis-acid catalyzed Diels-Alder reactions of acyclic (Z)-1,3-dienes. J Am Chem Soc 119:7402–7403

Rzasa RM, Romo D, Stirling DJ, Blunt JW, Munro HMG (1995) Structural and synthetic studies of the pateamines: synthesis and absolute configuration of the hydroxydienoate fragment. Tetrahedron Lett 36:5307–5310

Rzasa RM, Shea HA, Romo D (1998) Total synthesis of the novel, immunosuppressive agent (-)-pateamine A from Mycale sp. employing a β-lactam-based mycrocyclization. J Am Chem Soc 120:591–592

Scheuer PJ (ed) (1978) Marine natural products, chemical and biological perspectives, vol I. Academic, New York

Scheuer PJ (ed) (1983) Marine natural products, chemical and biological perspectives, vol V. Academic, New York

Scheuer PJ (ed) (1987) Bioorganic marine chemistry, vol I. Springer, Berlin Heidelberg New York

Schmitz WD, Messerschmidt NB, Romo D (1998) A β-lactone-based strategy applied to the total synthesis of (8*S*, 21*S*, 22*S*, 23*R*)- and (8*R*, 21*S*, 22*S*, 23*R*)-okinonellin B. J Org Chem 63:2058–2059

Schreiber SL (1991) Chemistry and biology of the immunophilins and their immunosuppressive ligands. Science 251:283–287

Schreiber SL (1992) Using the principles of organic chemistry to explore cell biology. Chem Eng News 70:22–32

Seebach D, Sutter MA, Weber RH, Zuger MF (1993) Poly(hydroxyalkanoates): a fifth class of physiologically important organic biopolymers? Org Syn Coll VII:215–220

Seki T, Masayuki S, Mackenzie L, Kaspar HF, Yasumoto T (1995) Gymnodimine, a new marine toxin of unprecedented structure isolated from New Zealand oysters and the dinoflagellate, Gymnodinium sp. Tetrahedron Lett 36:7093–7096

Sonogashira K, Tohda Y, Hagihara N (1975) A convenient synthesis of acetylenes: catalytic substitutions of acetylenic hydrogen with bromoalkanes, iodoarenes, and bromopyridines. Tetrahedron Lett 4467–4470

Soriente A, De Rosa M, Scettri A, Sodano G, Terencio MC, Paya M, Alcaraz MJ (1999) Manoalide. Curr Med Chem 6:415–431

Stewart M, Blunt JW, Munro M HG, Robinson WT, Hannah DJ (1997) The absolute stereochemistry of the New Zealand shellfish toxin gymnodimine. Tetrahedron Lett 38:4889–4890

Su B, Jacinto E, Hibi M, Kallunki T, Karin M, Ben-Neriah Y (1994) JNK is involved in signal integration during costimulation of T lymphocytes. Cell 77:727–736

Sugimoto T, Ishihara J, Murai A (1999) Synthesis of the B,C,D,E,F-ring fragment of pinnatoxins. Synlett 541–544

Suzuki A (1998) In: Diedrich F, Stang PJ (eds) Metal-catalyzed cross-coupling reactions. Wiley-VCH, Weinheim, chap 2

Taber DF, Silverberg LJ (1991) Enantioselective reduction of β-keto esters. Tetrahedron Lett 32:4227–4230

Taber DF, Deker PB, Silverberg LJ (1992) Enantioselective Ru-mediated synthesis of (-)-indolizidine-223AB. J Org Chem 57:5990–5994

Takai K, Nitta K, Utimoto K (1986) Simple and selective method for RCHO (E)-RCH=CHX conversion by means of a CHX_3-$CrCl_2$ system. J Am Chem Soc 108:7408–7410

Tamai Y, Someya M, Fukumoto J, Miyano S (1994a) Asymmetric [2+2] cycloaddition of ketene with aldehydes catalyzed by Me_3Al complexes of axially chiral 1,1'-binaphthalene-2,2'diol derivatives. J Chem Soc Chem Commun 1549–1550

Tamai Y, Yoshiwara H, Someya M, Fukumoto J, Miyano S (1994b) Asymmetric [2+2] cycloaddition of ketene with aldehydes catalyzed by chiral bissulfonamide-trialkylaluminium complexes. J Chem Soc Chem Commun 2281–2282

Taunton J, Colins JL, Schreiber SL (1996) Synthesis of natural and modified trapoxins, useful reagents for exploring histone deacetylase function. J Am Chem Soc 118:10412–10422

Tebbe FN, Parshall GW, Reddy GS (1978) Olefin homologation with titanium methylene compounds. J Am Chem Soc 100:3611–3613

Thomson AW, Starzl WE (eds) (1994) Immunosuppressive drugs: developments in anti-rejection therapy. Arnold, London

Torisawa Y, Nakagawa M, Takami H, Nagata T, Ali MA, Hino T (1994) Diels-Alder reaction of the N-protected 3-phenylthio-2(1H)-dihydropyridinone derivatives. Heterocycles 39:277–292

Torisawa Y, Hosaka T, Tanabe K, Suzuki N, Motohashi Y, Hino T, Nakagawa M (1996) Synthesis of a tetracyclic substructure of manzamine A via the Diels-Alder reaction of dihydropyridinones. Tetrahedron 52:10597–10608

Uemura D, Chou T, Haino T, Nagatsu A, Fukuzawa S, Zheng S, Chem H (1995) Pinnatoxin A – a toxic amphoteric macrocycle from the Okinawan bivalve Pinna-Muricata. J Am Chem Soc 117:1155–115

Walker M, Heathcock CH (1991) Extending the scope of the Evans asymmetric aldol reaction: preparation of anti and "non-Evans" syn aldols. J Org Chem 56:5457–5750

Wang Y, Zhao C, Romo D (1999) A stereocomplementary approach to β-lactones: highly diastereoselective synthesis of *cis*-β-lactones, a β-chloro acid, and a tetrahydrofuran. Org Lett 1:1197–1199

Wasserman HH (1987) New methods in the formation of macrocyclic lactams and lactones of biological interest. Aldrichimica Acta 20:63–74

Weinreb SM, Nahm S (1981) N-Methoxy-N-methylamides as effective acylating agents. Tetrahedron Lett 22:3815–3818

Wenthold RJ, Hampson DR (1988) A kainic acid receptor from frog brain purified using domoic acid affinity chromatography. J Biol Chem 263:2500–2505

White JD, Kawasaki M (1992) Total synthesis of (+)-latrunculin-A, an ichthyotoxic metabolite of the sponge Latrunculia magnifica, and its C-15 epimer. J Org Chem 57:5292–5300

White JD, Johnson AT (1994) Synthesis of the aliphatic depside (+)-bourgeanic acid. J Org Chem 59:3347

Wipf P, Lim S (1993) Rapid carboalumination of alkynes in the presence of water. Angew Chem Int Ed Engl 32:1068–1071

Yang HW, Romo D (1997) A highly diastereoselective, tandem Mukaiyama aldol-lactonization route to β-lactones: application to a concise synthesis of the potent pancreatic lipase inhibitor, (-)-panclicin D. J Org Chem 38:4–5

Yang HW, Romo D (1998) Practical, one-step synthesis of optically active β-lactones via the tandem Mukaiyama aldol-lactonization (TMAL) reaction. J Org Chem 63:1344–1347

Yang HW, Romo D (1999) A single-pot, mild conversion of β-lactones to β-lactams. J Org Chem 64:7657–7660

Yang HW, Zhao C, Romo D (1997) Studies of the tandem Mukaiyama aldol-lactonization (TMAL) reaction: a concise and highly diastereoselective route to β-lactones applied to the total synthesis of the potent pancreatic lipase inhibitor, (-)-panclicin D. Tetrahedron 53:16471–16488

Yang J, Cohn ST, Romo D (2000) Studies toward (-)-gymnodimine: concise routes to the spirocyclic and tetrahydrofuran moieties. Org Lett 2:763–766

Zemribo R, Romo D (1995) Highly diastereoselective [2+2] cycloadditions via chelation control: Asymmetric synthesis of β-lactones. Tetrahedron Lett 36:4159–4162

Zemribo R, Champ MS, Romo D (1996) Tandem transacylation debenzylation of β-lactones mediated by $FeCl_3$ leading to γ- and δ-lactones: application to the synthesis of (-)-grandinolide. Synlett 278–280

Zhao C, Romo D (2000) Mechanistic investigations of the tandem Mukaiyama aldol-lactonization (TMAL) reaction: evidence for a cyclic transition state arrangement. (submitted to J Org Chem)

5 Total Synthesis of Antifungal Natural Products

A.G.M. Barrett, W.W. Doubleday, T. Gross, D. Hamprecht,
J.P. Henschke, R.A. James, K. Kasdorf, M. Ohkubo,
P.A. Procopiou, G.J. Tustin, A.J.P. White, D.J. Williams

5.1 Introduction

There is great concern amongst the medical profession regarding fungal disease, with dermatophyte infections such as tinea pedis and candidiasis, although usually not fatal, being prevalent throughout the world (Kwon-Chung and Bennett 1992). Pathogens such as *Candida albicans, Cryptococcus neoformans, Pneumocystis carinii* and *Aspergillus fumigatus* have a far more gruesome reputation and are the cause of considerable mortality in immunocompromised patients. Populations at risk from these opportunistic fungal infections include AIDS patients, recipients of cancer chemotherapy and persons with genetically impaired or drug-suppressed immune function. Fungal pneumonia, induced by

Pneumocystis carinii, is the cause of death in the majority of AIDS patients. Invasive pulmonary aspergillosis is the scourge of patients subject to cancer chemotherapeutic regimes. Current treatments for serious systemic fungal infection are deficient, and the gold standard amphotericin is acutely toxic (Kwon-Chung and Bennett 1992). There are now resistance problems with azole fungistatic agents such as fluconazole. Novel therapies are therefore needed for serious fungal disease and for the management of the multitude of topical fungal infections. In consequence, several pharmaceutical companies worldwide are seeking to develop superior fungicidal agents. Due to evolutionary pressures of microbial antagonism, fermentation broths are rich sources of novel anti-infective agents. In consequence, natural product chemistry continues to play a major role in antibiotic discovery.

In 1990, Yoshida and co-workers in the Fujisawa laboratories in Tsukuba, Japan, reported the partial structural elucidation of a remarkable natural product (Yoshida et al. 1990)*. Fractionation of the fermentation broth from *Streptoverticillium fervens* and extensive chromatography led to the isolation of a structurally unique compound. The structure of this novel nucleoside was determined by extensive NMR spectroscopy and partial degradation. Although little excitement was generated upon realization of the unusual 5'-amino-5'-deoxy-5,6-dihydrouridine-derived moiety contained in FR-900848 (**1**), the unassuming Fujisawa file number for the new natural product, the fact that it possessed an unusual fatty acid side chain was most notable. This C_{23} fatty acid residue is endowed with five cyclopropanes, four of which are contiguous. Although Yoshida and co-workers' initial degradation studies led to a determination of the constitution of the molecule (Yoshida et al. 1990), there remained 11 elements of ambiguity in the structure: the geometry of Δ^{18}, the stereochemistry of the terminal cyclopropyl unit, and the stereochemistry of the quatercyclopropane moiety. Tanaka et al., however, established (H. Tanaka 1991, personal communication) that the central quatercyclopropane unit **2**, obtained by ozonolysis, reductive workup and acetylation, was C_2-symmetric.

FR-900848 (**1**) displays powerful, selective activity against filamentous fungi including *Aspergillus niger*, *Mucor rouxianus*, *Aureo-*

* For an earlier report of a natural product that may be identical with FR-900848 see Das et al. (1987).

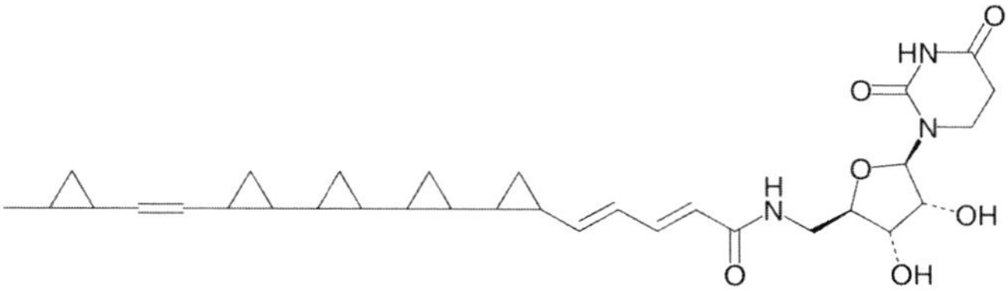

1

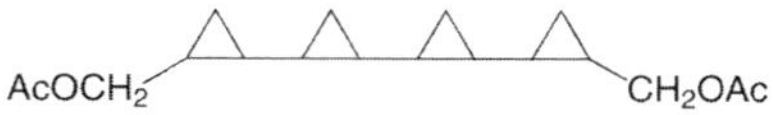

$AcOCH_2$ CH_2OAc

2

basidium pullulans, various *Trichophyton* sp., etc. It is essentially inactive, however, against nonfilamentous fungi such as *Candida albicans* and gram-positive and -negative bacteria. It displays in vivo activity and is not significantly toxic (Yoshida et al. 1990; Yoshida and Horikoshi 1989). Thus FR-900848 (**1**) represents an important new lead for the design of nucleoside antifungal agents active against the major human pathogen *Aspergillus fumigatus.* It is known that the fatty acid side chain of FR-900848 (**1**) is considerably restricted conformationally, but the influence of this on bioactivity and mode of antifungal action is uncertain. It is not yet clear what the exact biosynthetic origin of FR-900848 (**1**) is, nor is it obvious as to what evolutionary advantage there is in endowing a fatty acid with five cyclopropane ring systems with the attendant strain energy of about 130 kcal mol $^{-1}$. Finally, FR-900848 (**1**) provides an intriguing model upon which to design a new class of materials, namely stereoregular polycyclopropene, which might have unique and useful physical properties. All these factors underscore the potential importance of synthetic chemistry on FR-900848 (**1**) and related multicyclopropane arrays.

5.2 Stereoselective Syntheses of Bicyclopropanes

Extensive reporting on the synthesis and reactions of bicyclopropane arrays exists in the literature. Buchert and Reissig (1992), for example, have reported the synthesis of highly substituted bicyclopropanes, and Nijveldt and Vos have carried out an X-ray crystallographic study of

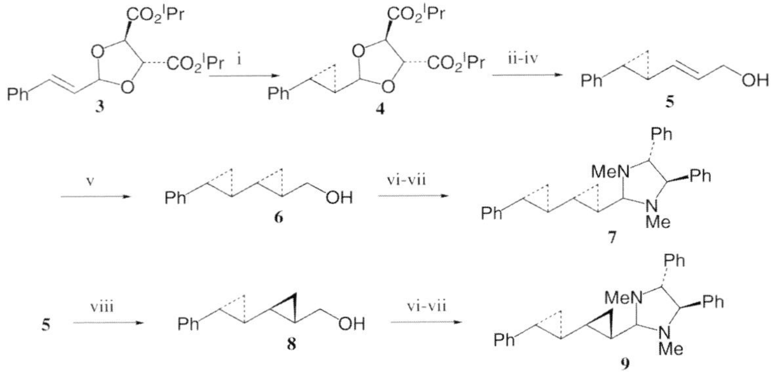

Reagents and conditions:(i) Et_2Zn, CH_2I_2, PhMe, -20 °C, 91%; (ii) TsOH, H_2O, THF, 60 °C, 93%; (iii) $(EtO)_2P(O)CH_2CO_2Et$, NaH, THF, 0 °C, 95%; (iv) DIBAL-H, CH_2Cl_2, -78 °C, 89%; (v) L-(+)-diethyl tartrate, Et_2Zn, CH_2I_2, $(CH_2Cl)_2$, -12 °C, 72%; (vi) PCC, NaOAc, SiO_2, CH_2Cl_2, 0 °C; (vii) (1*R*,2*R*)-*N*,*N*'-dimethyl-1,2-diphenyl-ethanediamine, Et_2O, 4Å sieves, 25 °C; (viii) D-(-)-diethyl tartrate, Et_2Zn, CH_2I_2, $(CH_2Cl)_2$, -12 °C, 84%.

Scheme 1. Reagent-controlled stereoselective synthesis of bicyclopropane derivatives

bicyclopropane (Nijveldt and Vos 1988). Prior to the discovery of FR-900848 (**1**), little attention was paid to issues of stereochemistry in bicyclopropane chemistry. We ourselves (Barrett and Tustin 1995; Barrett et al. 1994a, 1995a) and the Zercher (Theberge and Zercher 1994; Cebula et al. 1995; Theberge et al. 1996) and Armstrong and Maurer (1995) groups have independently reported stereoselective methods for the preparation of bicyclopropane systems relevant to the total synthesis of FR-900848 (**1**). These methodologies all utilize known asymmetric Simmons-Smith reactions to control all four stereocenters in the assembly of 1,6-disubstituted bicyclopropanes.

The *syn*- and *anti*-bicyclopropanes **6** and **8** were stereoselectively prepared by us using both Yamamoto (Arai et al. 1985; Mori et al. 1986) and Fujisawa (Ukaji et al. 1992) asymmetric cyclopropanation chemistry. Thus cyclopropanation of the chiral acetal **3** using Yamamoto methodology gave cyclopropane **4** (Arai et al. 1985) as a single diastereoisomer (Scheme 1) following chromatographic purification. The enantiomerically pure allylic alcohol **5** was obtained by subsequent acid hydrolysis, Horner-Emmons homologation and DIBAL-H reduction. Cyclopropanation of the alcohol **5** with diethylzinc and diiodomethane in the presence of l-(+)-diethyl tartrate or d-(-)-diethyl tartrate, respectively, according to the Fujisawa protocol (Ukaji et al. 1992), gave the *syn*- and *anti*-bicyclopropane derivatives **6** (ds 6:1) and **8** (ds 6:1). Thus, *syn*- or *anti*-bicyclopropanes can be prepared via reagent control. Treatment of alcohol **5** with diethylzinc and diiodomethane in the absence of tartrate esters, however, provided a mixture of both bicyclopropanes **6** and **8** (~1:1) in 82% yield. It is apparent from these observations that the pre-existing cyclopropane ring in alkene **5** does not lead to the induction of stereochemistry of the second cyclopropanation reaction. Zercher has observed similar low stereoselectivity on the cyclopropanation of racemic alkene **5** (Theberge and Zercher 1994). The structures of the bicyclopropanes **6** and **8** were proven by oxidation and conversion (Manganey et al. 1988a,b) into the imidazolidines **7** and **9**, respectively, and an X-ray crystal structure analysis (Barrett et al. 1994a) of isomer **7**.

Zercher has applied Charette's asymmetric cyclopropanation reaction (Charette and Juteau 1994; Charette et al. 1995) for the highly stereoselective synthesis of both *syn*- and *anti*-bicyclopropane methanol derivatives (exemplified in Scheme 2; Theberge and Zercher 1994). Thus the allylic alcohol **10** was transformed into the bicyclopropanes **11** (67%) and **12** (72%) using the tartramide additives **13** and **14**, respectively, to control absolute stereochemistry (ds >12:1). In addition, the Zercher group has applied double Charette asymmetric cyclopropanation chemistry and sulfur ylide methodologies to elaborate various bicyclopropanes with the *cis-trans* and *cis-cis* ring stereochemistries

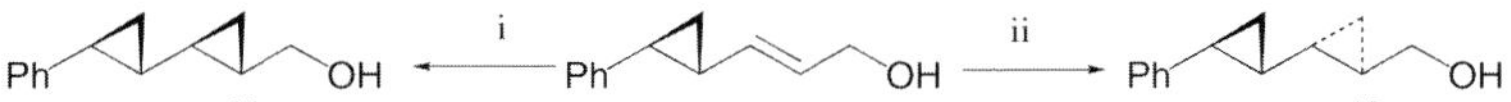

Reagents and conditions: (i) Et_2Zn, CH_2I_2, **13**, 67%; (ii) Et_2Zn, CH_2I_2, **14**, 72%.

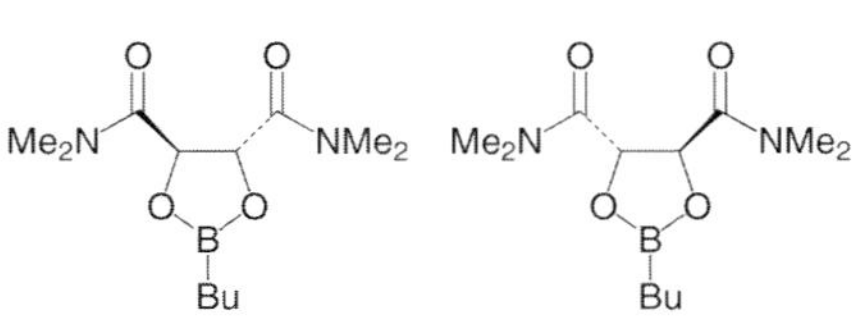

Scheme 2. Diastereoselective synthesis of bicyclopropanes

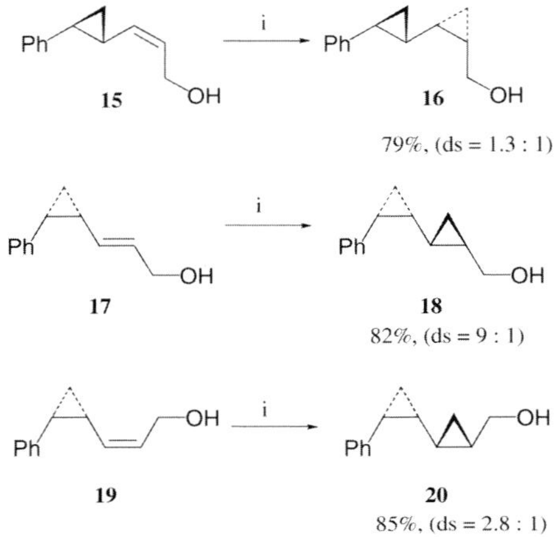

Ph 15 OH → i → Ph 16 OH

79%, (ds = 1.3 : 1)

Ph 17 OH → i → Ph 18 OH

82%, (ds = 9 : 1)

Ph 19 OH → i → Ph 20 OH

85%, (ds = 2.8 : 1)

Reagents and conditions: (i) Et_2Zn, CH_2I_2, 0 °C.

Scheme 3. Stereocontrol in the Simmons-Smith cyclopropanation of alkenylcyclopropanes

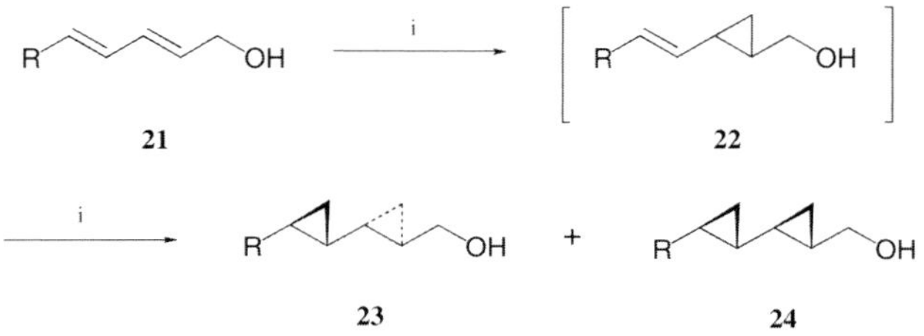

R 21 OH → i → [R 22 OH]

→ i → R 23 OH + R 24 OH

Reagents and conditions: (i) Et_2Zn, CH_2I_2, $(CH_2Cl)_2$, -20 °C.
Structures refer to racemic modifications
[R, %, ds: Me, 68, 5 : 1; Ph, 80, 5 : 1; ^{i}Pr, 72, 6 : 1;
c-C_6H_{11}, 78, 7 : 1; ^{t}BuMe$_2$SiOCH$_2$, 72, >95 : 5]

Scheme 4. Double cyclopropanation of 2,4-propadien-1-ol derivatives

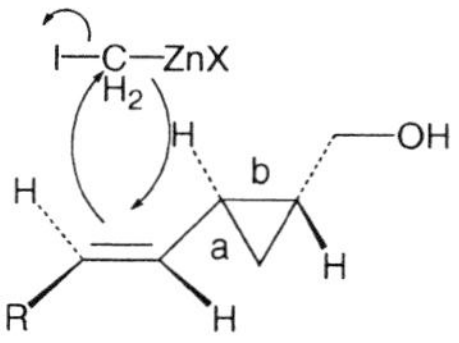

25

Fig. 1. *Anti*-stereocontrol in a Simmons-Smith cyclopropanation reaction

(Theberge and Zercher 1994; Cebula et al. 1995; Theberge et al. 1996). In some cases it was observed that the pre-existing cyclopropane ring dramatically influenced the stereochemistry of the second cyclopropanation reaction, as exemplified by the high *anti*-selectivity in the cyclopropanation of alkene **17** to provide bicyclopropane **18**. In contrast, lower stereoselectivities were observed on the cyclopropanation of alkenes **15** and **19** (Scheme 3). Generally in these systems, except when substrate and reagent stereochemical biases are mismatched, Simmons-Smith cyclopropanation in the presence of the Charette additives **13** and **14** proceeds with good to excellent stereoselectivity. It is probable that the diastereoselectivity in the conversion of alkene **17** into the bicyclopropane **18** has its origin in steric approach control and the minimization of allylic 1,3-strain (Hoffman 1989). The use of the Charette chemistry (Charette and Juteau 1994; Charette et al. 1995) will feature repeatedly in this paper, and there is no question that this methodology represents a significant advance in enantioselective synthesis.

We have reported that racemic bicyclopropane derivatives **23** and **24** are formed in good yield (68%–80%) and with good diastereoselectivity (5:1 to >95:5), favoring the *anti*-isomers **23**, upon reaction of 2,4-dienols **21** with diethylzinc and diiodomethane in 1,2-dichloroethane at –20°C (Scheme 4; Barrett and Tustin 1995). We speculate that the reaction involves the intermediacy of the monocyclopropane **22** and stereoelectronic control of the second cyclopropanation step (see Fig. 1). In this analysis, overlap of the most electron-rich cyclopropane σ-bond (bond a, not bond b) with the alkene π-system should enhance its nucleophilicity and favor *anti*-delivery of the zinc carbenoid electrophile.

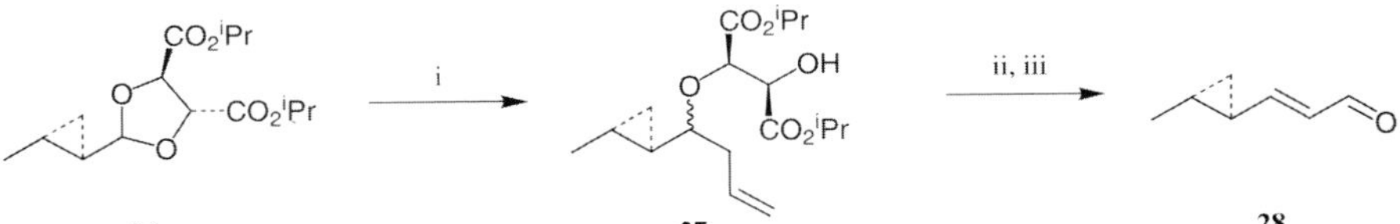

26 27 28

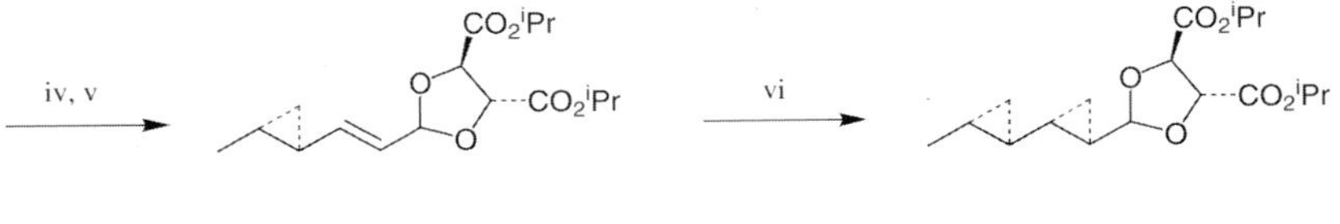

29 30

Reagents and conditions: (i) $Me_3SiCH_2CH{=}CH_2$, $TiCl_4$, CH_2Cl_2, -78 °C, 59%; (ii) O_3; Ph_3P work-up, 90%; (iii) Et_3N, Δ, 90%; (iv) EtOH, PPTS, $HC(OEt)_3$; (v) L-(+)-di-isopropyl tartrate, EtOH, 35%; (vi) CH_2I_2, Et_2Zn, 65%.

Scheme 5. Synthesis of bicyclopropane acetals

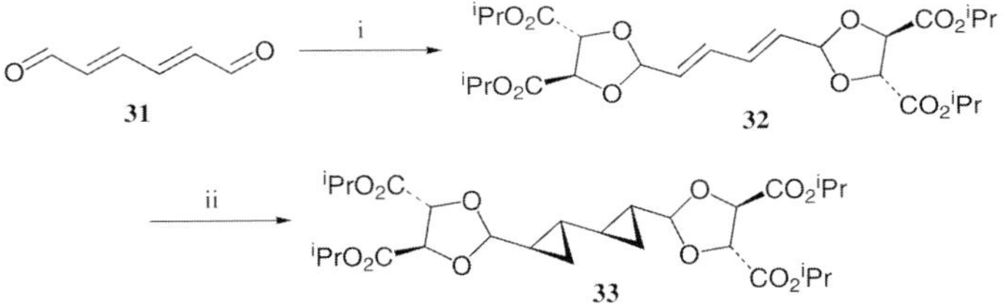

31 32 33

Reagents and conditions: (i) **34**, Me_3SiOTf (cat), $MeC(OSiMe_3){=}NSiMe_3$, CH_2Cl_2, -78 to 25 °C, 73%; (ii) Et_2Zn, CH_2I_2, $(CH_2Cl)_2$, -20 °C, 78%.

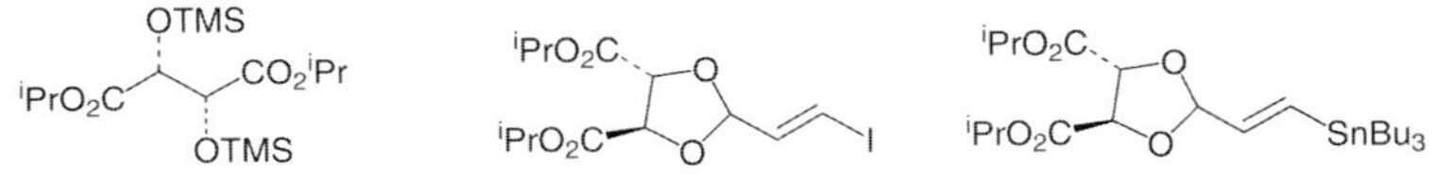

34 35 36

Scheme 6. Synthesis of a bicyclopropane diacetal

Armstrong prepared the *syn*-bicyclopropane **30** by application of both allylsilane and Yamamoto cyclopropanation (Arai et al. 1985) chemistry (Scheme 5; Armstrong and Maurer 1995). Thus, reaction of the acetal **26** with allylsilane in the presence of titanium chloride, and ozonolysis, followed by treatment with triethylamine gave the enal **28**. The corresponding acetal **29**, which was formed by condensation of the diethyl acetal derivative of enal **28** with l-(+)-diisopropyl tartrate, was further cyclopropanated to produce the bicyclopropane **30** (65%, >90% de). We used a double Yamamoto cyclopropanation approach to prepare a *syn*-bicyclopropane (Scheme 6; Barrett et al. 1995a; A.G.M. Barrett, W.W. Doubleday, G.J. Tustin, unpublished observations). Thus, Noyori acetalization (Tsunoda et al. 1980) of muconaldehyde (Davies and Whitham 1977) (**31**) or, alternatively, the Stille coupling (Stille 1986) of vinyl iodide **35** with the vinyl stannane **36** gave the dienyl diacetal **32**. In turn, both alkenes **35** and **36** were prepared (A.G.M. Barrett, W.W. Doubleday, G.J. Tustin, unpublished observations) from propargylic alcohol via radical hydrostannylation, PCC oxidation, and Noyori acetalization (Tsunoda et al. 1980). The diene **32** was cyclopropanated to furnish the bicyclopropane **33**, in good overall yield (56% from dial **31**) and essentially as a single diastereoisomer, by use of Yamamoto's adaptation of the Simmons-Smith reaction (Arai et al. 1985). The relative and absolute stereochemistry of the four new chiral centers present in the bicyclopropane **33** were determined by single crystal X-ray structure analysis (Barrett et al. 1995a). It is apparent from this result that diene double cyclopropanation provides facile entry to *syn*-bicyclopropane systems. Clearly, following the Yamamoto mechanistic model, each Lewis basic acetal oxygen should independently direct the Simmons-Smith zinc carbenoid to one face of the adjacent alkene.

Recently, Taylor and co-workers (1997, 1999) reported a novel iterative approach to *syn*- and *anti*-bicyclopropanes whose stereochemistries were purely substrate controlled (Scheme 7). Homoallylic alcohol **37**, prepared from 2- (phenoxymethyl)-oxirane and propargyltrimethylsilane followed by hydrogenation in the presence of Lindlar catalyst, was activated with trifluoromethanesulfonic anhydride and cyclized via a silicon-stabilized intermediate cyclopropylcarbinyl cation providing exclusively the *trans*-vinylcyclopropane **38** (89%). Epoxidation of the vinylcyclopropane **38** and reiteration of the acetylide elongation process provided the diastereomeric alcohols **39a** and **39b** which were cyclized

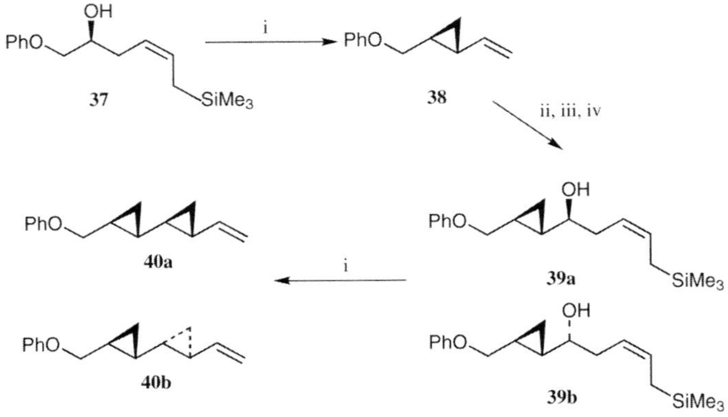

Reagents and conditions: (i) Tf_2O, CH_2Cl_2, 2,6-lutidine, -78°C, then Et_3N, -78 to 25 °C, 89%; (ii) mCPBA, CH_2Cl_2, 0°C; (iii) $HCCCH_2SiMe_3$, BuLi, $BF_3.OEt_2$, THF, -78 °C then epoxide prepared from **38**; (iv) H_2, Lindlar, MeOH, 50% from **38**.

Scheme 7. Allylsilanes in bicyclopropane synthesis

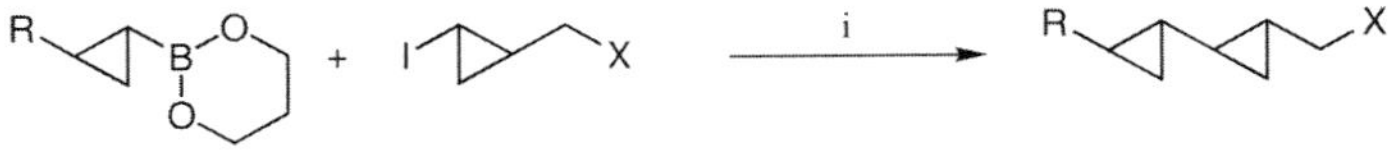

R = Alkyl, cyclopropyl X = OH, OBn

Reagents and conditions: (i) $Pd(OAc)_2$ (10 mol%), Ph_3P (50 mol%), DME, KO^{t}Bu, ^{t}BuOH, 80°C.

Scheme 8. Palladium[0]-catalyzed construction of bicyclopropane derivatives

upon treatment with trifluoromethanesulfonic anhydride to give diastereomerically pure *syn*- and *anti*-bicyclopropanes **40a** and **40b**, respectively. Surpisingly, cyclization of the benzyl ethers corresponding to the phenyl ethers **39a** and **39b** were also cyclized on trifluoromethanesulfonylation, but each gave rise to a mixture of the corresponding bicyclopropane derivatives (1:1 *syn*:*anti*; Tayler et al. 1999). Charette and De Freitas-Gil have shown that symmetric and unsymmetric contiguous bicyclopropanes **43** can be synthesized by cross-coupling of cyclopropylboronate esters **41** and iodocyclopropanes **42** under modified

Suzuki conditions (Charette and De Freitas-Gil 1997). This methodology is exemplified by the chemistry in Scheme 8 and is also applied to the generation of a tercyclopropane.

Recently, Itoh and co-workers have reported the use of lipase mediated resolution to produce *syn*-tetrafluorobicyclopropanedimethanol derivatives in high enantiomeric excess (Mitsukura et al. 1999).

5.3 Stereoselective Syntheses of Quatercyclopropanes and the Stereochemical Elucidation of FR-900848 (1)

We have examined three methodologies for the stereoselective assembly of quatercyclopropane arrays: (a) a double Yamamoto cyclopropanation followed by a double Charette cyclopropanation (Barrett et al. 1995a,b), (b) a double Charette cyclopropanation followed by a second double Charette cyclopropanation (Barrett and Kasdorf 1996a,b), and (c) a tetraene tetracyclopropanation reaction (A.G.M. Barrett, W.W. Doubleday, G.J. Tustin, unpublished observations). In the first approach (Scheme 9), acid-catalyzed hydrolysis of the diacetal **33** and direct homologation of the resulting dialdehyde by a double Wittig reaction gave a mixture of the *E,E*-diester **44** and the *E,Z*-diester **45** (3.7:1). The requisite diol **46** was produced in high yield (91%) by reduction of the pure corresponding diester **44** using DIBAL-H. Pre-mixing of diol **46** with dioxaborolane **14** followed by treatment with preformed bis (iodomethyl)zinc gave quatercyclopropane **47** (94%) essentially as a single diastereoisomer. Similarly, treatment with the antipodal dioxaborolane **13** provided the quatercyclopropane **48** (100%). It was evident on inspection of the ^{1}H and ^{13}C NMR spectra that the quatercyclopropanes **47** and **48** were two different C_2-symmetric isomers. We indirectly confirmed these structures by X-ray crystallographic analysis (Charette and De Freitas-Gil 1997) of the di-(4-bromobenzoate) ester derivative of **47**, which showed the four cyclopropyl units to be arranged helically (see Fig. 2, p. 161).

We also prepared the quatercyclopropane **47** from mucondiol using two sequential double Charette cyclopropanation reactions (Barrett and Kasdorf 1996a,b). Since this strategy was crucial in our total synthesis of FR-900848 (**1**) (Barrett and Kasdorf 1996a,b), discussion of this approach is delayed until the next section. The preparation of a quater-

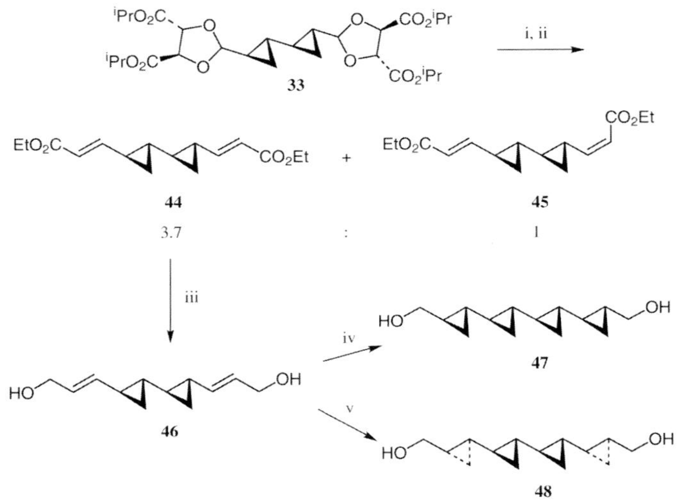

Reagents and conditions: (i) TsOH, THF/H_2O, 55 °C; (ii) $Ph_3P=CHCO_2Et$, CH_2Cl_2, 61% from **33**; (iii) DIBAL-H, CH_2Cl_2, -78 °C, 91%; (iv) **14**, $Zn(CH_2I)_2$, CH_2Cl_2, 0 to 25 °C, 94%; (v) **13**, $Zn(CH_2I)_2$, CH_2Cl_2, 0 to 25 °C, 100%.

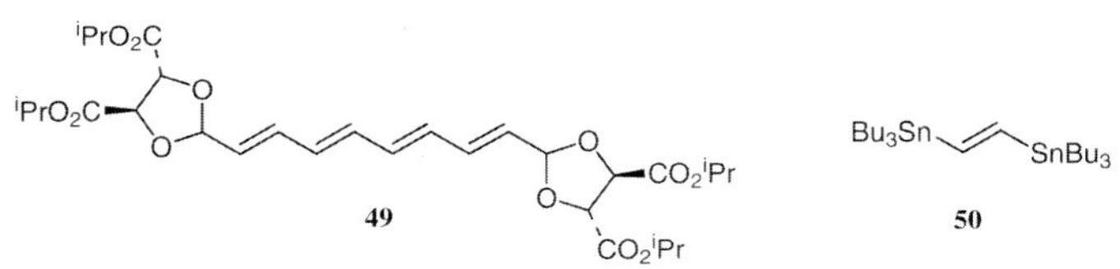

Scheme 9. Synthesis of quatercyclopropane derivatives

cyclopropane using direct tetracyclopropanation with diiodomethane and diethylzinc of the tetraene **49**, prepared from the iodoalkene **35** and alkene **50** using a sequence of Stille coupling (Stille 1986), iododestannylation and a second Stille coupling, was briefly examined; however, the stereochemical identity of this substance is not yet known.

In our structural elucidation of FR-900848 (**1**) our planning was driven by speculations as to the biosynthetic origin of the natural product (Barrett 1994). Since the carbon count of **1** is odd at C_{23}, we speculated that all the cyclopropanes are probably introduced during the fatty acid biosynthesis from acetate and a C_1 source such as *S*-adenosyl

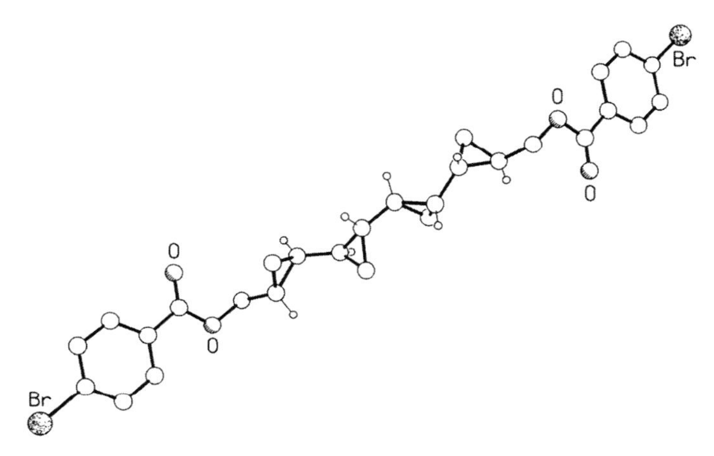

Fig. 2. The molecular structure of the di-(4-bromobenzoate) ester of diol **47** showing the absolute stereochemistry

methionine, rather than directly from the C_{18}-polyene **51**. As such, and since $\Delta^{2,4}$ are unequivocally *trans*, the geometry of all cyclopropanes and Δ^{18} are most likely to be all *trans*. If these suppositions are correct, then FR-900848 (**1**) should be either stereoisomer **52** or **53**, given that each enzymatic cyclopropanation would be expected to proceed with conservation of the alkene geometry and show the same absolute stereochemical bias. In consequence of these considerations, we sought to prepare model cyclopropane systems including quatercyclopropanes **47** and **48** to establish the stereochemistry of FR-900848 by correlation with key fragments from degradation. Falck and co-workers (1995) also embarked on the synthesis of polycyclopropanated compounds as precursors to FR-900848 (**1**), based on their speculations of the biosynthetic origin of the natural product. In 1995 the isolation and biosynthesis of U-106305, a natural product related to FR-900848 (**1**), was reported by Upjohn scientists (Kuo et al. 1995). This compound was found to be biosynthetically derived from both acetate and methionine (see below).

The experiments described below were used to confirm that the isolated olefinic unit of FR-900848 (**1**) is in fact *trans* (Scheme 10; Barrett et al. 1994b, 1995c, 1996a). Double Simmons-Smith cyclo-

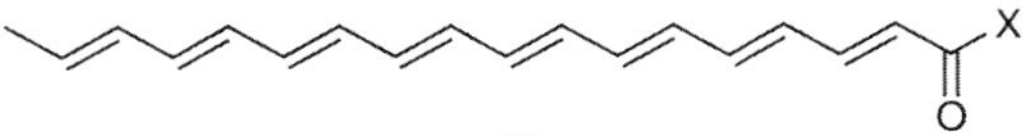

X

O

51

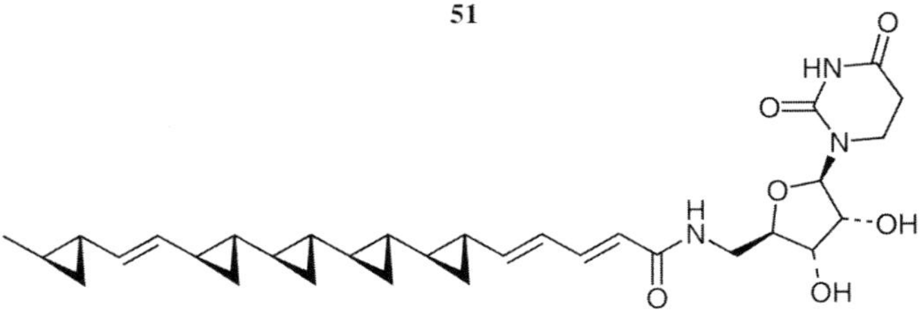

O

HN

O

N

O

H

N

OH

O

OH

52

O

HN

O

N

O

H

N

OH

O

OH

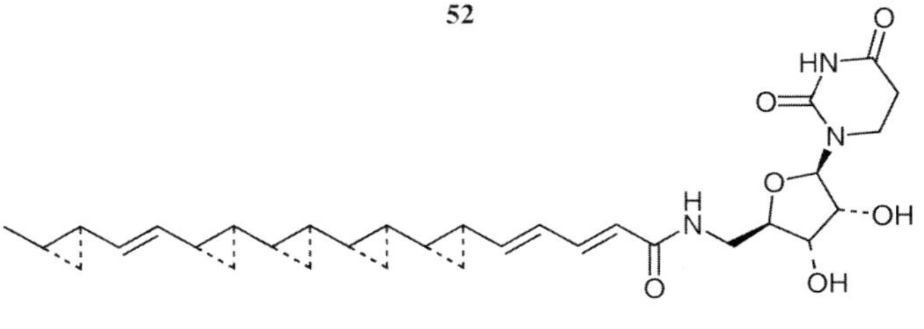

53

propanation of alkene **54** proceeded with high diastereoselectivity, furnishing only a single C_2-symmetric dicyclopropane **55** (89%). Acid-catalyzed hydrolysis of the isopropylidene group (63%) and sequential benzylidination gave the acetal **57** (64%). Whitham elimination (Hines et al. 1973) gave the geometrically pure *E*-alkene **58** (60%). Epimerization of diol **56** was achieved by consecutive Swern oxidation (Banwell and Onrust 1985) and sodium borohydride reduction, giving diol **59**. Whitham elimination of the corresponding benzylidene acetal **60** provided the geometrically pure *Z*-alkene **61**. Comparison of the ^{1}H NMR spectra of alkenes **58** and **61** with FR-900848 (**1**) was consistent with the assignment of the natural Δ^{18} geometry as *trans*. X-ray crystallographic studies (Barrett et al. 1994b, 1995c) for both 3,5-dinitrobenzoate diesters derived from diols **56** and **59** confirmed the structures of the intermediates.

The stereochemistry of the quatercyclopropane unit of FR-900848 (**1**) was established (Barrett et al. 1995b, 1996a) by comparisons of the acetates prepared from the model diols **47** and **48** with the corresponding degradation product **2** (Tanaka 1991) obtained from the natural product. We were delighted to find that quatercyclopropane **2** was

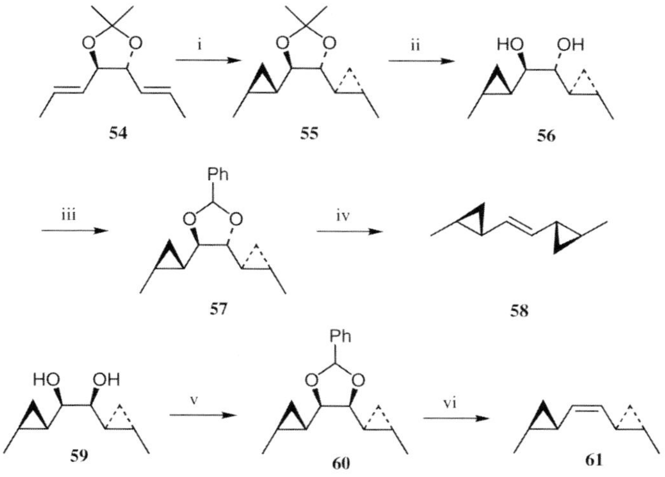

54 55 56 57 58 59 60 61

Reagents and conditions: (i) Et_2Zn, CH_2I_2, $(CH_2Cl)_2$, -20 °C, 89%; (ii) TsOH (cat), THF/H_2O (5/1), 70 °C, 63%; (iii) PhCHO, H_2SO_4 (cat), PhMe, Δ, 64%; (iv) BuLi, pentane, 60%; (v) PhCHO, CSA, 96%; (vi) BuLi, pentane.

Scheme 10. Stereospecific synthesis of dicyclopropylethene derivatives

identical in all respects, including the absolute stereochemistry with the diacetate derived from diol **47**. The imidazolidine cyclopropane derivatives **63** and **64** were synthesized from crotonaldehyde via acetal **62**

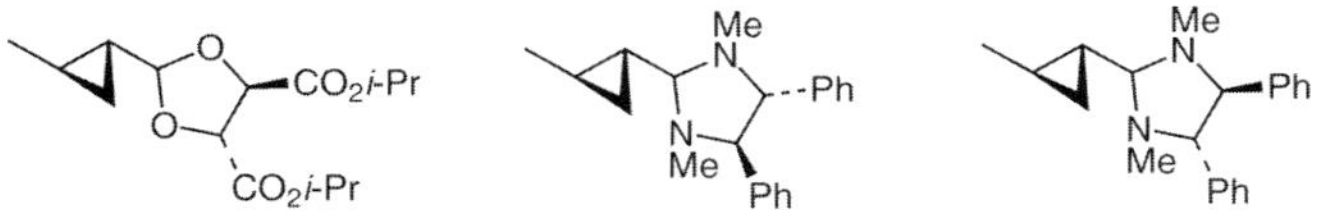

62 63 64

using Yamamoto asymmetric cyclopropanation (Arai et al. 1985; Mori et al. 1986), hydrolysis, and condensation with the corresponding chiral diamines (Manganey et al. 1988b). Reaction of the ozonolysis product obtained from an authentic sample of FR-900848 (**1**) with (1*R*, 2*R*)-*N*,*N*'-dimethyl-1,2-diphenylethanediamine provided an imidazolidine derivative which was identical with the synthetic adduct **63**. These facts together are consistent with the assignment of structure **52** to FR-900848.

5.4 Total Synthesis of FR-900848 (52)

We have reported the total synthesis of FR-900848 (**52**) using the sequence of Charette cyclopropanation reactions shown in Schemes 11, 12, and 13 (Barrett and Kasdorf 1996a,b). Thus, mucondiol (**65**) was bicyclopropanated in the presence of the chiral auxiliary **14** (Charette and Juteau 1994) giving the bicyclopropane **66** (89%), predominantly as a single enantiomer. Oxidation and subsequent direct homologation without purification of the intermediate dialdehyde provided a separable mixture of *E,E*-diester **44** and the *E,Z*-isomer **45** (8.7:1, 67% from **66**). Diester **45** was readily isomerized into the desired diester **44** under Hunter's conditions (Hunter 1995). Reduction of diester **44** (94%) with DIBAL-H and bicyclopropanation of the diol product, under Charette conditions (Charette and Juteau 1994), furnished the diol **47** (93%), predominantly as a single diastereoisomer. Monoprotection, oxidation, and Horner-Emmons homologation provided esters **68** and **69**. Hunter isomerization (Hunter 1995) was again used for converting the un-

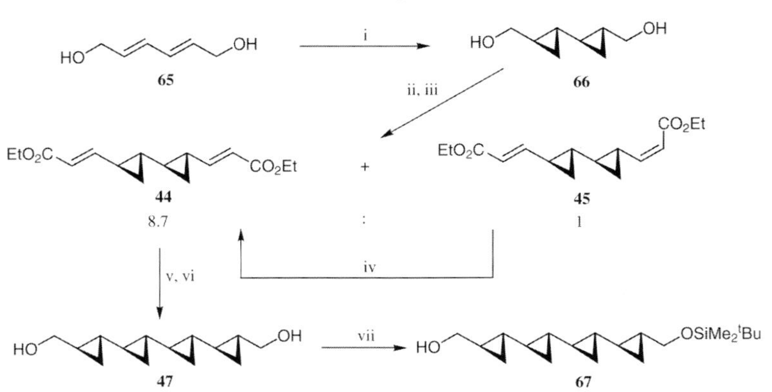

Reagents and conditions: (i) **14**, Et_2Zn, CH_2I_2, CH_2Cl_2, 0 to 25 °C, 89%; (ii) PCC, NaOAc, silica, CH_2Cl_2, 0 to 25 °C; (iii) $Ph_3P{=}CHCO_2Et$, CH_2Cl_2, 67% (from **66**); (iv) PhSH, *n*-BuLi, $Ti(OiPr)_4$, THF, 0 to 25 °C, 50%; (v) DIBAL-H, CH_2Cl_2, -78 °C, 94%; (vi) **14**, Et_2Zn, DME, CH_2I_2, CH_2Cl_2, -15 to 25 °C, 93%; (vii) NaH, $^tBuMe_2SiCl$, THF, 44%.

Scheme 11. Total synthesis of FR-900848: the quatercyclopropane unit

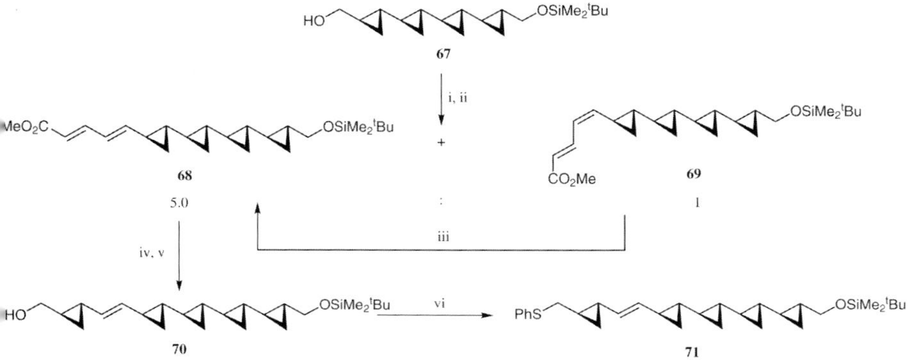

Reagents and conditions: (i) PCC, NaOAc, silica, CH_2Cl_2, 0 to 25 °C; (ii) (*E*)-$MeO_2CCH{=}CHCH_2P(O)(OMe)_2$, NaH, THF, 0 to 25 °C, 71% (from **67**); (iii) PhSH, *n*-BuLi, $Ti(OiPr)_4$, THF, 0 to 25 °C, 63%; (iv) DIBAL-H, CH_2Cl_2, -78 °C, 91%; (v) **14**, Et_2Zn, DME, CH_2I_2, CH_2Cl_2, -40 °C, 90%; (vi) *N*-(phenylsulfenyl)succinimide, *n*-Bu_3P, PhH, 89%.

Scheme 12. Elaboration of the complete multicyclopropane domain of FR-900848

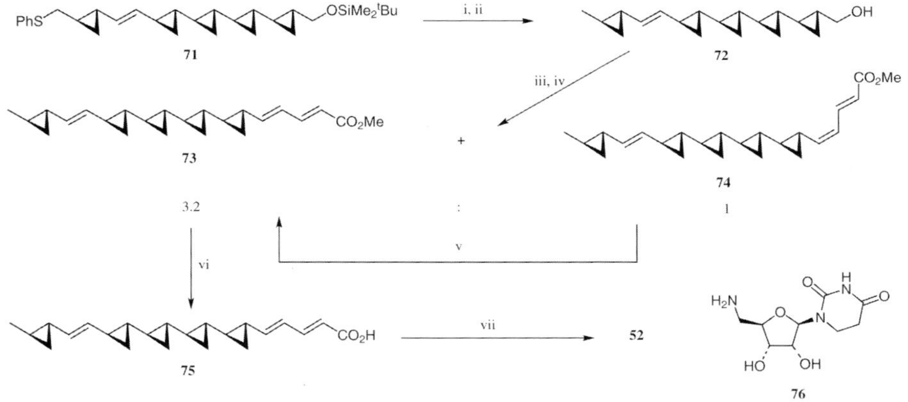

Reagents and conditions: (i) Raney Ni, EtOH, -40 °C; (ii) NH_4F, EtOH, 65 °C, 49% (from **71**); (iii) PCC, NaOAc, silica, CH_2Cl_2, 0 to 25 °C; (iv) (*E*)-$MeO_2CCH{=}CHCH_2P(O)(OMe)_2$, NaH, THF, 0 to 25 °C, 63% (from **72**); (v) PhSH, *n*-BuLi, $Ti(OiPr)_4$, THF, 0 to 25 °C, 51%; (vi) $KOSiMe_3$, CH_2Cl_2, 85%; (vii) **76**, BOP-Cl, Et_3N, DMA, 69%.

Scheme 13. Completion of the total synthesis of FR-900848 wanted isomer **69** into additional *E,E*-ester **68**. Finally, DIBAL-H reduction of *E,E*-ester **68** (91%), followed by a final Charette asymmetric cyclopropanation (Charette et al. 1995), gave the pentacyclopropane olefinic alcohol **70** in excellent yield (90%). Charette has also reported related monocyclopropanation reactions on related dienols (Charette et al. 1996).

Reaction of alcohol **70** with *N*- (phenylsulfenyl)succinimide and tributylphosphine (Walker 1977) provided the sulfide **71** (89%) which, upon treatment with Raney nickel followed by ammonium fluoride, gave alcohol **72** without significant alkene hydrogenation or cyclopropane degradation. Oxidation of the alcohol **72**, Horner-Emmons homologation with Hunter isomerization (Hunter 1995) of the unwanted *E,Z*-ester **74**, potassium trimethylsilanolate (Laganis and Chenard 1984)-mediated hydrolysis, and BOP-Cl (Cabre and Palomo 1984)-mediated coupling with the nucleoside amine **76** (Skaric et al. 1982) furnished FR-900848 (**52**). Much to our delight, the synthetic material was identical with an authentic sample of the natural product.

Falck and co-workers (1996) have also reported a total synthesis of FR-900848 (**52**) which clearly shows the benefit of Horeau amplification of enantiomeric purity (Scheme 14). The vinyl stannane **77** was transformed into the corresponding cyclopropane **78** using Charette cyclopropanation (Charette et al. 1995) followed by protection. Tin to lithium to copper exchange and oxygenation gave the corresponding bicyclopropane **79**, which was in turn converted via the carboxylic acid **80**, Barton bromodecarboxylation (Barton et al. 1983), and a second dimerization reaction to reveal the quatercyclopropane **82**. These two oxidative dimerization reactions were accompanied by an enhancement in enantiomeric purity from 88%–90% ee to >99% ee in a process that is a variant of the Horeau amplification principle (Rautenstrauch 1994). The diether **82** was converted into FR-900848 using, among others, a Julia-Peterson olefination reaction, Horner-Emmons homologation, and acylation via *p*-nitrophenyl ester activation.

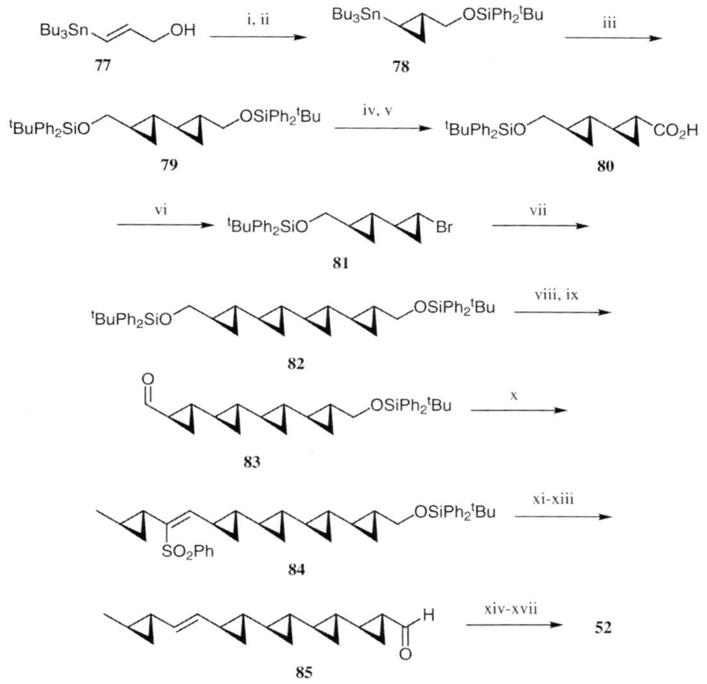

Bu_3Sn OH **77** —i, ii→ Bu_3Sn $OSiPh_2{}^tBu$ **78** —iii→

${}^tBuPh_2SiO$ $OSiPh_2{}^tBu$ **79** —iv, v→ ${}^tBuPh_2SiO$ CO_2H **80**

—vi→ ${}^tBuPh_2SiO$ Br **81** —vii→

${}^tBuPh_2SiO$ $OSiPh_2{}^tBu$ **82** —viii, ix→

O $OSiPh_2{}^tBu$ **83** —x→

$OSiPh_2{}^tBu$ SO_2Ph **84** —xi-xiii→

H O **85** —xiv-xvii→ **52**

Reagents and conditions: (i) **14**, Et_2Zn, CH_2I_2, CH_2Cl_2, 23 °C, 98%; (ii) ${}^tBuPh_2SiCl$, imidazole, DMF, 23 °C, 88%; (iii) s-BuLi, THF, -40 °C; $[ICuPBu_3]_4$, -78 °C; O_2, -78 °C, 73%; (iv) Bu_4NF, THF, 23 °C, 72%; (v) $RuCl_3$, $NaIO_4$, CCl_4, MeCN, H_2O, 23 °C, 91%; (vi) *N*-hydroxypyridine-2-thione, DCC, DMAP, $BrCCl_3$, 23 °C; hυ, 0 °C, 77%; (vii) t-BuLi, THF, -78 °C; $[ICuPBu_3]_4$, -78 °C; O_2, -78 °C, 75%; (viii) Bu_4NF, THF, 23 °C, 72%; (ix) TPAP, NMO, 4Å molecular sieves, CH_2Cl_2, 23 °C, 91%; (x) **86**, n-BuLi, THF, -78 °C, 65%; (xi) Li, naphthalene, THF, -78 °C, 70%; (xii) Bu_4NF, THF, 23 °C, 95%; (xiii) TPAP, NMO, 4Å molecular sieves, CH_2Cl_2, 23 °C, 91%; (xiv) (*E*)-$EtO_2CCH{=}CHCH_2P(O)(OEt)_2$, $LiN(SiMe_3)_2$, THF, -78 °C, 89%; (xv) LiOH, MeOH, H_2O, 23 °C, 90%; (xvi) 4-nitrophenol, DCC, DMAP, CH_2Cl_2, 23°C, 73%; (xvii) **76**, DMF, 23°C, 76%.

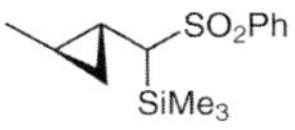

SO_2Ph $SiMe_3$ **86**

Scheme 14. Falck total synthesis of FR-900848

5.5 Total Synthesis and Structural Elucidation of U-106305 (87)

In 1995 scientists at Upjohn reported the isolation of U-106305 (**87**) from the fermentation broth of *Streptomyces* sp. UC 11136 (Kuo et al. 1995). U-106305 (**87**) is biologically active and is an inhibitor of cholesteryl ester transfer protein (CETP), which is of consequence in coronary heart disease. Although the full structure was not determined, it was established that all the cyclopropanes are *trans*-disubstituted and that both alkenes are *E*. Their statement that "U-106305 represents a structural class of compounds not previously reported from microbial fermentations" is inaccurate; FR-900848 (**52**) and U-106305 (**87**) are clearly very closely related, and, prior to our total synthesis, we assumed that U-106305 (**87**) had the same absolute stereochemistry as **88**.

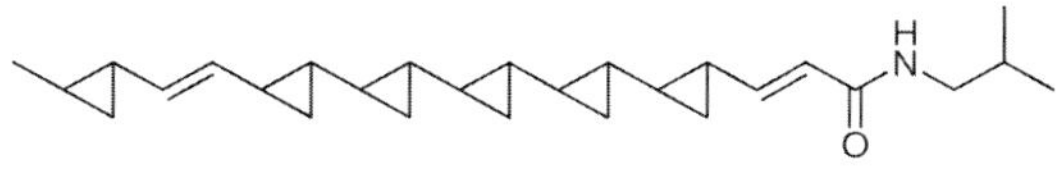

87

88

In our synthesis of U-106305 (**88**) we used a bi-directional approach, very similar to that utilized for FR-900848 (**52**), to assemble the C_2-symmetric quinquecyclopropane unit **94** (Scheme 15; Barrett et al. 1996b). Thus, Charette cyclopropanation (Charette et al. 1995) of diol **89** gave **90** in 89% ee. Diene diester **91** was prepared from **90** by Dess-Martin oxidation (Dess and Martin 1991) and Wittig olefination. Enantiomerically pure (*E,E*)-diester **91** was obtained by fractional recrystallization from Et_2O:hexanes and its structure was confirmed by X-ray crystallography (Barrett et al. 1997). The quinquecyclopropane **94** was prepared from diester **91** via tercyclopropane **93** using our usual chain-extension methodology and two double Charette cyclopropanation reactions (Charette et al. 1995). The structure of the quinquecyclopropane **94** was analyzed by X-ray crystallography (Barrett et al.

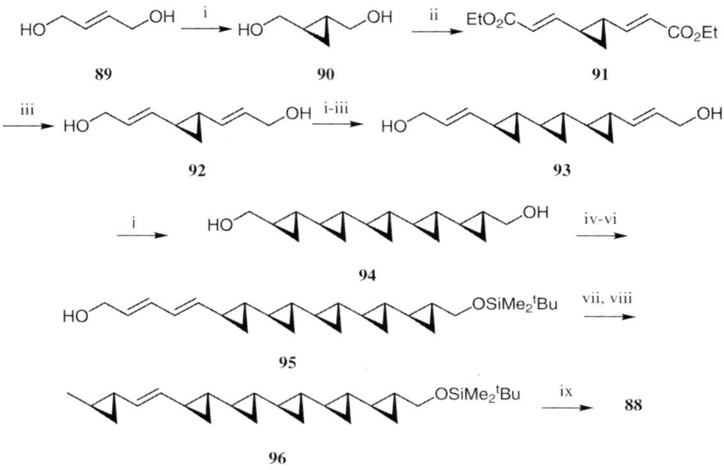

HO OH 89 —i→ HO OH 90 —ii→ EtO_2C CO_2Et 91

—iii→ HO OH 92 —i-iii→ HO OH 93

—i→ HO OH 94 —iv-vi→

HO $OSiMe_2{}^tBu$ 95 —vii, viii→

$OSiMe_2{}^tBu$ 96 —ix→ 88

Reagents and conditions: (i) **14**, 4Å molecular sieves, $Zn(CH_2I)_2$·DME, CH_2Cl_2, -40 to 25 °C, 83 - 91%; (ii) Dess-Martin periodinane, pyridine, CH_2Cl_2 or DMSO, 25 °C; PPh_3, ca. 10 °C; $Ph_3P{=}CHCO_2Et$, 75 - 81%; (iii) DIBAL-H, CH_2Cl_2, hexanes, -78 °C, 96 - 97%; (iv) ${}^tBuMe_2SiCl$, imidazole, CH_2Cl_2, 25 °C, 75% calcd. at 78% conversion; (v) Dess-Martin periodinane, pyridine, DMSO, 25 °C; PPh_3, 25 °C; (*E*)-$(MeO)_2P(O)CH_2CH{=}CHCO_2Me$, NaH, DBU, 25 °C, 88%; (vi) DIBAL-H, CH_2Cl_2, hexanes, -78 °C, 95%; (vii) **14**, 4Å molecular sieves, $Zn(CH_2I)_2$·DME, CH_2Cl_2, -50 °C to -25 °C, 72%; (viii) *N*-(phenylsulfenyl)succinimide, PBu_3, C_6H_6, 25 °C, 91%; Raney nickel, THF, 25 °C, 44%; (ix) Bu_4NF, THF, 25 °C; DMSO, pyridine, 25 °C; Dess-Martin periodinane, 25 °C; PPh_3, ca. 15 °C; $Cl^-Ph_3P^+{-}CH_2CONHCH_2{}^iPr$, DBU, 25 °C, 91%.

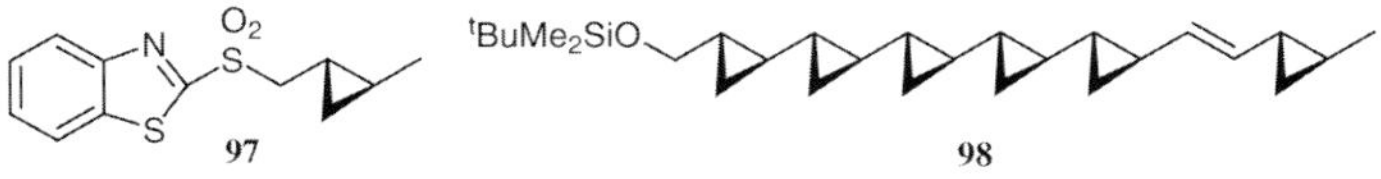

N O_2 S S 97 ${}^tBuMe_2SiO$ 98

Scheme 15. Total synthesis of U-106305

1996b) and found to be near-helical, as can be seen from Fig. 3. This compound was transformed into U-106305 (**88**) using fundamentally the same strategy as in our synthesis of FR-900848 (**52**). We were delighted to find that our synthetic material was identical in all respects to an authentic sample of U-106305, clearly establishing the full structure and absolute stereochemistry of U-106305 (**88**). Subsequent to our

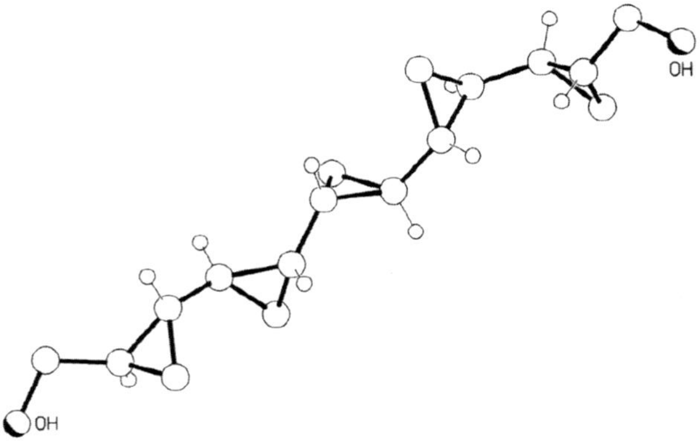

Fig. 3. The molecular structure of the quinquecyclopropane diol **94** showing the absolute stereochemistry

work, Charette and Lebel (1996) reported a total synthesis of the antipode of U-106305 (**88**) using a bi-directional strategy that was closely related to our work. Their synthesis differed in the use of a Julia-Lythgo coupling (Baudin et al. 1993; Smith et al. 1996) reaction to link the quinquecyclopropane and monocyclopropane units in the later stages. The use of sulfone **97** provided the corresponding alkene **98** with 4:1 *E*:*Z* geometric selectivity. Zercher (McDonald et al. 1997) also used a very similar bi-directional approach to the enantiomer of quinquecyclopropane **94** but has not yet reported the synthesis of the antipode of U-106305 (**88**).

5.6 Syntheses of Polycyclopropane Containing Coronanes

Recently we reported the synthesis of two coronanes, namely **105** and **106**, as prototypes for a novel class of macrocyclic compounds (Barrett et al. 1998). The larger, 22-membered ring coronane **106** showed a remarkable affinity for diethyl ether (NMR spectroscopy) as well as methylamine and 4-dimethylaminopyridine (complexes were observed by ES mass spectrometry). The structure of **105** was confirmed by an X-ray crystallographic study (Fig. 4). The smaller ring coronane **105**

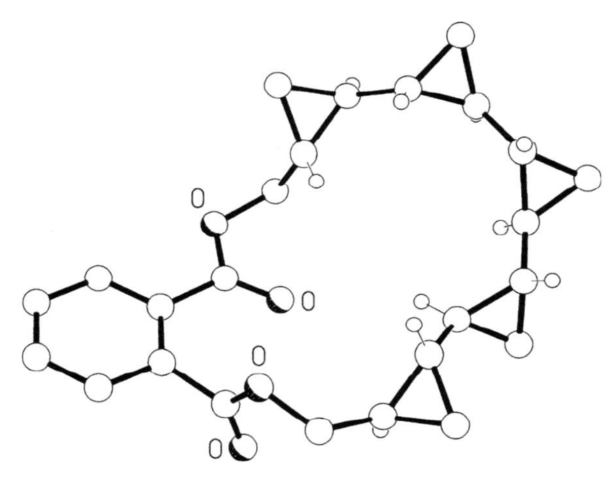

Fig. 4. The molecular structure of coronane **105** showing the absolute stereochemistry

was synthesized (Scheme 16) by treatment of diol **94** with *t*-butyldimethylsilyl chloride, giving the monoprotected alcohol **102** (54%) along with unreacted starting diol **94** (30%). Phthalic anhydride **99** was condensed with trimethylsilylethanol to provide monoester **100**, which was coupled under Steglich-esterification (Neises and Steglich 1978) conditions with the protected alcohol **102** to provide the unsymmetric diester **103** in quantitative yield. Double deprotection using tetrabutylammonium fluoride gave hydroxy acid **104** which, without isolation, was subjected to a Yamaguchi macrocyclization reaction (Inanaga et al. 1979) to provide coronane **105** in a high overall yield. Coronane **106** was similarly synthesized using this stepwise approach from the larger heptacyclopropanedimethanol (**101**) (Barrett et al. 1997). Using the same methodology, the *meta*- and *para*-isomers **107** and **108** of coronane **105** were prepared from the corresponding *meta*- and *para*-isomers of acid **100** and the monoprotected alcohol **102** in high overall yield (A.G.M. Barrett, M. Ohkubo, unpublished results). The structures of all these coronanes were confirmed by X-ray crystallography.

Synthesis of the related *bis*-aryl coronane **113** was undertaken (Scheme 17; James 1999) utilizing a methodology similar to that described above for the synthesis of coronane **105**. The monoester **111** was

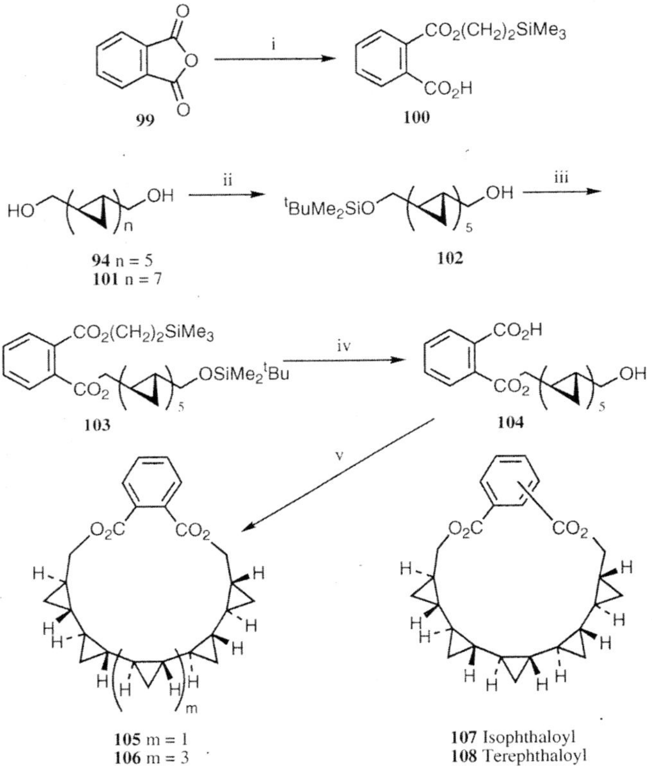

O O O **99** —i→ $CO_2(CH_2)_2SiMe_3$ CO_2H **100**

HO OH n **94** n = 5 **101** n = 7 —ii→ $^tBuMe_2SiO$ OH 5 **102** —iii→

$CO_2(CH_2)_2SiMe_3$ CO_2 $OSiMe_2{}^tBu$ 5 **103** —iv→ CO_2H CO_2 OH 5 **104**

v

O_2C CO_2 H H H H H H H H H H m

105 m = 1
106 m = 3

O_2C CO_2 H H H H H H H H H H H H

107 Isophthaloyl
108 Terephthaloyl

Reagents and conditions: (i) $Me_3Si(CH_2)_2OH$, NaH, THF, 81%; (ii) $^tBuMe_2SiCl$, imidazole, CH_2Cl_2, 54% for n = 5; (iii) **100**, DCC, DMAP, CH_2Cl_2, 100%; (iv) Bu_4NF, THF; (v) 2,4,6-$Cl_3C_6H_2COCl$, Et_3N, THF, 25 °C; DMAP, PhMe, Δ, 74% for m = 1; 67% overall for m =3.

Scheme 16. Synthesis of coronane prototypes

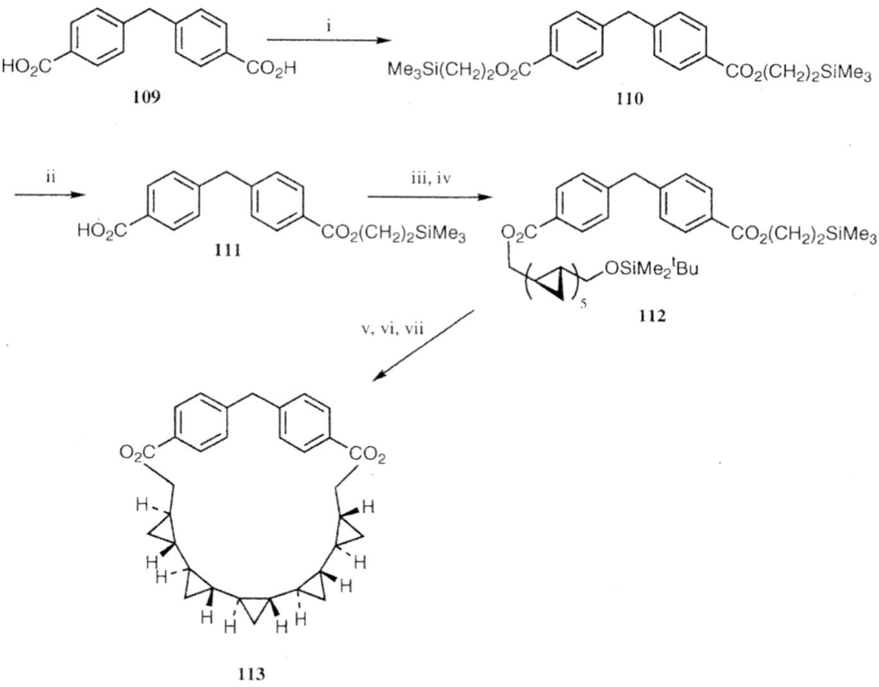

Reagents and conditions: (i) $Me_3Si(CH_2)_2OH$, DCC, DMAP, THF, 57%; (ii) Bu_4NF, THF, 49%; (iii) 2,4,6-$Cl_3C_6H_2COCl$, Et_3N, PhMe; (iv) **102**, DMAP, PhMe, 63% from **111**; (v) Bu_4NF, THF; (vi) 2,4,6-$Cl_3C_6H_2COCl$, Et_3N, THF; (vii) DMAP, PhMe, 110°C, 90% from **112**.

Scheme 17. Synthesis of a coronane with a methylenedibenzoic acid spacer

obtained by partial desilylation of diester **110** with tetrabutylammonium fluoride and was coupled with alcohol **102**, via the 2,4,6-trichlorobenzoyl mixed anhydride, providing the macrocyclization precursor **112** (63%). Deprotection of ester **112** provided the hydroxy acid which was used without purification in the subsequent Yamaguchi macrocyclization to provide the desired dilactone **113** in excellent yield. The structure of this coronane was also confirmed by X-ray crystallography (Fig. 5).

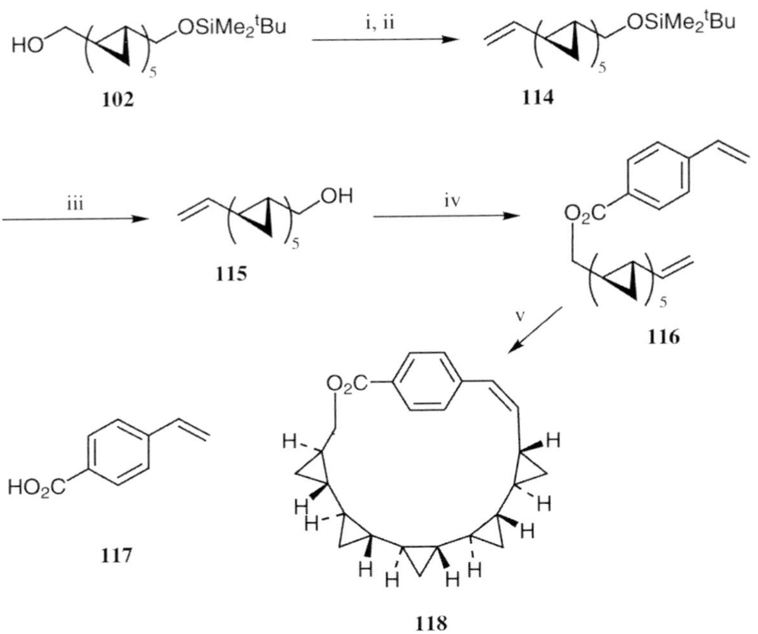

Reagents and conditions: (i) Dess-Martin periodinane, CH_2Cl_2; (ii) Ph_3PCH_3Br, ^{t}BuOK, THF, 76% from **102**; (iii) Bu_4NF, THF, 97%; (iv) **117**, DCC, DMAP, CH_2Cl_2, 96%; (v) $Cl_2(Cy_3P)_2Ru{=}CHPh$ (**119**), PhMe, Δ, 45%.

Scheme 18. Alkene metathesis and coronane construction

Ring closing metathesis (RCM) has proven itself to be a powerful tool for the synthesis of medium to large heterocycles (Grubbs and Chang 1998), and we have shown that it is applicable to the generation of coronanes, as exemplified in the synthesis of coronane **118** (Scheme 18; James 1999). Alcohol **115** was synthesized from monoprotected alcohol **102** by Dess-Martin oxidation (Dess and Martin 1991) and Wittig homologation, followed by desilylation to provide the target alcohol **115** in high overall yield. This alcohol **115** was subjected to DCC-mediated coupling with 4-vinylbenzoic acid **117** (Leebrick and Ramsden 1958) to provide the diene **116** (96%). Diene **116** was then

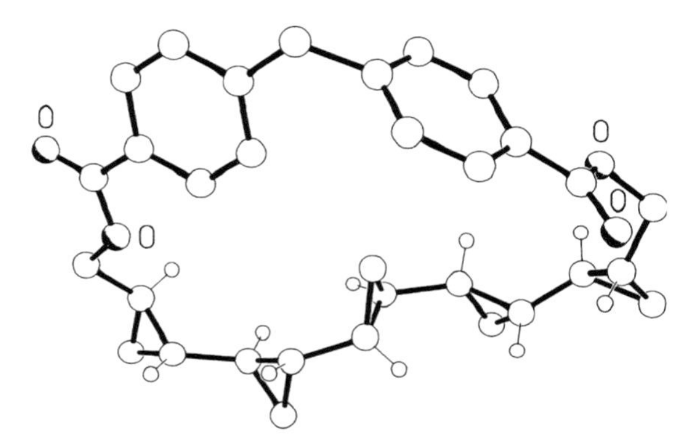

Fig. 5. The molecular structure of *bis*-aryl coronane **113** showing the absolute stereochemistry

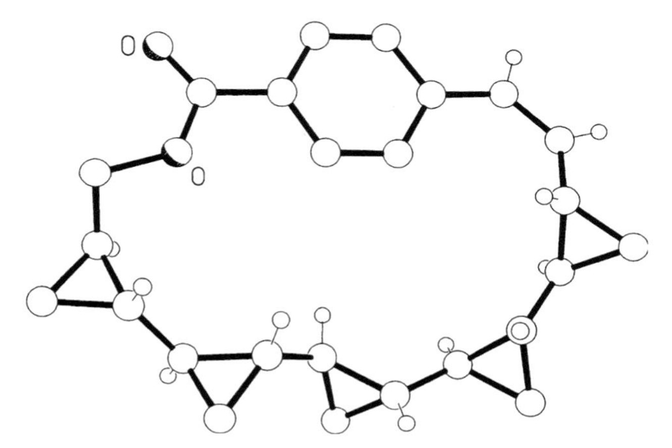

Fig. 6. The molecular structure of coronane **118** showing the absolute stereochemistry

subjected to RCM (20 mol%, **119**) providing the desired macrocycle **118** (45%), with the remaining mass balance in this reaction composed of unreacted diene **116**. Macrocyclic lactone **118** was obtained as a crystalline solid and its structure was confirmed by X-ray crystallographic analysis, showing that the olefin had *Z*-geometry (Fig. 6).

5.7 Future Directions

Much remains to be done on the chemistry and biology of multiple cyclopropane arrays. It is clear from the Upjohn work that the biological effects mediated by multiple cyclopropane fatty amides should be much more general than merely anti-infective. Further aspects of the synthesis, reactions, and properties of cyclopropane oligomers will be reported in due course

Acknowledgements. We thank GlaxoWellcome Research & Development for their most generous endowment (to A.G.M.B.) and for general support of our research, the Wolfson Foundation for establishing the Wolfson Centre for Organic Chemistry in Medical Science at Imperial College, the Engineering and Physical Sciences Research Council, the U.S. Department of Education under the Graduate Assistance in Areas of National Need Program, the Overseas Research Students Program for fellowship support (to K.K.), the National Institutes of Health (AI-22252) for support when this work started in the USA, the European Commission for a TMR Research Training Grant (to D.H.), the Fujisawa Pharmaceutical Company Ltd. for generous donations of samples of FR-900848 (**1**) and key spectroscopic data, Pharmacia & Upjohn, Inc. for a sample of U-106305 (**87**), ChemGenics Pharmaceutical Inc. for support of our research on antifungal agents, and G.D. Searle & Company for their generous unrestricted support.

References

Arai I, Mori A, Yamamoto H (1985) An asymmetric Simmons-Smith reaction. J Am Chem Soc 107:8524–8256

Armstrong RW, Maurer KW (1995) Synthesis of vicinal cyclopropanes. Tetrahedron Lett 36:357–360

Banwell MG, Onrust R (1985) A versatile new strategy for the synthesis of tropolones. Tetrahedron Lett 26:4543–4546

Barrett AGM (1994) 6th international Kyoto conference on new aspects of organic chemistry, Kyoto, Japan, 7–11 Nov

Barrett AGM, Kasdorf K (1996a) Total synthesis of the antifungal agent FR-900848. J Chem Soc Chem Commun 325–326

Barrett AGM, Kasdorf K (1996b) Total synthesis of the pentacyclopropane antifungal agent FR-900848. J Am Chem Soc 118:11030–11037

Barrett AGM, Tustin GJ (1995) Studies towards the synthesis of FR-900848 – stereoselective preparation of *anti*-bicyclopropane derivatives. J Chem Soc Chem Commun 355–356

Barrett AGM, Doubleday WW, Tustin GJ, White AJP, Williams DJ (1994a) Approaches to the assembly of the antifungal agent FR-900848 – studies on the asymmetric synthesis of bicyclopropanes and an X-ray crystallographic analysis of (4*R*,5*R*)-2-[(1*R*,3*S*,4*S*,6*R*)-6-phenyl-1-bicyclopropyl]-1,3-dimethyl-4,5-diphenylimidazolidine. J Chem Soc Chem Commun 1783–1784

Barrett AGM, Kasdorf K, Williams DJ (1994b) Approaches to the assembly of the antifungal agent FR-900848 – studies on double asymmetric cyclopropanation and an X-ray crystallographic study of (1*R*,2*R*)-1,2-bis-[(1*S*,2*S*)-2-methylcyclopropyl]-1,2-ethanediyl 3,5-dinitrobenzoate. J Chem Soc Chem Commun 1781–1782

Barrett AGM, Doubleday WW, Kasdorf K, Tustin GJ, White AJP, Williams DJ (1995a) Approaches to the assembly of the antifungal agent FR-900848 – studies on the synthesis of C-2 symmetrical tetracyclopropane derivatives and an x-ray crystallographic study of (1*R*,3*S*,4*S*,6*R*)-bicyclopropyl-1,6-di-(2-[(4*R*,5*R*)-di(isopropyloxycarbonyl)-1,3-dioxolane]). J Chem Soc Chem Commun 407–408

Barrett AGM, Kasdorf K, Tustin GJ, Williams DJ (1995b) Determination of the full structure and absolute stereochemistry of the antifungal agent FR-900848 – an X-ray crystallographic study of (1*R*,3*S*,4*R*,6*S*,7*S*,9*R*,10*S*,12*R*)-quatercyclopropyl-1,12-dimethanediyl di-4-bromobenzoate. J Chem Soc Chem Commun 1143–1144

Barrett AGM, Kasdorf K, White AJP, Williams DJ (1995c) Approaches to the assembly of the antifungal agent FR-900848 – determination of the geometry of the dicyclopropylethene unit and an X-ray crystallographic study of (1*R*,2*S*)-1,2-bis[(1*S*,2*S*)-2-methylcyclopropyl]-1,2-ethanediyl 3,5-dinitrobenzoate. J Chem Soc Chem Commun 649–650

Barrett AGM, Doubleday WW, Kasdorf K, Tustin GJ (1996a) Stereochemical elucidation of the pentacyclopropane antifungal agent FR-900848. J Org Chem 61:3280–3288

Barrett AGM, Hamprecht D, White AJP, Williams DJ (1996b) Total synthesis and stereochemical assignment of the quinquecyclopropane-containing cholesteryl ester transfer protein inhibitor U-106305. J Am Chem Soc 118:7863–7864

Barrett AGM, Hamprecht D, White AJP, Williams DJ (1997) Iterative cyclopropanation: a concise strategy for the total synthesis of the hexacyclopropane cholesteryl ester transfer protein inhibitor U-106305. J Am Chem Soc 119:8608–8615

Barrett AGM, Gross T, Hamprecht D, Ohkubo M, White AJP, Williams DJ (1998) Multiple asymmetric cyclopropanation: synthesis and X-ray crystallographic studies of a prototype coronane and all *anti-trans*-1,15-quinquecyclopropanedimethanol. Synthesis 490–494

Barton DHR, Crich DC, Motherwell WB (1983) A practical alternative to the Hunsdiecker reaction. Tetrahedron Lett 24:4979–4982

Baudin JB, Hareau G, Julia SA, Lorne R, Ruel O (1993) Stereochemistry of direct olefin formation from carbonyl compounds and lithiated heterocyclic sulfones. Bull Chim Soc Fr 130:856–878

Buchert M, Reissig HU (1992) Highly functionalized vinylcyclopropane derivatives by regioselective and stereoselective reaction of Fischer carbene complexes with 1,4-disubstituted electron-deficient 1,3-dienes. Chem Ber 125:2723–2729

Cabre J, Palomo AL (1984) New experimental strategies in amide synthesis using *N,N*-bis[2-oxo-3-oxazolidinyl] phosphorodiamidic chloride. Synthesis 413–417

Charette AB, De Freitas-Gil RP (1997) Synthesis of contiguous cyclopropanes by palladium-catalyzed Suzuki-type cross-coupling reactions. Tetrahedron Lett 38:2809–2812

Charette AB, Juteau H (1994) Design of amphoteric bifunctional ligands – application to the enantioselective Simmons-Smith cyclopropanation of allylic alcohols. J Am Chem Soc 116:2651–2652

Charette AB, Lebel H (1996) Enantioselective total synthesis of (+)-U-106305. J Am Chem Soc 118:10327–10328

Charette AB, Marcoux J-F (1995) The asymmetric cyclopropanation of acyclic allylic alcohols – efficient stereocontrol with iodomethylzinc reagents. SynLett 1197

Charette AB, Prescott S, Brochu C (1995) Improved procedure for the synthesis of enantiomerically enriched cyclopropylmethanol derivatives. J Org Chem 60:1081–1083

Charette AB, Juteau H, Lebel H, Deschênes D (1996) The chemo- and enantioselective cyclopropanation of polyenes: chiral auxiliary vs chiral reagent-based approach. Tetrahedron Lett 37:7925–7928

Clauss R, Hinz W, Hunter R (1997) Low temperature isomerisation of (*Z*)-α,β-unsaturated esters into their (*E*)-isomers by LiTi $(O^iPr)_4SPh$ and LiSPh. Synlett 57–59

Das BC, Cosson JP, Guittet E, Hassani M, Staron T, DeMaack F (1987) Book of abstracts, 35 ASMS conference on mass spectrometry and allied topics, 24–29 May, Denver, Colorado, pp 952–953

Davies SG, Whitham GH (1977) Benzene oxide-oxepin. Oxidation to muconaldehyde. J Chem Soc Perkin Trans I:1346–1347

Dess DB, Martin JC (1991) A useful 12-I-5 triacetoxyperiodinane (the Dess-Martin periodinane) for the selective oxidation of primary or secondary alcohols and a variety of related 12-I-5 species. J Am Chem Soc 113:7277–7287

Falck JR, Mekonnen B, Yu J (1995) 209th American Chemical Society National Meeting, Anaheim, Calif., April 2–6, abstract 057

Falck JR, Mekonnen B, Yu J, Lai J-Y (1996) Synthesis of the polycyclopropane antibiotic FR-900848 via the Horeau gambit. J Am Chem Soc 118:6096–6097

Grubbs RH, Chang S (1998) Recent advances in olefin metathesis and its application in organic synthesis. Tetrahedron 54:4413–4450

Hines JN, Peagram MJ, Thomas EJ, Whitham GH (1973) Some reactions of benzaldehyde acetals with alkyl-lithium reagents: a stereospecific olefin synthesis from 1,2-diols. J Chem Soc Perkin Trans I:2332–2337

Hoffman RW (1989) Allylic 1,3-strain as a controlling factor in stereoselective transformations. Chem Rev 89:1841–1860

Hunter R (1995) Frank Warren Conference, Orange Free State, South Africa, April 4–7

Inanaga J, Hirata K, Saeki H, Katsuki T, Yamaguchi M (1979) A rapid esterification by means of mixed anhydride and its application to large-ring lactonization. Bull Chem Soc Jpn 52:1989–1993

James RA (1999) Cyclopropane methodology. PhD thesis, Imperial College, chap 6, pp 136–151

Kuo MS, Zielinski RJ, Cialdella JI, Marschke CK, Dupuis M, Li GP, Kloosterman DA, Spilman CH, Marshall VP (1995) Discovery, isolation, structure elucidation, and biosynthesis of U-106305, a cholesteryl ester transfer protein inhibitor from UC-11136. J Am Chem Soc 117:10629–10634

Kwon-Chung JK, Bennett JE (1992) Medical mycology. Lee and Febiger, Philadelphia

Laganis ED, Chenard BL (1984) Metal silanolates – organic soluble equivalents for O^{-2}. Tetrahedron Lett 25:5831–5834

Leebrick JR, Ramsden HE (1958) Synthesis and reactions of *p*-vinylphenylmagnesium chloride. J Org Chem 23:935–936

Manganey P, Grojea F, Alexakis A, Normant JR (1988a) Improved optical resolution of R-star, R-star *N,N'*-dimethyl 1,2-diphenyl ethylene diamine. Tetrahedron Lett 29:2675–2676

Manganey P, Grojea F, Alexakis A, Normant JR (1988b) Resolution and determination of enantiomeric excesses of chiral aldehydes via chiral imidazolidines. Tetrahedron Lett 29:2677–2680
McDonald WS, Verbicky CA, Zercher CK (1997) Two-directional synthesis of polycyclopropanes. An approach to the quinquecyclopropane fragment of U-106305. J Org Chem 62:1215–1222
Mitsukura K, Korekiyo S, Itoh T (1999) Synthesis of optically active bisdifluorocyclopropanes through a chemo-enzymatic reaction strategy. Tetrahedron Lett 40:5739–5742
Mori A, Arai I, Yamamoto H (1986) Asymmetric Simmons-Smith reactions using homochiral protecting groups. Tetrahedron 42:6447–6458
Neises B, Steglich W (1978) Simple method for the esterification of carbocyclic acids. Angew Chem Int Ed Engl 17:522
Nijveldt D, Vos A (1988) Single-crystal X-ray geometries and electron-density distributions of cyclopropane, bicyclopropyl and vinylcyclopropane. 1. Data collection, structure determination and conventional refinements. 2. Multipole refinements and dynamic electron-density distributions. 3. Evidence for conjugation in vinylcyclopropane and bicyclopropyl. Acta Crystallogr Sect B Struct Sci B44:281, 289, 296
Rautenstrauch V (1994) The two expressions of the Horeau principle, nth-order Horeau amplifications, and scales for the resulting very high enantiopurities. Bull Soc Chim Fr 131:515–524
Skaric V, Katalenic D, Skaric D, Salaj I (1982) Stereochemical transformations of 5'-amino-5'-deoxypuridine and its 5,6-dihydro analogue – 5'-*N*-aminoacyl derivatives of 5'-amino-5'-deoxy-5,6-dihydrouridine. J Chem Soc Perkin Trans 1:2091
Smith ND, Kocienski PJ, Street SDA (1996) A synthesis of (+)-herboxidiene A. Synthesis 652
Stille JK (1986) The palladium-catalyzed cross-coupling reactions of organotin reagents with organic electrophiles. Angew Chem Int Ed Engl 25:508–523
Taylor RE, Ameriks MK, LaMarche MJ (1997) A novel approach to oligocyclopropane structural units. Tetrahedron Lett 38:2057–2060
Taylor RE, Engelhardt FC, Yuan H (1999) Oligocyclopropane structural units from cationic intermediates. Org Lett 1:1257–1260
Theberge CR, Zercher CK (1994) Diastereoselective synthesis of bicyclopropanes. Tetrahedron Lett 35:9181–9184
Theberge CR, Zercher CK (1995) A divergent diastereoselective approach to bicyclopropanes. Tetrahedron Lett 36:5495–5498
Theberge CR, Verbicky CA, Zercher CK (1996) Studies on the diastereoselective preparation of bis-cyclopropanes. J Org Chem 61:8792–8798
Tsunoda T, Suzuki M, Noyori R (1980) A facile procedure for acetalization under aprotic conditions. Tetrahedron Lett 21:1357–1358

Ukaji Y, Nishimura M, Fujisawa T (1992) Enantioselective construction of cyclopropane rings via asymmetric Simmons-Smith reaction of allylic alcohols. Chem Lett 61–64

Walker KAM (1977) A convenient preparation of thioethers from alcohols. Tetrahedron Lett 4475–4478

Yoshida M, Horikoshi K (1989) FR-900848 substance and preparation thereof. United States Patent 4,803,074; Fujisawa Pharmaceutical, Japan, 7 Feb

Yoshida M, Ezaki M, Hashimoto M, Yamashita M, Shigematsu N, Okuhara M, Kohsaka M, Horikoshi K (1990) A novel antifungal antibiotic, FR-900848. 1. Production, isolation, physicochemical and biological properties. J Antibiot 43:748–754

6 Combinatorial Methods to Engineer Small Molecules for Functional Genomics

J.A. Ellman

6.1 Introduction

The complete sequencing of the human genome will result in the identification of a huge number of coded proteins. However, in order for this information to be useful, the biological function of the coded proteins must be determined. Combinatorial small-molecule libraries will play a critical role in elucidating the function of these proteins. Combinatorial libraries can be used to rapidly assess the natural substrate specificity of newly identified enzymes. In addition, small-molecule libraries can be used to identify cell-permeable ligands that selectively activate or inactivate a protein target and therefore serve as powerful tools for understanding the function of the protein in cells and in animals.

Combinatorial libraries that target protein families will be particularly powerful for establishing protein function. Ligands to any member of a protein family can potentially be identified by designing combinatorial scaffolds and displays based upon common structural and mechanistic features of that protein family. To illustrate the power of this targeted library approach, we will give an overview of targeted libraries that we have developed to identify cell-permeable inhibitors and to establish the substrate specificity of proteases, for which several thousand are predicted to be coded by the human structural gene pool (Neurath 1999).

6.2 Proteases Are Important Targets for the Treatment of Human Disease

Proteases, which catalyze the hydrolysis of amide bonds in peptides and proteins, are involved in the regulation of most physiological processes in addition to their digestive role of protein degradation (Neurath 1999). For example, proteases play a central role in blood coagulation, fibrinolysis, the complement system, the processing of protein hormone precursors, and apoptosis. Proteases are also essential for the replication, nutrition, and host invasion of bacterial, viral, and parasitic pathogens.

Not surprisingly, proteases have served as important therapeutic targets in a multitude of diseases, including inflammation, cancer, and cardiovascular disease and viral, parasitic, and bacterial infections. The immense potential of protease inhibitors in the treatment of human disease is demonstrated by the huge impact of HIV protease inhibitors on the treatment of AIDS and of angiotensin-converting enzyme inhibitors on the treatment of hypertension.

6.3 Protease Substrate Specificity

Substrate specificity, or the ability to discriminate among many potential substrates, is central to the function of proteases. Knowledge of a protease's substrate specificity may not only give valuable insights into its biological function but also provide the basis for potent substrate and inhibitor design. The use of synthetic substrates is typically employed to define substrate specificity. However, the synthesis and assay of single substrates is tedious and impractical for proteases with specificity beyond P1 and often results in a limited substrate-specificity profile.

6.4 Combinatorial Libraries of Protease Substrates

Combinatorial approaches have recently been used to address the identification of substrate-specificity profiles for proteases. All of these combinatorial methods involve the generation of libraries of substrates, proteolysis of the substrates, and identification of the optimal substrate sequence. Substrate libraries can be broken down into two categories: those that are biologically generated and those that are synthetically generated. Biological library methods include the display of peptide libraries on filamentous phages (Matthews and Wells 1993), the randomization of amino acids at physiological cleavage sites (Bevan et al. 1993), and the identification of macromolecular cleavage sites in in vitro transcription/translation cDNA libraries (Kothakota et al. 1997). The diversity of these libraries is often constrained to the transformation efficiency of the host organism and can only contain naturally occurring amino acids.

Synthetic substrate libraries circumvent the dependence on transformation efficiency and naturally occurring amino acids. While combinatorial synthesis allows for the creation of millions of compounds, these methods are useful only when coupled with powerful analytical assays that allow for the identification of the preferred substrate. Discontinuous analysis of the cleavage products through Edman degradation (Birkett et al. 1991), mass spectroscopy (Berman et al. 1992), and chromatography (Schellenberger et al. 1993) has proven useful for qualitative assessment of optimal substrates from soluble or support-bound peptides. Substrate consensus sequences have also been obtained using support-bound fluorescence-quenched substrate libraries prepared by the process of split synthesis, which results in single substrate sequences on each of the resin beads (Lam and Lebl 1998). Partial proteolysis of the support-bound libraries and subsequent sequence determination of the substrates on the most fluorescent beads provides the consensus sequences (Meldal et al. 1994). Unfortunately, this method suffers from two major limitations. First, the kinetics of support-bound substrates can differ greatly from those of soluble substrates (Hilaire et al. 1999). Second, like the other methods previously mentioned, identification of the substrate occurs after the cleavage event, making the kinetic analysis more cumbersome. A method that avoids these limitations and gives a quantitative assessment of protease substrate preference is the use of positional scanning-synthetic combinatorial libraries (PS-SCL).

6.5 Positional Scanning Libraries of Fluorogenic Substrates

Positional scanning-synthetic combinatorial libraries (PS-SCL) of fluorogenic peptide substrates have the potential to be a very powerful tool for determining protease specificity. In contrast to other combinatorial libraries, this library format provides rapid and continuous information on each of the varied substituents in the substrate. A positional scanning library with the general structure Ac-X-X-X-Asp-AMC was prepared previously by Rano et al. to rapidly and accurately assess the P4-P2 specificity for caspases that require Asp in the P1* position (Rano et al. 1997). Specific cleavage of the amide bond after the Asp residue liberates a fluorescent 7-amino-4-methylcoumarin (AMC) leaving group, thus allowing for the simple determination of cleavage rates for a library of substrates. The P1 Asp-coumarin substrate was conveniently linked to an insoluble polymer through the Asp carboxylic acid side chain, which allowed for library synthesis by standard peptide synthesis. However, the method used to synthesize the library was specific for an aspartic acid at P1. The employment of strategies to link P1 amino

* Nomenclature for the substrate amino acid preference is Pn, Pn-1,...P2, P1, P1', P2',..., Pm-1', Pm'. Amide bond hydrolysis occurs between P1 and P1'. Sn, Sn-1,..., S2, S1, S1', S2',..., Sm-1', Sm' denotes the corresponding enzyme binding sites (18)

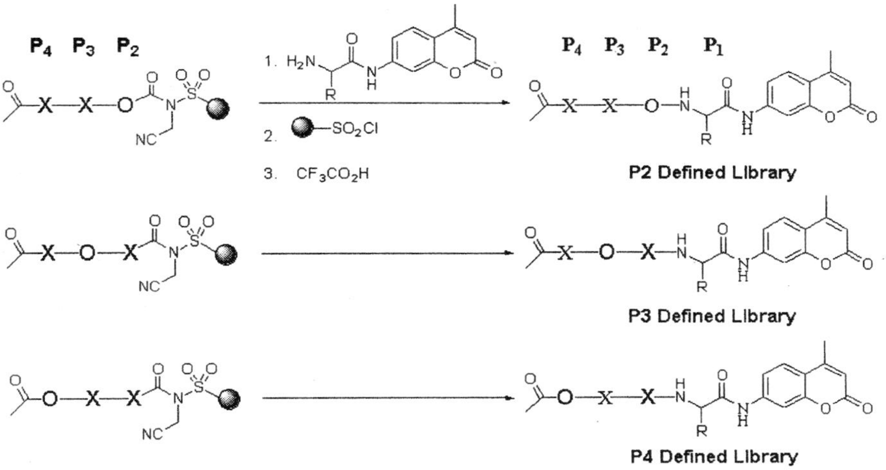

Fig. 1. Synthesis of positional scanning substrate libraries with diversity at the P1 position

acid-coumarin derivatives through side chain functionality may prove viable for some residues. However, linkage through hydrophobic side chain functionalities (Leu, Phe, Val, etc.) will prove difficult, and therefore, the utility of this strategy is limited. By developing a general strategy to incorporate all 20 proteinogenic amino acids at the P1 position of a PS-SCL, the extended P4–P1 specificity of virtually any protease could be rapidly determined.

We have designed and developed a general method for the preparation of positional scanning-synthetic combinatorial substrate libraries that is free from the limitations of previous approaches and enables incorporation of any amino acid at the P1 position. To incorporate diversity at the P1 position, fluorogenic AMC amides of amino acids are coupled with a support-bound PS-SCL to provide library compounds in solution (Fig. 1). In this approach the PS-SCL are prepared employing an alkanesulfonamide "safety-catch" linker and solid-phase peptide synthesis (Backes and Ellman 1999; Backes et al. 1996).

Several key factors needed to be addressed to prepare the support-bound tripeptide libraries (Fig. 2). First, conditions were identified for

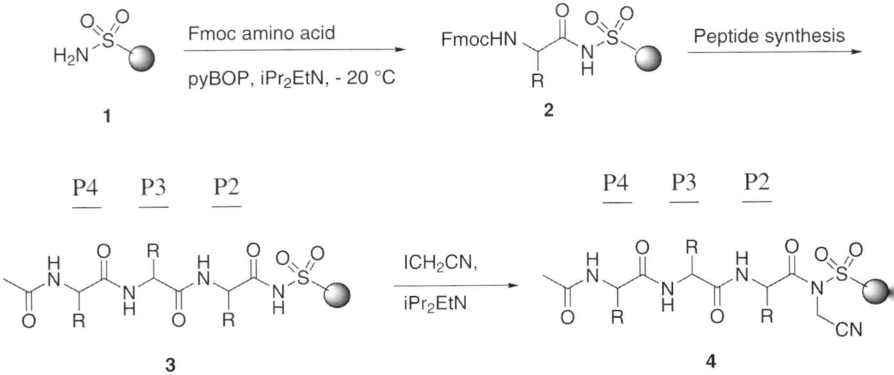

Fig. 2. Preparation of support-bound tripeptide libraries

coupling all 20 proteinogenic Fmoc-protected amino acids onto the sulfonamide resin **1** with good efficiency and with less than 1% racemization. For each amino acid, suitable side-chain protecting groups were also identified that are compatible with the resin coupling step and the cyanomethyl activation step of the sulfonamide linker of **3**.

In addition, in order to achieve complete release of the positional scanning library from the support, it is necessary to employ an excess of the AMC amide (Fig. 1). This requires a purification step to remove excess AMC amide from the fluorogenic peptide library, since the AMC amide increases background fluorescence, thereby complicating library evaluation. Numerous scavengers were evaluated. Only polymer-bound sulfonyl chloride provided complete removal of excess reagent. Potassium bicarbonate is also added as an easily filtered HCl scavenger.

We have demonstrated the utility of this method by the preparation of three positional scanning libraries Ac-O-X-X-Lys-AMC (P4), Ac-X-O-X-Lys-AMC (P3), and Ac-X-X-O-Lys-AMC (P2), where X represents a mixture of amino acids and O represents a defined amino acid (Backes et al. 1999). According to the positional scanning method, each library is divided into 20 sublibraries. In each sublibrary a single amino acid is present at the defined position.

We have used these libraries to map out the P4–P2 substrate specificity of the well-studied proteases involved in the fibrinolytic cascade,

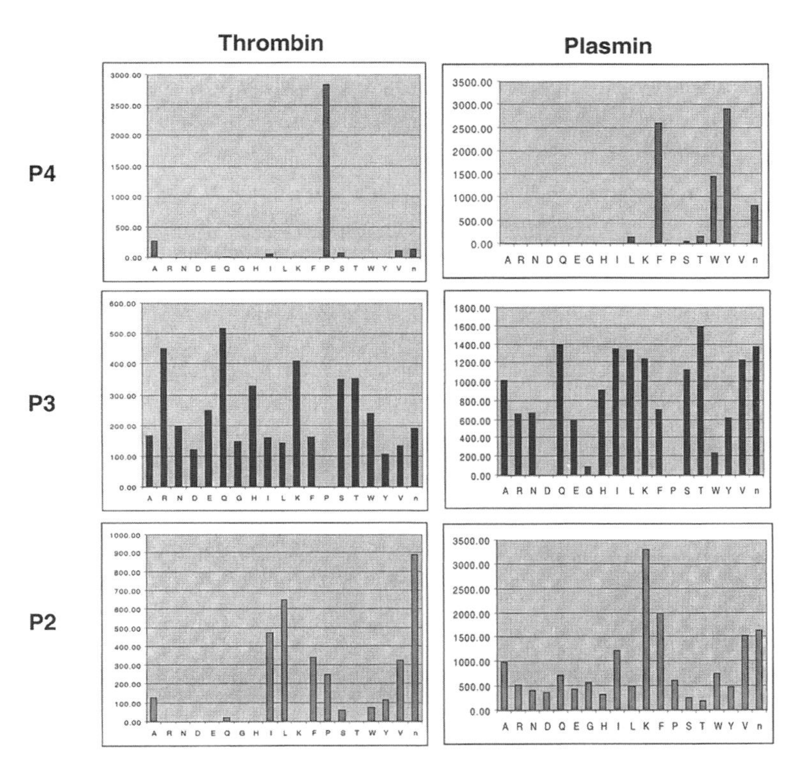

Fig. 3. Thrombin and plasmin P4–P2 substrate specificity profiles (P1 defined as Lys)

Table 1. PS-CSL specificity analysis confirmed by kinetic analysis on pure, single fluorogenic substrates

P4 P3 P2 P1	k_{cat} (s^{-1})	K_m (μM)	k_{cat}/K_m ($M^{-1}s^{-1}$)
K T Y K	11.3±0.4	20.8±3.3	543.8±71.2
K T F K	20.1±0.6	29.7±3.6	677.2±68.7
K T S K	8.8±1.3	440±100	20.1±2.5
F T Y K	17.5±0.9	51.0±10.2	342.4±53.6
L T F K	33.2±3.9	296±70	111.9±14.9
L E F K	5.5±0.3	74.6±9.9	72.5±6.3

plasmin and thrombin (Fig. 3), which are known to prefer basic residues at the P1 position (Lottenberg et al. 1983). For each enzyme, the resulting enzymatic profiles from the library were verified by single-substrate kinetic analysis, as shown for plasmin in Table 1. In addition, the profiles resemble the known physiological cleavage sites of the two enzymes.

In current efforts, fluorogenic PS-SCL substrate libraries are being prepared that will enable the rapid determination of the extended P4–P1 specificity for almost any protease.

6.6 Mechanism-based Design of Protease Inhibitor Libraries

One of the most powerful strategies for designing inhibitors of proteases is to employ a mechanism-based pharmacophore as a key binding element that interacts with the active site of the protease. Pharmacophores have been developed that target each of the four major protease classes, which are defined according to the mechanism of protease-catalyzed amide bond hydrolysis. Examples of mechanism-based pharmacophores used to target proteases include secondary alcohols for aspartyl proteases, ketone and aldehyde carbonyls for cysteine and serine proteases, and hydroxamic acids for zinc-metalloproteases. The most straightforward approach to obtaining a potent inhibitor is to directly incorporate a minimal mechanism-based pharmacophore into the peptide substrate. Unfortunately, the utility of peptide-based inhibitors as chemical genetic (pharmacological) tools is limited because peptides generally have very poor cell permeability and almost always have poor pharmacokinetics in animals due to rapid serum-clearing times, and/or rapid liver clearance and biliary excretion.

Relatively small inhibitors (<700 amu) that display *nonpeptide* functionality about an appropriate isostere have therefore been the focus of efforts to improve cell permeability and pharmacokinetic properties. The immense power of this approach is clearly demonstrated by the successful development of approved AIDS drugs targeting HIV-1 protease. However, this approach requires the preparation of many different compounds in a labor-intensive and time-consuming iterative process before small, cell-permeable inhibitors can be identified that have high affinity.

The synthesis and evaluation of small-molecule libraries employing mechanism-based pharmacophores have begun to have a significant impact upon the development of protease inhibitors (for a review see Whittaker 1998). Early efforts by Bastos et al. (1985), Owen et al. (1991) and Campbell et al. (1995) demonstrated the utility of incorporating ketoamide-based serine protease pharmacophores, the statine aspartyl protease pharmacophore, and phosphonate-based metalloprotease pharmacophores into peptides. We further advanced the utility of mechanism-based libraries by developing strategies that allow a full range of diverse, nonpeptide functionality to be displayed at *all* variable sites about the mechanism-based pharmacophore (Kick and Ellman 1995).

6.7 Aspartyl Protease Inhibitor Libraries

We first demonstrated this conceptual strategy in targeting aspartyl proteases, which are a ubiquitous class of enzymes that play an important role in mammals, plants, fungi, parasites, and retroviruses. The aspartyl proteases are endopeptidases that use two aspartic acid residues to catalyze the hydrolysis of amide bonds. Potent inhibitors of the aspartyl proteases have been developed that utilize as the minimal pharmacophore a secondary alcohol that serves as a stable mimetic of the tetrahedral intermediate (Fig. 4). We specifically chose to display functionality about the hydroxyethylamine-based mimetics **5a** and **5b** (Kick and Ellman 1995), since this isostere is amenable to the introduction of a wide variety of side chains about both sides of the secondary alcohol.

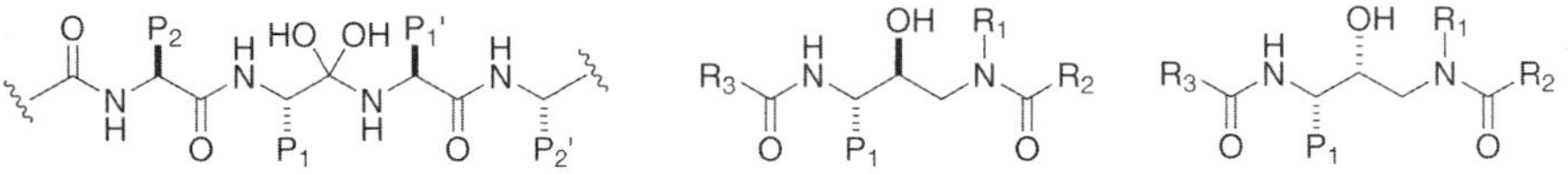

Fig. 4. Tetrahedral intermediate of hydrolysis. Hydroxyethylamine inhibitor structures **5a** and **5b**

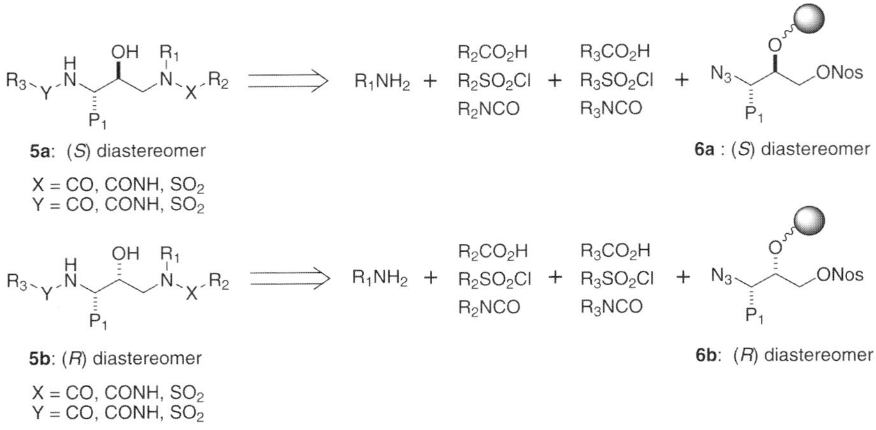

R_1NH_2 + R_2CO_2H / R_2SO_2Cl / R_2NCO + R_3CO_2H / R_3SO_2Cl / R_3NCO +

5a: (*S*) diastereomer

X = CO, CONH, SO_2
Y = CO, CONH, SO_2

6a : (*S*) diastereomer

R_1NH_2 + R_2CO_2H / R_2SO_2Cl / R_2NCO + R_3CO_2H / R_3SO_2Cl / R_3NCO +

5b: (*R*) diastereomer

X = CO, CONH, SO_2
Y = CO, CONH, SO_2

6b: (*R*) diastereomer

Fig. 5. Synthesis strategy to incorporate R_1, R_2, and R diversity elements

The aspartyl protease inhibitors are prepared by the introduction of three readily available building blocks upon the minimal scaffolds **6a** and **6b** (Fig. 5). Amine nucleophiles are then employed to introduce the R_1 substituent and acylating agents serve to introduce the R_2 and R_3 substituents. Both the *S* and *R* secondary alcohol diastereomers must be accessed, since the preferred alcohol stereochemistry depends on both the targeted aspartyl protease and the overall inhibitor structure. After trifluoroacetic acid-mediated cleavage from the solid support, the final inhibitors are usually obtained in 70%–80% overall yields for the solid-phase synthesis sequence (Kick and Ellman 1995).

In an important extension of the approach (Lee et al. 1998), multiple side chains can also be introduced at the P_1 position using Grignard reagents (Fig. 6). Chelation-controlled reduction of **7a** with $Zn(BH_4)_2$ proceeds with 85:15 diastereoselectivity. Standard functional group transformations then provide **6a**. Reduction of **7b** (Fig. 6) under electronic control employing l-selectride proceeds in 90:10 diastereoselectivity to provide the other diastereomer **6b** after the same series of functional group transformations. Using these sequences, the final inhibitors **5a** and **5b** are obtained in 45%–65% overall yields for the 12-step process.

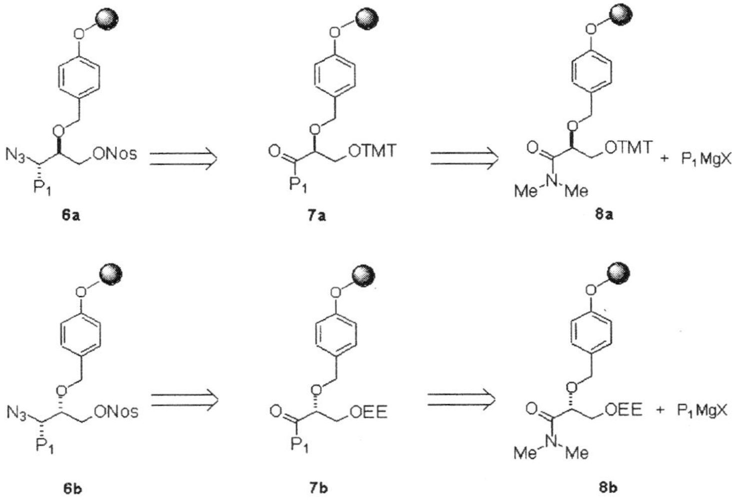

Fig. 6. Synthesis strategy to incorporate P1 side chain diversity

We have collaborated extensively with Professor Irwin D. Kuntz at the University of California at San Francisco to select the building blocks to be incorporated into the inhibitor libraries (Kick et al. 1997; Haque et al. 1999). Two complementary building block selection methods have been used. First, when the structures of the protease targets were available, we have integrated combinatorial chemistry and structure-based design through use of the DOCK cassette of programs. This set of programs, which was developed in the Kuntz group, was the first and remains one of the most popular structure-based design programs. These programs enable the efficient structure-based selection of side chains to be incorporated into the library. Second, we have used diversity metrics to maximize the display of diverse side chain functionality in our library methods. This second approach is clearly necessary when the structure of the protease target is not available. In addition, this approach can result in the identification of side chains that would not be anticipated by structure-based methods.

Based upon the above solid-phase synthesis approach and library design strategy, we have prepared multiple inhibitor libraries (Lee et al.

1998; Kick et al. 1997; Haque et al. 1999). Several of the libraries contained hundreds of potential inhibitors and two libraries each contained 1000 potential inhibitors. The inhibitor libraries were prepared in a spatially addressable microtiter-based format, such that the inhibitors were prepared as discrete compounds rather than as compound mixtures.

The libraries targeting cathepsin D and plasmepsins I and II were assayed using a high-throughput fluorescence-based assay configured in a microtiter-based format (Lee et al. 1998; Kick et al. 1997; Haque et al. 1999). This assay procedure enables the screening of hundreds of compounds in a single day.

6.8 Inhibitors of Cathepsin D

The initial successful report of the identification of low nanomolar inhibitors of cathepsin D was one of the first examples of integrating structure-based with combinatorial chemistry (Kick et al. 1997). Subsequent library design, synthesis, and screening iterations towards cathepsin D have resulted in the development of extremely potent small-molecule inhibitors of cathepsin D with single-digit nanomolar to high picomolar K_i values (Fig. 7). These inhibitors, which do not incorporate any amino acid units, are the most potent small-molecule inhibitors to have been reported to date against cathepsin D. In more recent efforts, we have optimized for other desirable characteristics in addition to potency, including minimal binding to serum proteins, molecular weights under 650 amu and good calculated log P values (e.g., inhibitor **10**, Fig. 7).

The identified cathepsin D inhibitors have served as important pharmacological tools in several collaborations. For example, Professor Gary Lynch at the University of California at Irvine has studied the effects of inhibiting cathepsin D on Alzheimer's disease as well as other neurodegenerative diseases. Neurofibrillary tangles, a defining feature of Alzheimer's disease, are composed of hyperphosphorylated tau protein and fragments of tau (Matsuo et al. 1994). While age-related changes in kinases and phosphatases are involved in generating the precursors, the discovery that hyperphosphorylated tau and neurofibrillary tangles are found in subjects with Niemann-Pick's type C disease

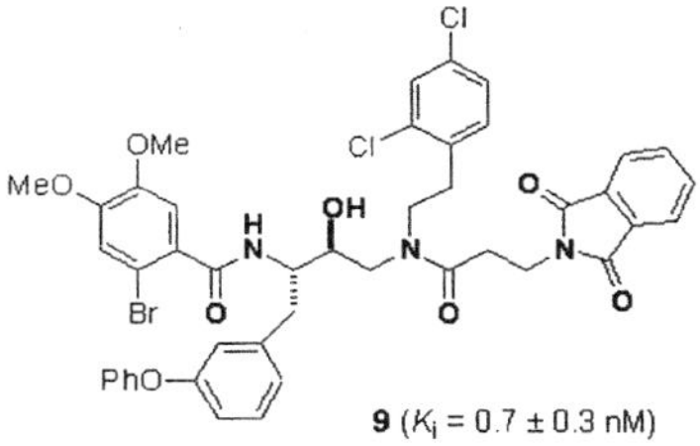

Fig. 7. Representative cathepsin D inhibitors identified using combinatorial libraries

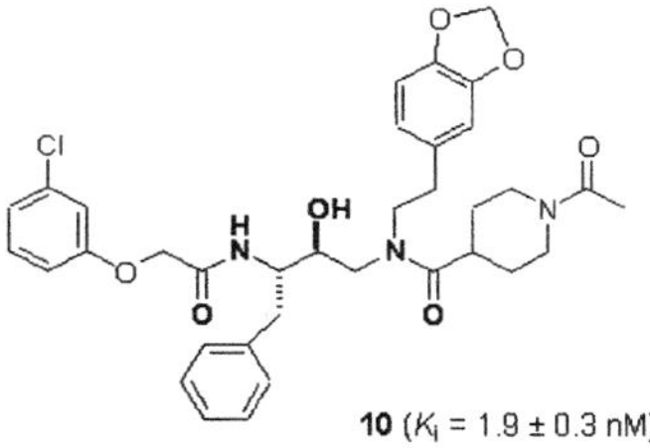

(Love et al. 1995) raised the possibility that lysosomal disturbances are also a contributing factor. Also pointing to this conclusion is the observation that inhibition of select lysosomal proteases in cultured rat brain slices results in the extralysosomal formation of hyperphosphorylated tau fragments recognized by antibodies against human tangles (Bi et al. 1999b). Efforts to link lysosomal dysfunction to tangles have focused on cathepsin D. This protease, which is elevated in AD vulnerable neurons in advance of pathology (Troncoso et al. 1998), was recently shown to cleave tau at neutral (cytoplasmic) pH, resulting in fragments corresponding in mass to those found in tangles (Bednarski and Lynch 1996). These results suggest that selective inhibitors of cathepsin D could be useful in regulating the formation of the precursors to neurofibrillary tangles.

Based upon this premise, Professor Lynch has tested whether the cathepsin D inhibitors can block production of known precursors to neurofibrillary tangles. Partial lysosomal dysfunction was induced in cultured hippocampal slices with N-Cbz-l-phenylalanyl-l-alanine-diazomethylketone (ZPAD), an inhibitor of the cysteine proteases, cathepsins B and L (Troncoso et al. 1998). This leads within 48 h to hyperphosphorylated tau protein fragments recognized by antibodies against human tangles (Bednarski et al. 1997). Inhibitor **10** (inhibitor EA1 in Fig. 8) blocked production of the fragments in a dose-dependent fashion (Bi et al. 1999a). The threshold was in the submicromolar range, with higher concentrations producing complete suppression. The effects were selective and not accompanied by pathophysiology. In addition, EA1 (inhibitor **10**) does not affect tau processing unless lysosomal

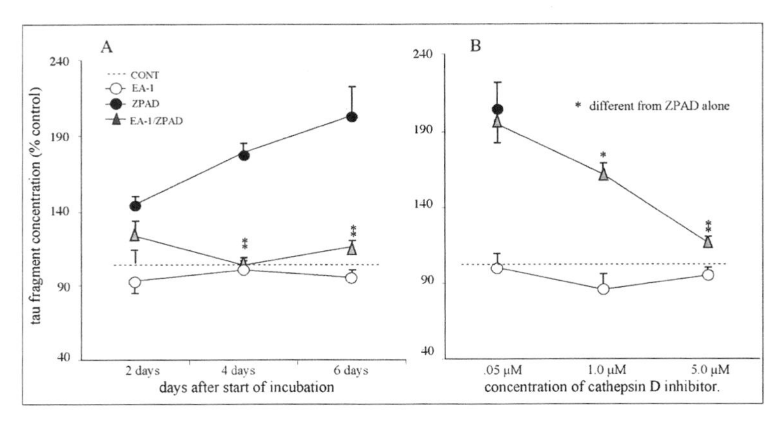

Fig. 8. Inhibition of tau protein proteolysis in human hippocampal cell model of Alzheimer's disease

function is induced. Comparable results were obtained with three structurally distinct inhibitors that were also obtained from the libraries. These results support the hypothesis that cathepsin D links lysosomal dysfunction to the etiology of Alzheimer's and suggest a new approach to treating the disease.

6.9 Inhibitors of Malarial Aspartyl Proteases

Malaria is a parasitic disease that afflicts 300–500 million people worldwide, killing 1–2 million annually (Wyler 1993). *Plasmodium falciparum* is the most dangerous form of the four malaria parasites that infect human beings and is responsible for more than 95% of malaria-related deaths. Increasing resistance of *Plasmodium falciparum* to existing therapies has heightened concerns about malaria in the international health community. Indeed, some *P. falciparum* strains have been identified that are resistant to all known antimalarial drugs (Murray and Perkins 1996). Unfortunately, there has been little economic incentive for the pharmaceutical industry to develop new therapies for this third world disease.

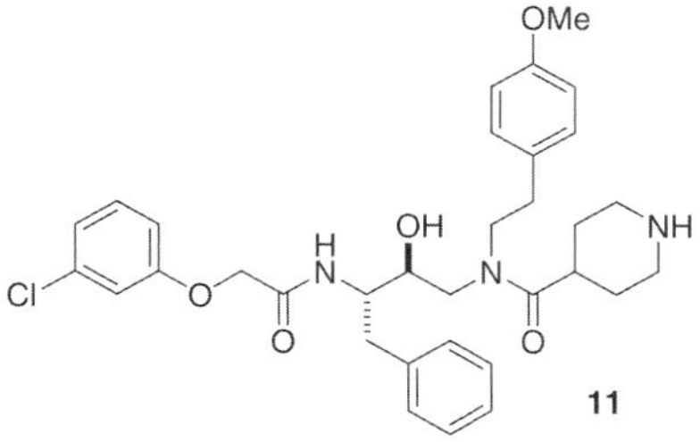

Plasmepsin I	**~2 nM (K_i)**
Plasmepsin II	**4.3 nM (K_i)**
Cell Culture	**1 μM (IC_{50})**

11

Fig. 9. Representative inhibitors of plasmepsins I and II identified using combinatorial libraries

The *Plasmodium* parasite invades human erythrocytes and consumes up to 75% of the hemoglobin present (Goldberg 1993). Three enzymes that have been identified to digest hemoglobin in the acidic parasite food vacuole are a cysteine protease, falcipain, and two aspartyl proteases, plasmepsins I and II (Plm I and II) (Luker et al. 1996). Daniel Goldberg, who first identified and characterized the plasmepsins, has previously demonstrated that peptidic inhibitors of these enzymes show modest activity in cell culture (IC_{50} >20 μ*M*). However, much more potent, cell-permeable inhibitors with improved pharmacokinetics are required. We have therefore entered into a multicollaborative effort to identify potent inhibitors of the essential malarial proteases, the plasmepsins I and II.

Several library design, synthesis, and screening iterations towards Plm I and II have resulted in the development of a number of extremely potent small-molecule inhibitors with single-digit nanomolar K_i values to Plm I and Plm II (Haque et al. 1999). We have optimized for other desirable characteristics in addition to potency, including minimal binding to serum proteins, molecular weights under 600 amu, good calculated log P values, and greater than 15-fold selectivity over the most closely related human aspartyl protease, cathepsin D (representative inhibitor **11**, Fig. 9). The most promising inhibitors have single-digit micromolar to submicromolar IC_{50} values against the parasite in cell culture.

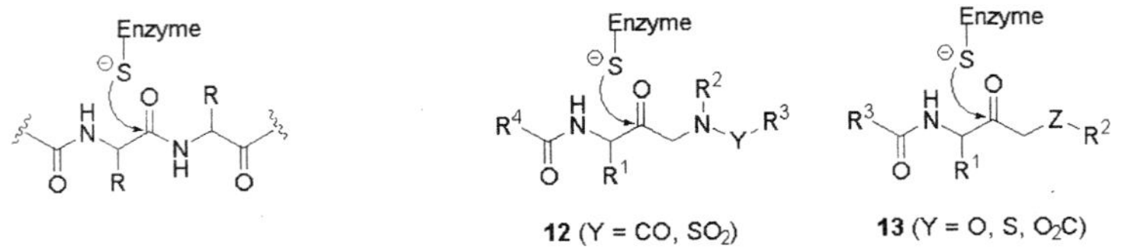

12 (Y = CO, SO_2) **13** (Y = O, S, O_2C)

Fig. 10. Attack of active site cysteine upon scissile amide. Ketone-based inhibitor structures **12** and **13**

6.10 Cysteine Protease Inhibitor Libraries

We have also developed targeted library approaches towards cysteine proteases (Lee et al. 1999), which play a role in a multitude of biological processes and have been implicated in the pathogenesis of many diseases (for a review on cysteine proteases and their inhibitors: Otto and Schirmeister 1997). Characterized by a conserved cysteine residue in the active site, this class of proteases includes the calpains, which have been implicated in neurodegenerative disorders, cathepsin K, which has been linked to osteoporosis (Bossard et al. 1996), and the caspase family of proteases, recently shown to be involved in programmed cell death (Thornberry and Lazebnik 1998). Additionally, cysteine proteases are the primary class of proteases found in parasites (McKerrow et al. 1993), which are among the most prevalent pathogens worldwide.

Cysteine proteases produce a biological response by catalyzing the hydrolysis of amide bonds in peptides and proteins through nucleophilic attack of the active-site cysteine residue upon the amide carbonyl (Fig. 10). A common feature of virtually all cysteine protease inhibitors is an electrophilic functionality, such as a carbonyl or a Michael acceptor that can react with the nucleophilic cysteine residue (Otto and Schirmeister 1997). Peptidyl aldehydes were the first class of reversible inhibitors to be reported. However, the inherent reactivity of the aldehyde pharmacophore to nucleophilic attack and oxidation is a considerable liability for attaining good pharmacokinetics. Additionally, aldehyde-based inhibitors are limited, since functionality can be displayed on only one side of the carbonyl. In contrast, ketone-based pharmacophores **12** and **13** (Fig. 10) are more stable in vivo than the corresponding aldehyde-based inhibitors. In addition, the display of functionality on both sides of the carbonyl provides the potential to achieve enhanced specificity through multiple interactions with the active site. Indeed, Veber and co-workers recently reported several examples of reversible peptidyl ketones with nanomolar inhibition towards cathepsin K (Yamashita et al. 1997). A number of researchers have developed irreversible ketone-based inhibitors targeting several different cysteine proteases (Otto and Schirmeister 1997).

We have therefore chosen to employ the ketone carbonyl as the minimal pharmacophore, since it enables the display of functionality on both sides of the carbonyl and provides the potential to achieve specificity through multiple interactions with the active site. A general solid-phase synthesis approach to display diverse functionality about a ketone carbonyl would therefore enable the preparation of libraries of potential inhibitors targeting the cysteine protease class (Fig. 11).

Chloromethyl ketone **15**, which introduces the P_1 side chain and provides sites for further functionalization on both sides of the ketone carbonyl, can be prepared in enantiomerically pure form in one pot from the corresponding protected amino acid according to well-precedented methods (Fig. 11). Linking to support through the ketone carbonyl is ideal because the carbonyl functionality is the only invariant part of a ketone-based inhibitor regardless of the cysteine protease that is targeted. The hydrazone linkage allows for nucleophilic substitution at the α-position while simultaneously preventing nucleophilic attack at the carbonyl. The hydrazone also prevents racemization, which is problematic for the corresponding enolizable α-acylamino-substituted chiral ketone. Nucleophilic displacement of the support-bound α-chloro hydrazones **16** with amines followed by acylation provides entry to the amidomethyl ketone class of reversible cysteine protease inhibitors **12**. Displacement with carboxylates or thiolates provides entry to the acyloxymethyl and mercaptomethyl ketone classes of cysteine protease inhibitors **13**. Removal of the Alloc group, acylation, and subsequent acidic cleavage from support provides the fully substituted ketone products **12** and **13** without racemization in 40%–100% overall yields.

A key challenge in this synthesis effort was the identification of the appropriate hydrazone linkage that would allow the previously described chemistry to be performed. First, a precedent did not exist for performing this series of transformations on α-halohydrazones, and the known conversion of α-halohydrazones to azodienes provided signifi-

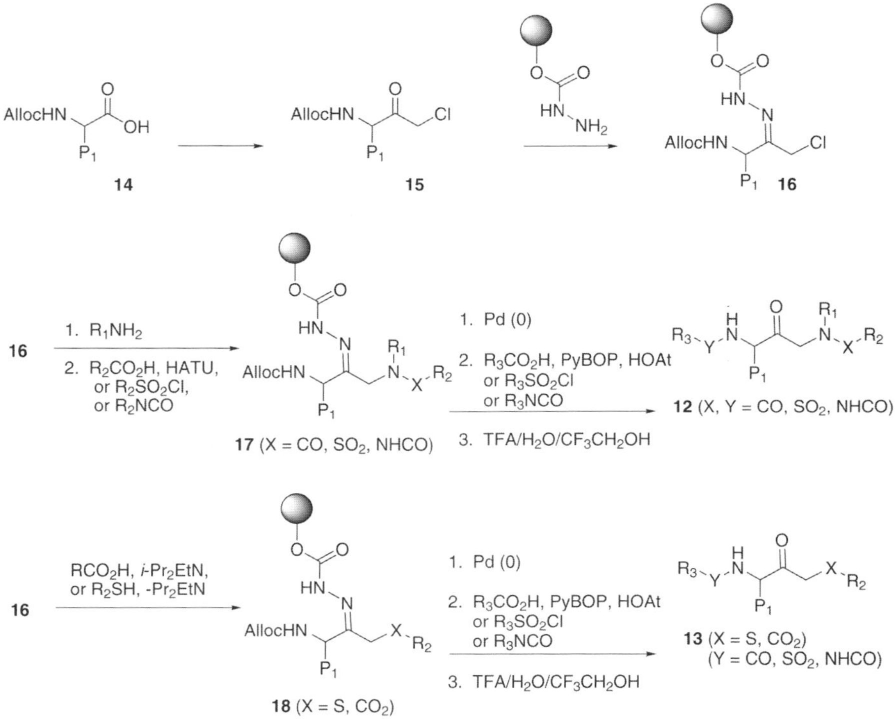

Fig. 11. Synthesis of amidomethyl, carboxymethyl, and mercaptomethyl ketone inhibitors

cant potential for complications (for a review on azodiene chemistry: Attanasi and Filippone 1997). In addition, although hydrazine linkers had previously been employed to couple peptidyl aldehydes and trifluoromethyl ketones to solid supports, these pre-formed handle strategies required the preparation of the hydrazone derivative in solution before loading onto solid support (Murphy et al. 1992). Using the developed chemistry, the synthesis and screening of ketone-based cysteine protease inhibitor libraries is currently in progress.

6.11 Conclusion

Combinatorial libraries can provide a great deal of insight into protein function. In this regard, combinatorial libraries that target protein families will be particularly powerful. In this chapter we have given an overview of our research on libraries designed to target proteases. First, fluorogenic substrate libraries have been described that enable the complete P4–P1 substrate profile to be determined for a majority of proteases. This method should greatly facilitate the elucidation of the natural substrates of proteases. Second, library strategies have been described that enable the rapid identification of potent, cell-permeable inhibitors of the aspartyl and cysteine protease classes. The libraries are designed to display diverse functionality about minimal tight-binding fragments based upon the mechanism of each of the protease classes. The identified cell-permeable, small-molecule inhibitors have proven to be powerful "chemical genetic" tools for understanding the function of the targeted proteases in cells and in animals.

References

Attanasi OA, Filippone P (1997) Working twenty years on conjugated azo-alkenes (and environs) to find new entries in organic synthesis. Synlett 1128–1140

Backes BJ, Ellman JA (1999) An alkanesulfonamide "safety-catch" linker for solid-phase synthesis. J Org Chem 64:2322–2330

Backes BJ, Virgilio AA, Ellman JA (1996) Activation method to prepare a highly reactive acylsulfonamide safety-catch linker for solid-phase synthesis. J Am Chem Soc 118:3055–3056

Backes BJ, Harris JL, Leonetti F, Ellman JA, Craik C (1999) Strategy to prepare positional-scanning libraries of fluorogenic peptide substrates that incorporate diverse P1 substituents: facile and accurate specificity determination of thrombin and plasmin. Nature Biotech 18:187–193

Bastos M, Maeji NJ, Abeles RH (1995) Inhibitors of human heart chymase based on a peptide library. Proc Natl Acad Sci U S A 92:6738–6742

Bednarski E, Lynch G (1996) Cytosolic proteolysis of tau by cathepsin D in hippocampus following suppression of cathepsins B and L. J Neurochem 67:1846–1855

Bednarski E, Ribak CE, Lynch G (1997) Suppression of cathepsins B and L causes a proliferation of lysosomes and the formation of meganeurites in hippocampus. J Neurosci 17:4006–4021

Berman J, Green M, Sugg E, Andergegg R (1992) Rapid optimization of enzyme substrates using defined substrate mixtures. J Biol Chem 267:1434–1437

Bevan A, Brenner C, Fuller RS (1993) Quantitative assessment of enzyme specificity in vivo: P2 recognition by Kex2 protease defined in a genetic system. Proc Natl Acad Sci USA 95:10384–10389

Bi X, Lin B, Haque T, Lee CE, Skillman AG, Kuntz ID, Ellman JA, Lynch G (1999a) Novel cathepsin D inhibitors block the formation of hyperphosphorylated tau fragments in hippocampus. J Neurochem 74:1469–1477

Bi X, Zhou J, Lynch G (1999b) Lysosomal protease inhibitors induce meganeurites and tangle-like structures in entorhinohippo-campal regions vulnerable to Alzheimer's disease. Exp Neurol 158:312–327

Birkett AJ, Yelamos B, Rodriguez-Crespo I, Gavilanes F, Peterson DL (1991) Determination of enzyme specificity in a complex mixture of peptide substrates by N-terminal sequence analysis. Anal Biochem 196:137–143

Bossard MJ, Tomaszek TA, Thompson SK, Amegadzie BY, Hanning CR, Jones C, Kurdyla JT, McNulty DE, Drake FH, Gowen M, Levy MA (1996) Proteolytic activity of human osteoclast cathepsin K – expression, purification, activation, and substrate identification. J Biol Chem 271:12517–12524

Campbell DA, Bermak JC, Burkoth TS, Patel DV (1995) A transition state analogue inhibitor combinatorial library. J Am Chem Soc 117:6738–6742

Goldberg DE (1993) Hemoglobin degradation in Plasmodium-infected red blood cells. Semin Cell Biol 4:355–361

Haque TS, Skillman AG, Lee CE, Habashita H, Gluzman IY, Ewing TJA, Goldberg DE, Kuntz ID, Ellman JA (1999) Single digit nanomolar, low molecular weight non-peptide inhibitors of malarial aspartyl protease plasmepsin II. J Med Chem 42:1428–1440

St Hilaire PM, Willert M, Juliano MA, Juliano L, Meldal M (1999) Fluorescence-quenched solid phase combinatorial libraries in the characterization of cysteine protease substrate specificity. J Comb Chem 1:509–523

Kick EK, Ellman JA (1995) Expedient method for the solid-phase synthesis of aspartic acid protease inhibitors directed toward the generation of libraries. J Med Chem 38:1427–1430

Kick EK, Roe DC, Skillman AG, Liu G, Ewing TJA, Sun Y, Kuntz ID, Ellman JA (1997) Structure-based design and combinatorial chemistry yield low nanomolar inhibitors of cathepsin D. Chem Biol 4:297–309

Kothakota S, Azuma T, Reinhard C, Klippel A, Tang J, Chu K, McGarry TJ, Kirschner MW, Koths K, Kwiatkowski DJ (1997) Caspase-3-generated

fragment of gelsolin: effector of morphological change in apoptosis. Science 278:294–298
Lam KS, Lebl M (1998) Synthesis of a one-bead one-compound combinatorial peptide library. Methods Mol Biol 87:1–6
Lee A, Huang L, Ellman JA (1999) General solid-phase method for the preparation of mechanism-based cysteine protease inhibitors. J Am Chem Soc 121:9907–9914 (available on the web: ASAP)
Lee CE, Kick EK, Ellman JA (1998) General solid-phase synthesis approach to prepare mechanism-based aspartyl protease inhibitor libraries. Identification of potent cathepsin D inhibitors. J Am Chem Soc 120:9735–9748
Lottenberg R, Hall JA, Blinder M, Binder EP, Jackson CM (1983) The action of thrombin on peptide p-nitroanilide substrates. Substrate selectivity and examination of hydrolysis under different reaction conditions. Biochim Biophys Acta 742:539–557
Love S, Bridges LR, Case CP (1995) Neurofibrillary tangles in Niemann-Pick disease type C. Brain 118:119–129
Luker KE, Francis SE, Gluzman IY, Goldberg DE (1996) Kinetic analysis of plasmepsins I and II, aspartic proteases of the Plasmodium falciparum digestive vacuole. Mol Biochem Parasitol 79:71–78
Matsuo ES, Shin RW, Billingsley ML, Vandevoorde A et al (1994) Biopsy-derived adult human brain tau is phosphorylated at many of the same sites as Alzheimer's disease paired helical filament tau. Neuron 13:989–1002
Matthews DJ, Wells JA (1993) Substrate phage: selection of protease substrates by monovalent phage display. Science 260:1113–1117
McKerrow JH, Sun E, Rosenthal PJ, Bouvier J (1993) The proteases and pathogenicity of parasitic protozoa. Annu Rev Microbiol 47:821–853
Meldal M, Svendsen I, Breddam K, Auzanneau FI (1994) Portion-mixing peptide libraries of quenched fluorogenic substrates for complete subsite mapping of endoprotease specificity. Proc Natl Acad Sci U S A 91:3314–3318
Murphy AM, Dagnino R, Vallar PL, Trippe AJ, Sherman SL, Lumpkin RH, Tamura SY, Webb TR (1992) Automated synthesis of peptide C-terminal aldehydes. J Am Chem Soc 114:3156–3157
Murray MC, Perkins ME (1996) Chemotherapy of malaria. Annu Rep Med Chem 31:141–150
Neurath H (1999) Proteolytic enzymes, past and future. Proc Natl Acad Sci U S A 96:10962–10963
Otto H-H, Schirmeister T (1997) Cysteine proteases and their inhibitors. Chem Rev 97:133–171
Owens RA, Gesellchen PD, Houchins BJ, DiMarchi RD (1991) The rapid identification of HIV protease inhibitors through the synthesis and screening of defined peptide mixtures. Biochem Biophys Res Commun 181:402–408

Rano TA, Timkey T, Peterson EP, Rotonda J, Nicholson DW, Becker JW, Chapman KT, Thornberry NA (1997) A combinatorial approach for determining protease specificities: application to interleukin-1 beta converting enzyme (ICE). Chem Biol 4:149–155

Schellenberger V, Turck CW, Hedstrom L, Rutter WJ (1993) Mapping the S' subsites of serine proteases using acyl transfer to mixtures of peptide nucleophiles. Biochemistry 32:4349–4353

Thornberry NA, Lazebnik Y (1998) Caspases: enemies within. Science 281:1312–1316

Troncoso JC, Cataldo AM, Nixon RA, Barnett JL, Lee MK, Checler F, Fowler DR, Smialek JE, Crain B, Martin LJ (1998) Neuropathology of preclinical and clinical late-onset Alzheimer's disease. Ann Neurol 43:673–676

Whittaker M (1998) Discovery of protease inhibitors using targeted libraries. Curr Opin Chem Biol 2(3):386–396

Wyler DJ (1993) Malaria-overview and update. Clin Infect Dis 16:449–458

Yamashita DS, Smith WW, Zhao B, Janson CA, Tomaszek TA, Bossard MJ, Levy MA, Oh H, Carr TJ, Thompson SK, Ijames CF, Carr SA, McQueney M, D'Alessio KJ, Amegadzie BY, Hanning CR, Abdel-Meguid S, DesJarlais RL, Gleason JG, Veber DF (1997) Structure and design of potent and selective cathepsin K inhibitors. J Am Chem Soc 119:11351–11352

7 Natural Products in Drug Discovery

H. Müller, O. Brackhagen, R. Brunne, T. Henkel, F. Reichel

The role and importance of natural products and natural products research has often been the subject of discussion in recent years. Some companies have curtailed their efforts in this research area, for reasons of cost, lack of success, or to make way for such upcoming new technologies as combinatorial synthesis. We therefore initiated a study to address the question: Is there a future for natural products in industrial drug discovery programs?

7.1 Drug Discovery in Pharmaceutical Industries – an Overview

Even today, after more than 100 years of research in pharmaceutical industries, there is still a great need for innovative drugs. Only one third of all diseases can be treated efficiently (Fig. 1). That means there is a need for novel lead structures, where low-molecular-weight compounds

Atherosclerosis
cardiovascular diseases: infarction, stroke; most frequent cause of death

Tumor diseases
second most frequent cause of death (23% in Germany)

Rheumatic diseases
high morbidity, intensive treatment costs and loss of employment

Alzheimer / senile dementia
5% - 10% of the population older than 65 years, high welfare costs

Allergies
asthma, neurodermatitis

Osteoporosis
appr. 1/3 of all women older than 50 years sustain fractures

Viral diseases
e.g., epidemic outbreak of AIDS

Other infectious diseases
problem of resistance, hospitalized strains

Fig. 1. Diseases urgently requiring innovative therapies

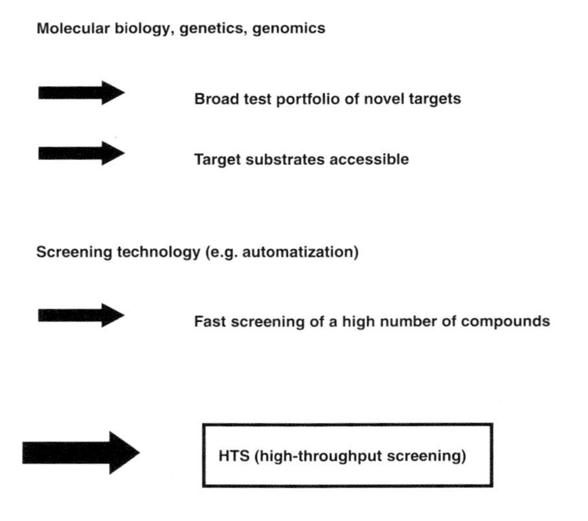

Fig. 2. Changes of paradigms in industrial drug discovery

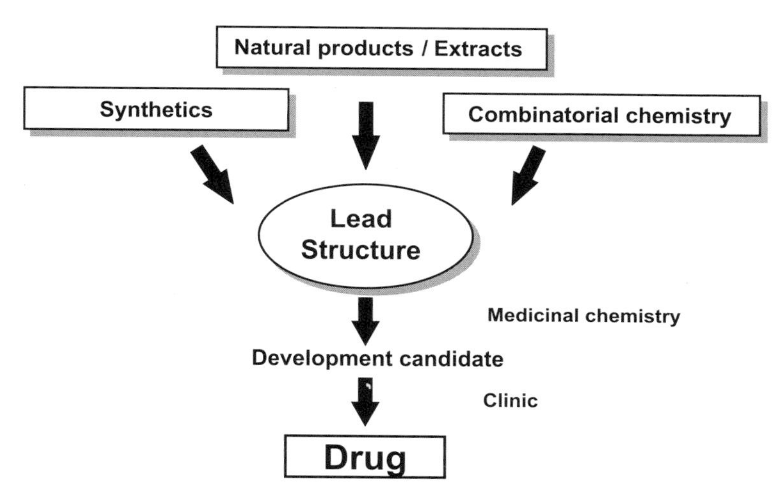

Fig. 3. Sources of compounds for drug discovery

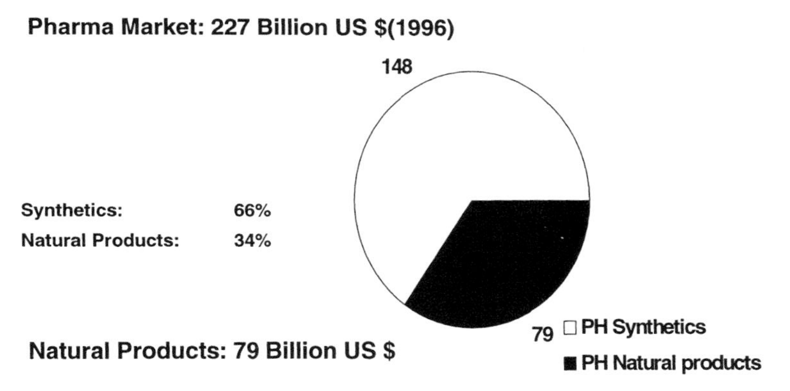

Fig. 4. Market share of natural products and drugs derived from natural products

are screened by more or less random screening approaches. However, the conditions for success in finding these new lead structures are better than ever before. Dramatic changes (Fig. 2) in industrial drug discovery programs through genetic engineering and modern screening technologies have enabled us to screen a high number of compounds in assay systems that were beyond our imagination only a few years ago. However, while running high-throughput screening (HTS) we have to keep in mind that a high number of screens is not the only guarantee of success. One also has to look carefully at the quality of the assay systems and of the substance pools dedicated to the screening programs.

The quality of a substance pool is defined by the sources of compounds that can be used (Fig. 3). However, considering that there are 15,000,000 synthetics as compared with 120,000 natural products, one may doubt whether it is still useful and responsible to invest further resources in this research area.

7.2 Benchmarking Results

The following data demonstrate the economic importance of natural product research in the recent past and present activities in this field: With a market share of about 80 billion US $, natural products and compounds derived from natural products play a very important role in the world pharmaceutical market (Fig. 4). An analysis of the top 20 brands shows that 40% are natural products or derived from natural products (Table 1).The analysis of the New Chemical Entities – newly launched drugs – 34% of which are natural products, shows the same trend. Thus, the economic importance of natural products seems to continue.

Only 5% of the NME (new molecular entities) – reflecting recent research activities – are natural products. However, this supposed decrease does not reflect a cutback in natural products research, since these data do not provide any information on congeners or compounds derived from natural products. We get almost the identical result if we look at the new "Pharmaceutical Projects".

Although these data strongly indicate the importance of natural products in drug discovery, they provide a more or less historical view and do not necessarily show the future role of natural product research.

Table 1. Economic importance of natural products in pharmaceuticals: top 20 brands, 1997 (from Pharma Quant Overview, January 1998)[a]

Product	Indication	Company	Sales 1997 (billions of dollars)
Losec/Prilosec	Proton pump inhibitor	Astra/Astra Merck	4.05
Zocor[b]	Hypolipidemic	Merck & Co.	3.39
Vasotec[b]	ACE inhibitor	Merck & Co.	2.57
Prozac	Antidepressant	Eli Lilly	2.51
Zantac	H2 antagonist	Glaxco Wellcome	2.50
Norvasc	Calcium antagonist	Pfizer	2.10
Claritin	Antihistamine	Schering Plough	1.75
Augmentin[b]	Penicillin antibiotic	Smith Kline Beecham	1.56
Seroxat/Paxil	Antidepressant	Smith Kline Beecham	1.52
Zoloft	Antidepressant	Pfizer	1.51
Ciprobay	Quinolone antibiotic	Bayer	1.46
Premarin family[b]	Hormone product	American Home Products	1.43
Mevalotin[b]	Hypolipidemic	Sankyo	1.41
Sandimmun/ Neoral[b]	Immunosuppressant	Novartis	1.36
Pravachol[b]	Hypolipidemic	Bristol Myers Squibb	1.35
Biaxin/Klaricid[b]	Macrolide antibiotic	Abbott	1.30
Novolin	Human insulin	Novo Nordisk	1.29
Adalat	Calcium antagonist	Bayer	1.20
Epogen	Colony-stimulating growth factor	Amgen	1.16
Voltaren	NSAID	Novartis	1.15

[a]Eight of 20 (40%) are natural products.
[b]Natural product or derived from natural product.

7.3 Molecular Diversity

The quality of substance pools is a key factor influencing the output and success of industrial HTS. Therefore, we wanted to know whether natural products represent another chemical quality than synthetics and if it is possible to enhance the molecular diversity of the substance pools through natural products.

Since there is no general definition of molecular diversity, we approached this problem with two studies: In the first study – the "statistical analysis" – we used ten pharmacophoric groups that are very often responsible for bioactivity as a basis for establishing criteria for molecular diversity (Fig. 5):

Acid,
Amine,
Arene,
Amide,
Alcohol,
Ketone/aldehyde,
Ether,
Ester,
Alkyl,
Polaric group.

- **Number of different pharmacophoric groups per compound**
- **Distribution of frequency of the single pharmacophoric groups**
- **Distribution of combinations of phamacophoric groups**
- **Number of molecules with head atoms**
- **Number of molecules with rotating C-C bonds**
- **Number of rings per molecule**
- **Number of rotatable bonds per molecule**
- **Number of chiral central atoms per molecule**

Fig. 5. Criteria of molecular diversity for statistical analysis

Table 2. Substance pools used for statistical and similarity analyses

Substance pool	No. of entries
ACD (Available Chemical Directory)	180,000 synthetic compounds
"Drugs" (e.g., Pharmaprojects, RD Focus)	15,000 compounds
DNP (Dictionary of Natural Products, Chapman & Hall)	80,000 natural products
Synthetic Compounds (Bayer Substance Pool)	250,000 synthetic compounds

Second, we performed a similarity analysis by 2-D fingerprinting, whereby the two-dimensional connectivities of every single compound in the investigated pool were compared with one another. For the two studies we used various substance pools (Table 2) with a total of more than 500,000 entries.

7.3.1 Statistical Analysis

A comparison of only the number of pharmacophoric groups of the substance pools shows no significant difference between synthetics and natural products. However, a look at the distribution of the pharmacophoric groups reveals many more alcoholic and ether groups among the natural products and many more aromatics, amines, and amides among the synthetics (Fig. 6). Thus, there are significant qualitative differences between synthetics and natural products.

A look at the selected group combinations and their abundance (Table 3) gives a very similar result. In the natural products pool there is a much higher percentage of combinations such as alcohol/ether, alcohol/ester, arene/alcohol, or arene, alcohol, ether. Combinations such as arene/amine, amine/amide, or amine/arene/amide are more frequent in the synthetics. Again you find distinct qualitative differences between natural products and synthetics. This is also true for most of the other selected pool properties (Table 4). Thus, the statistical analysis shows a remarkable complementarity of natural products and chemically synthesized compounds.

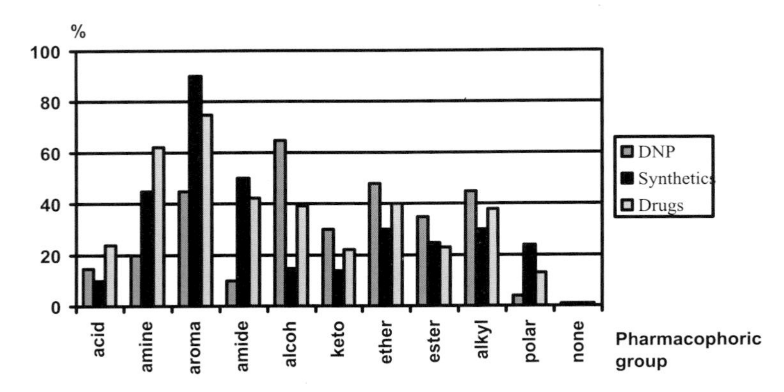

Fig. 6. Distribution of pharmacophoric groups

Table 3. Selected group combinations and their abundance (in percent): results of statistical analysis

Group	Drugs	Synthetic compounds	DNP[a] (Chapman & Hall)
Alcohol ether	19	5	41
Alcohol ester	10	3	30
Aromatic alcohol	24	13	40
Aromatic alcohol ether	12	5	27
Amine aromatic	50	40	15
Aromatic amide	31	43	12
Amine aromatic amide	20	15	5

[a]Dictionary of Natural Products includes list of known natural products that are not drugs or synthetic products.

Table 4. Selected pool properties and their abundance: results of statistical analysis

Property	Drugs	Synthetic compounds	DNP (C & H)
Bridgehead atoms (3-ring bonds), (%)	25	9	49
Bridgehead atoms (4-ring bonds), (%)	4	1.4	13
A-C-C-A (%)	74	58	66
Rings/molecule (n)	3	2.6	3.3
Chiral centers/molecule (n)	1.2	0.1	3.2
Rotatable bonds/molecule (n)	10.7	8	11.1

7.3.2 Similarity Analysis

In this study the molecular diversity of the substance pools is determined by "Chemspace" using a commercially available software, "Unity System". In this Chemspace the compounds are shown in an n-dimensional space. Structural qualities are defined by descriptors and fix the location in the Chemspace. The distance within this space is a measurement of the molecular diversity of the compared substances (Fig. 7).

~ 500,000 Substances (appr. 80,000 natural products)

? Compounds are shown in an *n*-dimensional space

? Structural qualities are described by descriptors and fix the location in this Chemspace

? The distance in this space is a measurement of diversity or relationship of these substances

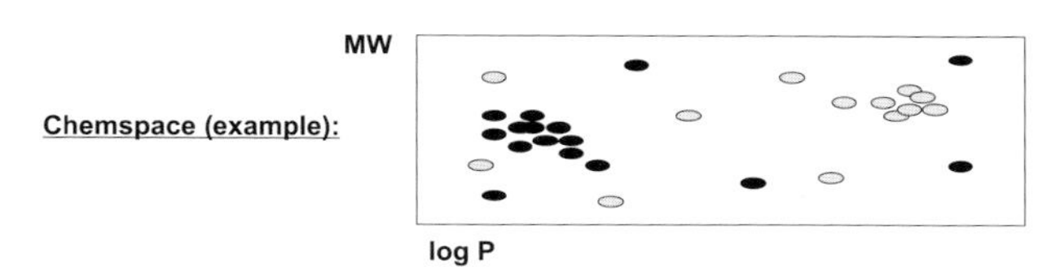

Fig. 7. Analysis of similarity to determine molecular diversity of substance pools by Chemspace (unity system)

The results of the similarity analysis show that more than 40% of the natural products do not overlap with the synthetics. The second study also reveals a significant complementarity between natural products and synthetics.

That means that natural products increase the molecular diversity of screening pools and thereby also the quality of the substance pools. In other words, we will lose molecular diversity and quality of the screening pools if we do not integrate natural products in our screening programs.

7.4 Further Approaches to Increasing Molecular Diversity of Screening Pools by Natural Product Research

Given that probably fewer than 5% of the biological sources are involved in screening programs, it is quite evident that *new biological sources* are still an important tool for the discovery of new natural products. Recently, more and more "unnatural" natural products have been synthesized via *combinatorial biosynthesis* through genetic engineering. This technique might also become a very interesting tool for producing analogues of complex natural products.

Besides the complementarity mentioned above, we think that because of their rigidity, higher stability, and unusual structural elements, natural products might be very interesting building blocks for automated synthesis, combinatorial chemistry, and classical synthesis. In particular, natural products prepared by fermentation or from plants which provide 1–10 g material are interesting materials. Two first examples – Teicoplanin and Balanol – were used as templates specifying ligands in combinatorial chemistry as early as 1996. The often supposed antagonism between combinatorial synthesis and natural product research has almost disappeared.

In order to create selected natural product pools for drug discovery programs *data bases* are very useful. In the following examples we used the BNPD data base. The contributions of biological sources, bacteria, fungi, and plants to the production of natural products are almost at the same level. At 13%, animals provide only a small portion of natural products.

Major contribution to current therapeutic successes in medicine

Worldwide intensive research activities

Natural products and synthetic compounds are complementary

Improvement of the quality of screening pools by natural product research

Fig. 8. Role of natural products in drug discovery

A closer look at the animals as a source of natural products shows that especially the marine organisms are good sources of natural products. The sponges, contributing 45%, are the most productive group.

Considering the time dependency of natural products and biological activity, fungi and marine macro-organisms seem to have been most popular during recent years. The same is true for natural products from algae.

The analysis also shows that the various biological sources provide different structural types. Thus, if we are interested in glycosides and macrocycles we should look into the bacteria, including rare actinomycetes. In insects, worms, bacteria, fungi, algae, and mollusks one will find peptides. Plants are a good source of N- and O-heterocycles, alicycles, and arenes.

7.5 Conclusion

Based on the evidence presented here, we conclude that *natural product research is and will be an integral part of drug research and development (Fig. 8).*

References

Henkel T (1997) IBC conference on combinatorial chemistry on natural products: statistical investigations of databases to assess molecular diversity of natural products and the importance of certain biological sources for the lead finding process. San Francisco, Calif., USA, December 1997

Henkel T, et al (1999) Statistical investigations into the structural complementarity of natural products and synthetic compounds. Angew Chem Int Ed 38:643–647

Müller H (1997) Dechema-Kolloquium: Konsequenzen des Rio-Abkommens für die chemisch-pharmazeutische Industrie; Wirkstofforschung: Naturstoffe und ihre Bedeutung für die Pharmaindustrie. Frankfurt, Germany, February 1997 (lecture)

Müller H (1998) 10. Irseer Naturstofftage der DECHEMA "Aktuelle Entwicklungen in der Naturstofforschung; Naturstoffe und ihre Bedeutung in der pharmazeutischen Wirkstofforschung. Irsee/Kaufbeuren, Germany, February 1998 (lecture)

8 Tools for Drug Discovery: Natural Product-based Libraries

S. Grabley, R. Thiericke, I. Sattler

8.1 Introduction

About 30% of the drugs sold worldwide are based on natural products. Though recombinant proteins and peptides account for an increasing market volume, the superiority of low-molecular-mass compounds in human disease therapy remains undisputed, due mainly to more favorable compliance and bioavailability properties. In the past, new therapeutic approaches have often evolved from research involving natural products. Numerous examples from medicine impressively demonstrate the innovative potential of natural compounds and their impact on progress in drug discovery and development. However, natural products are currently undergoing a phase of reduced attention because of the enormous effort which is necessary to isolate the active principles and to elucidate their chemical structures.

Approaches to improving and accelerating the joint drug discovery and development process are expected to arise mainly from innovation in drug target elucidation and from lead finding. Breakthroughs in molecular biology, cell biology, and genetic engineering in the 1980s gave access to understanding diseases on the molecular or on the gene level. Subsequently, novel target-directed screening assays of promising therapeutic significance, automation, and miniaturization resulted in high-throughput screening (HTS) approaches, thus changing the industrial drug discovery process fundamentally. Furthermore, elucidation of the human genome will provide access to a dramatically increased number of new potential drug targets that have to be evaluated for drug discovery. HTS enables the testing of an increasing number of drug targets and samples. Approaching up to 100,000 assay points per day is no longer futuristic. Therefore, for the accelerated identification of first-lead structures new concepts for generating large varieties of structurally diverse test samples are required.

In most cases, first hits are discovered by random screening approaches. Subsequent validation steps follow to find out whether a hit gives rise to a lead compound that proves beneficial for optimizing the structure/activity relationship by the well-established powerful tools of rational drug design. Issues of interest comprise not only bioactivity at the target of interest but also applicability, bioavailability, biostability, metabolism, toxicity, specificity of bioactivity, distribution, tissue selectivity, and cell penetration properties. As a key prerequisite for development, access to large quantities of the compound for clinical studies and commercialization has to be guaranteed.

In order to really take advantage of HTS, access to high sample numbers covering a broad range of low-molecular-mass diversity is essential. The answer of organic chemistry to high capacities in testing was combinatorial chemistry. The answer of natural product chemistry is still pending. However, in the past, compounds from nature have often opened up completely new therapeutic approaches. Moreover, natural compounds substantially contributed to the identification and understanding of novel biochemical pathways in vitro and in vivo, and consequently proved to make available not only valuable drugs but also essential tools in biochemistry and molecular cell biology (Grabley and

Thiericke 1999a,b). Therefore, it is worthwhile to study exhaustively the molecular basis of biological phenomena of new and/or unusual chemical structures which have evolved naturally.

Efforts to enhance molecular diversity from nature are addressed by combinatorial biochemistry approaches that aim at altering and combining biosynthetic genes (Hutchinson 1999). Although these promising strategies may probably contribute to HTS test sample supply in the long run, improvements are urgently required in the near future to accelerate innovation and to focus the attention of management boards in the pharmaceutical enterprises again towards natural product-based test sampling.

We presume that comprehensive compound collections comprising pure natural substances, their derivatives, and analogues as well as libraries generated by the standardized fractionation of extracts of natural origin represent most valuable tools for current drug discovery programs.

8.2 Natural Sources for the Development of Structurally Diverse Compound Libraries

8.2.1 Introduction

For thousands of years mankind has known about the benefits of drugs from nature. The drugs used by ancient civilizations were extracts of plants or animal products, with a few inorganic salts. Mainly plants have highly been estimated in medical use. In contrast, micro-organisms were not known to biosynthesize secondary metabolites useful for medicinal application until the discovery of the penicillins. However, the accidental discovery of penicillin from the culture broth of *Penicillium notatum* in 1928 by Alexander Fleming and its introduction in 1941/42 as an efficient antibacterial drug revolutionized medicinal chemistry and pharmaceutical research by stimulating completely new strategies in industrial drug discovery. In the following decades, micro-organisms attracted considerable attention as a new source of pharmaceuticals. From the screening of a huge number of microbial extracts an unexpected diversity of natural compounds performing a broad variety of biological activities became evident. Despite the superior role of phytogenic drugs in the past and the tremendous number of plant metabolites described in the literature today, secondary metabolites accessible from the culture broth of micro-organisms dominate the field of applied natural products research (for reviews see Grabley and Thiericke 1999a,b; Vandamme 1984; Goodfellow et al. 1988; Omura 1992; Gräfe 1992; Yarbrough et al. 1993; Grabley et al. 1994a; Kuhn and Fiedler 1995).

8.2.2 Molecular Diversity from Micro-organisms

8.2.2.1 Bacteria

Among the bacteria, streptomycetes and myxobacteria (Reichenbach and Höfle 1999) play a dominant role with respect to secondary metabolism. About 70% of the natural compounds from microbial sources derive from actinomycetes, mainly from strains of the genus *Streptomyces* that can easily be isolated from soil samples. The genus *Streptomyces* comprises more than 500 species that perform with outstanding diversity in secondary metabolism, yielding a yet increasing variety of new chemical structures. Actually, about 110 genera of the order *Actinomycetales* are known. So far, a number of these genera have been neglected with respect to the investigation of their potential for the biosynthesis of bioactive secondary metabolites.

Within our own research program on the exploitation of structural diversity from microbial diversity a major focus is on the isolation and investigation of streptomycetes and rare actinomycetes. During the past 3 years, we have been able to describe six novel genera of the order *Actinomycetales*. These isolates have been reported in the *International Journal of Systematic Bacteriology* and deposited in various international culture collections. So far, isolates from these groups of rare actinomycetes cannot be studied exhaustively for their secondary metabolism due to insufficient growth. Therefore, we are addressing the development of appropriate cultivation parameters. Together with other groups, we attempt to identify DNA sequences that may serve as probes for the biosynthesis of certain types of secondary metabolites in order to focus on strains that are putative producers of secondary metabolites. Based on the results obtained with streptomycetes and other bacteria such as myxobacteria it can be predicted that morphological differentiation is related to secondary metabolism. Therefore, in future we will emphasize the identification of rare actinomycetes undergoing morphological changes. However, cultures of *Streptomyces* remain our most important resource for accessing new natural products. In 1998, we succeeded in isolating 74 microbial metabolites. Their structure elucidation yielded 35 new compounds, more than 90% of these deriving from strains of *Streptomyces*.

8.2.2.2 Fungi

Besides the prokaryotic actinomycetes and myxobacteria, certainly, fungi are one of the most significant groups of organisms to be studied for drug discovery purposes. In particular, Fungi Imperfecti already have provided mankind with many different bioactive secondary metabolites, many of them having entered clinical applications, such as the β-lactam antibiotics, griseofulvin, cyclosporin A, or lovastatin. Furthermore, most new natural products described in the literature are currently isolated from fungi (Henkel et al. 1999). Therefore, fungi are an outstanding source for the isolation of structurally diverse small molecules that are highly qualified to supplement compound libraries for drug discovery.

Many fungi occupy thousands of unsuspected niches in nature, including cohabitation with larger life forms such as higher plants, or the marine environment. Given the huge number and variety of higher plants, the number of their associated microfungi, belonging mainly to the Ascomycotina and Deuteromycotina, is expected to be enormous (Petrini et al. 1992). It is estimated that only a few percent of the world's fungi are known today (Strobel et al. 1996). With respect to drug discovery, mycelium cultures of fungi are of major interest. Their metabolites are easily accessible in large quantities by fermentation processes, thus providing sufficient amounts for extended screening programs as well as preclinical and clinical studies.

8.2.2.3 Microalgae

Most kinds of micro-organisms investigated for their secondary metabolism are heterotrophic. However, microalgae, an assemblage of prokaryotic (cyanophyta) and eukaryotic micro-organisms with oxygenic photosynthesis, have to be considered as well. Microalgae abound in nature's aquatic habitats, both fresh-water and marine. They are also found in moist soils and artificial aquatic habitats, many of them living in symbiosis, e.g., with protozoae or fungi (Becker 1994).

Microalgae are typically cultivated under nonsterile conditions. Their biomass and metabolites, such as pigments, polyunsaturated fatty acids, and polysaccharides, are preferably used as additives in food manufacturing, cosmetics, and pharmaceuticals. However, microalgae are expected to be a rich source of new bioactive compounds with potential application as lead structures for the development of new drugs or agrochemicals. Until now, microalgae have been studied mainly according to their toxicogenic potential and their impact on poisoning of the animal and human food chain (Luckas 1996). In this context, most reported data are from dinoflagellates (Shimizu 1993) and cyanophyta (Attaway and Zaborsky 1993). So far, due to inadequate sterile in vitro cultivation conditions, microalgae have been neglected with respect to the systematic investigation of their potential to biosynthesize structurally diverse nontoxic bioactive secondary metabolites.

Therefore, in close cooperation with AnalytiCon AG (Potsdam), and the group of R. Buchholz at the TU Berlin, we have optimized in vitro culturing of microalgae by utilizing aseptic strains of various taxa in order to assess their biosynthetic potential by chemical and physicochemical screening techniques in combination with selected biological testing. To date, the metabolic patterns of approximately 500 different strains have been characterized (Ebert 1999).

For the isolation, taxonomic characterization, cultivation, and maintenance of microalgae strains specialized biological expertise, sophisticated bioengineering techniques, and special equipment are essential. In order to focus on the key issues of evaluating and estimating the chemical diversity of secondary metabolites from microalgae, the majority of strains for our studies were received from national and international culture collections. About 500 strains were selected, covering a maximum of biological diversity with respect to taxa variety. However, within the collections axenic isolates from marine sources are strongly underrepresented.

Compared with using heterotrophic micro-organisms, natural product screening with microalgae requires improved equipment in order to guarantee defined growth conditions with respect to light, medium, temperature, and carbon dioxide supply. In-house development and

Fig. 1. Instrumentation for the cultivation of 50 microalgae strains in 100-ml tubes in parallel

assembly yielded an apparatus for the cultivation of 50 isolates in 100-ml glass tubes in parallel (Fig. 1).

Culture conditions were optimized by considering both the enhancement of biomass and a promising metabolite pattern of the culture extracts detected by thin-layer chromatography (TLC) or HPLC analytics. For about 70% of the strains we succeeded in obtaining biomass yields of 1–2 g/l (Fig. 2). Secondary metabolite isolation for the generation of a microalgal compound library was achieved in our group by cultivation of promising producers in the 20-l scale, applying a glass vessel bioreactor equipped with an inside illumination device co-developed with Die Leuchtwerbung Ilmenau GmbH (Ebert et al. 1997). Further scale-up experiments have been realized by the TU Berlin utilizing either their photo loop-reactor with external illumination (10 l

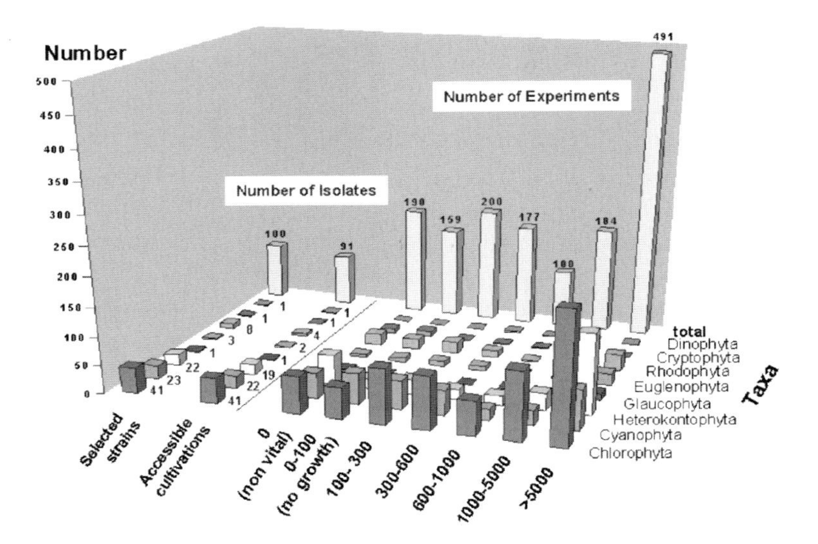

Fig. 2. Results from 1501 optimization experiments illustrate the correlation of biomass yields (by measuring OD 700; starting point was scheduled at 100) and microalgae taxa

and 70 l), or their photo-bioreactor with light-diffusing optical fibers (10 l and 100 l).

We selected more than 40 compounds deriving from 25 different strains for isolation, structure elucidation, and subsequent target-directed screening based on a microplate format library. The compounds isolated so far are representatives of steroids, phytols, glycolipids, glycoketones, purines, pyrimidines, indoles, diketopiperazines, and amino acids. Most of them are already described in the literature; however, their biological activities have not been studied in detail until now. Therefore, transfer to a compound library for drug discovery purposes is valuable. Our compounds will be supplied to the Natural Products Pool to be screened within the drug discovery programs of a number of industrial partners (see Sect. 8.6.2). In the future, in cooperation with others, we will focus our studies on strains freshly isolated from various terrestrial or marine samples. Furthermore, we will address the co-cultivation of microalgal strains with associated micro-organisms in order to consider their impact on microalgal secondary metabolism.

8.2.3 Molecular Diversity from Plants

Although nowadays micro-organisms, mainly bacteria and microfungi, have attracted superior interest for applied natural products research, higher plants remain a major source of new bioactive compounds due to the complexity and variability of their secondary metabolism, which presumably is enforced by defense strategies. Therefore, secondary metabolites isolated from plant extracts are essential with respect to the generation of valuable compound libraries for drug discovery. Studies addressing the variability of secondary metabolism in dependence on the place of origin have demonstrated the impact of the habitat. Therefore, current efforts focus on the investigation of plants from yet unexplored locations.

Plant metabolites such as alkaloids or terpenoids are structurally unique and modifications of the biosynthetic pathways yield a tremendous diversity of derivatives. However, in case of hit finding, a number of problems arise referring to back-tracing and accessing sufficient quantities for more detailed biological or pharmacological studies. Chemical synthesis cannot solve the problem in cases of complex structures, as exemplified by well-known phytogenic drugs such as morphine, codeine, reserpine, vincristine, the cardiac glycosides or, more recently, paclitaxel, and camptothecin.

However, in some cases valuable precursors can be made accessible from plants at moderate prices, thus contributing to improved manufacturing processes, supplemented by chemical synthesis and biocatalysis or biotransformation. Precursors sometimes are of considerable advantage because they also give access to unnatural analogues or derivatives valuable both for the generation of compound libraries for screening purposes and for the optimization of biological properties. For instance, the manufacturing of oral contraceptives and other steroidal hormones is based on starting materials of plant origin such as diosgenin and hecogenin.

8.2.4 The Marine Environment

Recent trends in drug discovery from natural sources emphasize investigation of the marine environment, yielding numerous, often highly complex chemical structures. So far, in most cases, in vitro cultivation techniques for the supply of sufficient quantities for biological activity profiling and clinical testing are missing. Focus on marine biotechnology is currently strengthened by results indicating that marine micro-organisms are substantially involved in the biosynthesis of marine natural products initially isolated from macro-organisms such as invertebrates (Attaway and Zaborsky 1993; de Vries and Beart 1995; König and Wright 1996, 1999; Jensen and Fenical 1996).

Therefore, within our efforts to exploit molecular diversity from microbial diversity, we started to investigate the potential of marine fungi. Taking the number and the chemical diversity of their terrestrial counterparts as an indicator, cultures of marine fungi promise to be a superior source for drug discovery. However, the true potential of marine fungi has not yet surfaced, as no unique secondary metabolites have been reported so far. This is possibly due to the predominant isolation and cultivation of ubiquitous fungi even from samples collected in the marine environment. Therefore, in cooperation with the Alfred-Wegener-Institute (Bremerhaven, Germany) and others our investigational focus is on obligate marine fungi that have undergone their evolution in the marine environment and thus, presumably, differ in their biosynthetic pathways from ubiquitous fungi. Defining appropriate conditions for strain isolation and cultivation on a large scale will be essential if we are to benefit from using marine fungi within industrial HTS programs.

The question of whether marine natural products will play a major role in drug discovery in the future remains open. Today, toxic principles dominate the spectrum of biological activities in products isolated from marine sources. This may be due partly to the major application of cytotoxicity-directed screening assays. However, it has to be considered that defense strategies are necessary to survive in the highly competitive marine environment, thus resulting in a tremendous diversity of highly toxic compounds affecting targets that are involved in eukaryotic cell-signaling processes. The strong toxic properties of marine metabolites often preclude their application in medicine. On the other hand, a number of metabolites have proven to be valuable tools in biochemistry and cell and molecular biology. For example, the water-soluble polyether type neurotoxin, maitotoxin, produced by the marine dinoflagellate *Gambierdiscus toxicus* heads the list of non-peptide toxins (Yasumoto et al. 1976). Maitotoxin can be obtained by in vitro cultivation techniques (isolation of 25 mg from 5000 l) and serves as a unique pharmacological tool for studying calcium transport (Gusovsky and Daly 1990). Currently, various other marine natural products that exhibit considerable toxic potency are hopeful candidates for clinical use, mainly in anticancer therapy. However, today clinical studies on marine metabolites and their testing within compound libraries are hampered by the insufficient supply of material.

8.3 Strategies for Exploiting Structural Diversity from Natural Sources

8.3.1 Introduction

Today, more than 30,000 diseases have been clinically described. Fewer than one third of these can be treated symptomatically, and only a few can be cured. New chemical entities are required to enable therapeutic innovations. The exploitation of structural diversity from natural sources is expected to contribute to improved lead discovery. In particular, low-molecular-mass natural products from bacteria, fungi, plants, and invertebrates, from either terrestrial or marine environments, represent unique structural diversity. In order to gain access to this outstanding molecular diversity, various strategies such as (target-directed) biological, physicochemical, or chemical screening have been developed.

In contrast to biological screening, the physicochemical, and chemical screening approaches a priori provide no correlation to a defined biological effect. Here, the selection of promising secondary metabolites from natural sources is based on physicochemical properties or on chemical reactivity, respectively. In both strategies, the first step is a chromatographic separation of compounds from the complex mixtures obtained from plants, bacteria, fungi, or animals. In a second (analytical) step, physicochemical properties or chemical reactivities of the separated secondary metabolites are analyzed. Both strategies have proven to be efficient supplemental and alternative methods, especially with the aim of discovering predominantly new secondary metabolites that can contribute to the development of valuable natural compound libraries (Grabley et al. 1999).

8.3.2 Physicochemical Screening Based on HPLC

Mycelium extracts, culture filtrates, or crude extracts of microbial broths as well as samples obtained from plant and animal extraction can be subjected to standardized reversed-phase HPLC by making use of various coupling techniques. Most common is HPLC coupled to a multi-wavelength UV/VIS monitor (diode array detection, DAD; Fiedler 1984). Comparison of the data (retention time and UV/VIS spectra) with those of reference substances acts as selection criteria. However, success of this strategy depends upon the amount and quality of pure references in the database. Based on the UV/VIS monitoring, the HPLC-DAD screening is well suited to screening for metabolites which bear significant chromophores. In combination with the efficient separation via HPLC, this screening procedure can be applied advantageously to plant material which contains numerous colored compounds. HPLC-DAD screening has already been successfully used for natural product screening and has led to the discovery of several new metabolites such as the naphthoquinone juglomycin Z (Fiedler et al. 1994), naphthgeranine F (Volkmann et al. 1995), the dioxolides (Blum et al. 1996), the antibiotically active fatty acid (*E*)-4-oxonon-2-enoic acid (Pfefferle et al. 1996), the insecticidal NK374200 (Morino et al. 1995), or the quinoxaline group metabolite echinoserine (Blum et al. 1995).

The data obtained from HPLC-DAD analysis are often helpful in de-replication, e.g., during high-throughput biological screening programs. However, the dependence on a UV/VIS detectable chromophore in the metabolite to be analyzed limits its possible application. Therefore, such alternative or supplemental detection methods as mass spectrometry (LC-MS) (Rodriguez et al. 1996; Gu et al. 1997) or nuclear magnetic resonance (LC-NMR) (Hostettmann et al. 1997) have been applied.

8.3.3 Chemical Screening Based on TLC

The TLC-based chemical screening approach has been developed for the investigation of metabolites from microbial cultures. In order to apply it in a reproducible way, standardized procedures for sample preparation and at least 50-fold concentration are required. The concentrates obtained from both the mycelium and the culture filtrate are analyzed by applying a defined amount to high-performance thin-layer chromatography (HPTLC) silica-gel plates, which then are chromatographed using different solvent systems. In a next step the metabolite pattern of each strain is analyzed by making use of visual detection (colored substances), UV extinction/fluorescence, and colorization reactions obtained by staining with different reagents (e.g., anisaldehyde/sulfuric acid, naphthoresorcin/sulfuric acid, orcinol, blue tetrazolium, and Ehrlich's reagent). The advantage of this combination of reagents lies in the broad structural spectrum of metabolites stained. The procedure focuses mainly on the chemical behavior and reactivity of the components and renders a good visualization of the secondary metabolite pattern (metabolic fingerprint) produced by each strain. On the basis of reference substances, significant spots on the chromatograms can be classified as: (a) a constituent of the nutrient broth, (b) a frequently formed and thus widely distributed microbial metabolite, or (c) a strain-specific compound that merits further attention.

Applied to hundreds of culture extracts from microbes (e.g., streptomycetes), these screening concepts resulted in various structurally new secondary metabolites that, in a following step, have to be assayed in a broad spectrum of biological test systems. Therefore, the availability of sufficient quantities of pure metabolites is crucial for success. However, in our experience, chemical screening typically gives rise to metabolites that are accessible in reasonable quantities. Furthermore, compounds identified by chemical screening represent almost all known structural classes (Burkhardt et al. 1996; Schneider et al. 1996; Göhrt et al. 1996, 1992; Grabley et al. 1992a,b, 1993, 1994b, 1996; Fuchser et al. 1994; Grote et al. 1988a,b; Mayer and Thiericke 1993a,b; Dräger et al. 1996; Rohr and Thiericke 1992; Drautz et al. 1985; Fiedler et al. 1986; Bach et al. 1993; Schönewolf et al. 1991; Schönewolf and Rohr 1991; Henkel and Zeeck 1991; Tanaka et al. 1993a,b; Omura et al. 1990; Henkel et al. 1991; Zerlin and Thiericke 1994; Hoff et al. 1992; Ritzau

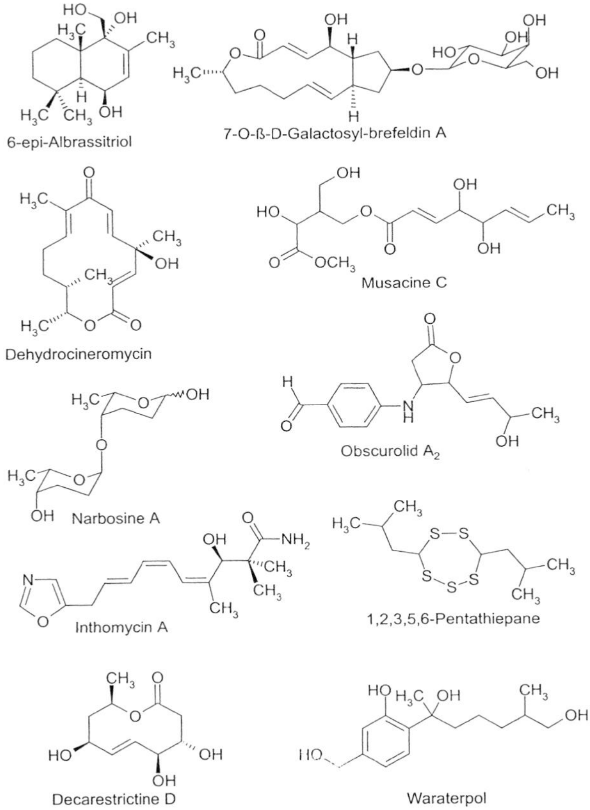

6-epi-Albrassitriol

7-O-ß-D-Galactosyl-brefeldin A

Dehydrocineromycin

Musacine C

Obscurolid A_2

Narbosine A

Inthomycin A

1,2,3,5,6-Pentathiepane

Decarestrictine D

Waraterpol

Fig. 3. Selected examples of the structural diversity discovered by chemical screening

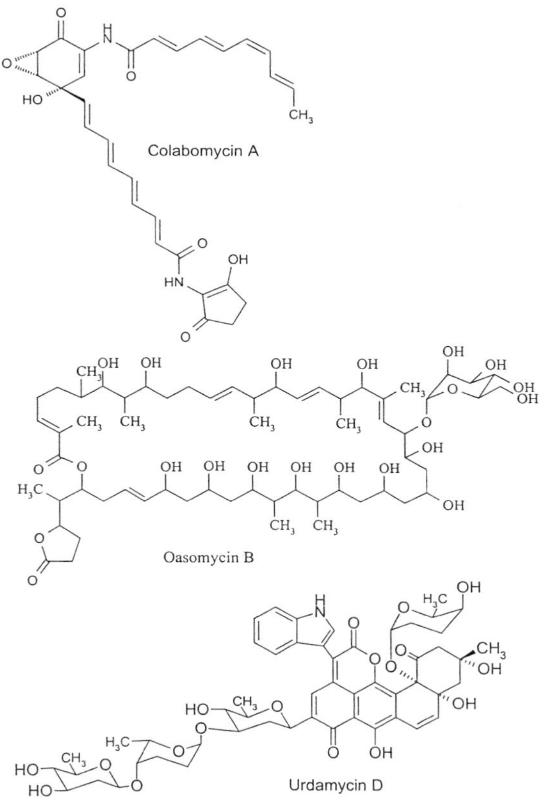

Colabomycin A

Oasomycin B

Urdamycin D

Fig. 3. Continued

et al. 1993; Thiericke and Zerlin 1996; Henne et al. 1997; Drautz et al. 1986; Rohr and Zeeck 1987; Henne et al. 1993). Therefore, chemical screening serves as a most valuable approach to the generation of natural compound libraries from microbial origin. Some selected examples are summarized in Fig. 3. So far, a certain percentage of secondary metabolites from chemical screening programs have shown striking biological effects and led to further studies.

Recently, in our laboratory, the sample preparation procedure (adsorption of the metabolites present in the culture filtrate) has been further developed towards automation and additional separation steps (see also Sect. 8.4.2). On the basis of new adsorber resins, high-quality sample preparation is now possible, allowing a more thorough TLC analysis of the metabolites produced. This leads to an improvement in the separation of overlapping spots and therefore to a higher reliability of the database-assisted analysis. On the other hand, sample application onto the TLC plates is possible via commercially available automated spotting stations. In consequence, chemical screening was adapted to the 96-well format, thus efficiently supplementing biological screening attempts.

In comparison to a TLC-based screening, the chromatographic resolution and sensitivity of HPLC-based physicochemical screening are of superior quality. On the other hand, TLC allows for parallel, quick, and cheap handling of samples and is superior in the mode of detection (UV/VIS and staining). As well as eluted compounds from HPLC separation, spots from TLC can easily be subjected to subsequent physicochemical analysis (MS, IR, NMR, etc.) via scraping off and elution from the silica gel materials.

8.3.4 Future Potential of Physicochemical and Chemical Screening Approaches

The future potential of physicochemical and chemical screening approaches lies in the possibility to tap the outstanding structural resources from nature and to build collections of pure natural products, new and known, which can be used advantageously for broad biological screening. Natural compound collections of substantial structural diversity contribute to improved lead discovery and efficiently supplement synthetic libraries (e.g., from classical or combinatorial synthesis). Often, it is better to run a biological screen with pure compounds from natural sources rather than with crude extracts. In order to extend the opportunities arising from testing pure compounds, a collection of natural products, their derivatives, and analogues was established under the leadership of our institute (see Sect. 8.6.2).

8.4. Methods and Technologies for Building High-Quality Test Sample Libraries

8.4.1 Introduction

In drug discovery via biological screening, the selection criterion usually is a desired biological effect aimed at a defined pharmaceutical application (target-directed biological screening; Omura 1992). Biological screening has been developed into a powerful concept, culminating in HTS, which integrates and makes use of recent findings in molecular and cell biology. Today, success depends on the therapeutic value of the bioassays running in the primary biological screening, and on the period required for the identification of first promising lead compounds in order to start with lead optimization procedures.

At present, libraries from combinatorial chemistry are the major source of compounds for HTS programs in drug discovery. On the other hand, nature has proven to be an outstanding source of new and innovative therapeutics. Due to the complexity of cellular metabolism, extracts from natural sources usually contain numerous different components in various amounts. Therefore, integration into automated drug-screening approaches requires additional efforts. Dealing with rough or enriched extracts from natural sources is highly cost-intensive but of remarkable interest. At present, extraction procedures for sample preparation are usually performed with a low automation rate. There exists a strong need in automation approaches for routinely performed, standardized procedures that involve additional steps of fractionation and concentration of highly diluted crude extracts originating from natural materials. For preparing fractionated natural extracts the following criteria have to be fulfilled: (a) wide scope of chemical adsorption and elution characteristics, (b) sufficient recovery rates of interesting metabolites, (c) satisfactory resolution of chromatographic separation, and (d) feasibility and reproducibility of the practical procedure.

The first commercially available apparatus for the automated extraction and separation of plant material in a preparative scale is the HPLC-based workstation SEPBOX by AnalytiCon AG (Potsdam, Germany) in cooperation with Merck KGaA (Darmstadt, Germany). The SEPBOX concept, which is also applicable to the separation of secondary metabolites from microbial culture broths, allows efficient fractionation of crude material under standardized conditions (Bindseil et al. 1997). Within less than 24 h, 1–5 g crude extract is fractionated into up to 300 fairly pure components that can be collected in a microplate-compatible format and subsequently used directly in screening programs.

8.4.2 Automated Chromatographic Solid-Phase Extraction

Facing the need for high-quality test sample preparation as the basis for building libraries that fulfill the requirements of HTS programs, we developed a novel automated and efficient sample preparation method based on a multistep fractionation method by chromatographic solid-phase extraction (SPE). Our approach, which has the advantage of not needing HPLC-techniques, evolved from a procedure for sample preparation from microbial broths with XAD-16 resins that had been developed for chemical screening. The advanced protocol with novel polystyrol-based resins allows mixture fractionation through variation of the organic solvent content in the eluent (Fig. 4). In order to evaluate different adsorption materials, and to assign experimental parameters, we have devised a defined mixture of natural products covering a wide range of structural features and chemical polarity. Analysis of the sample quality was performed with HPLC and TLC as well (Schmid et al. 1999).

For fractionation of culture filtrates in our standard procedures we are now using ENV^+ and Amberchrom 161 resins and methanol/water mixtures for elution (exemplified by Fig. 5). The protocol makes it possible to generate single-step fraction samples as well as multiple-step fractionations with different resins from a single source. The automated multistep procedure, which is performed with modified RapidTrace modules from Zymark GmbH (Idstein, Germany) shows highly reliable performance and requires only minor manual intervention. In analogy to the SEPBOX concept, our multistep SPE process can be adapted to the separation and purification of mixtures deriving from combinatorial chemistry or any chemical synthesis. In cooperation with CyBio Instruments GmbH (formerly OPAL Jena GmbH, Jena, Germany), we are currently working on the development of an apparatus that will provide samples in the 96-well microplate format.

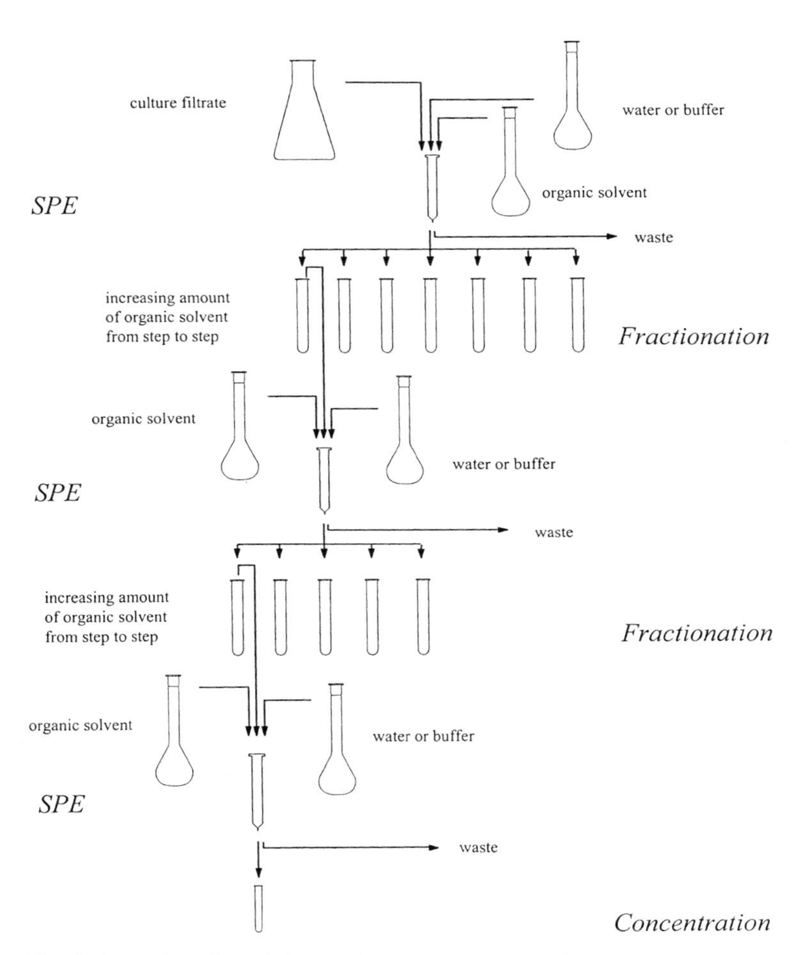

Fig. 4. General outline of the multistep SPE procedure for sample preparation from microbial broths

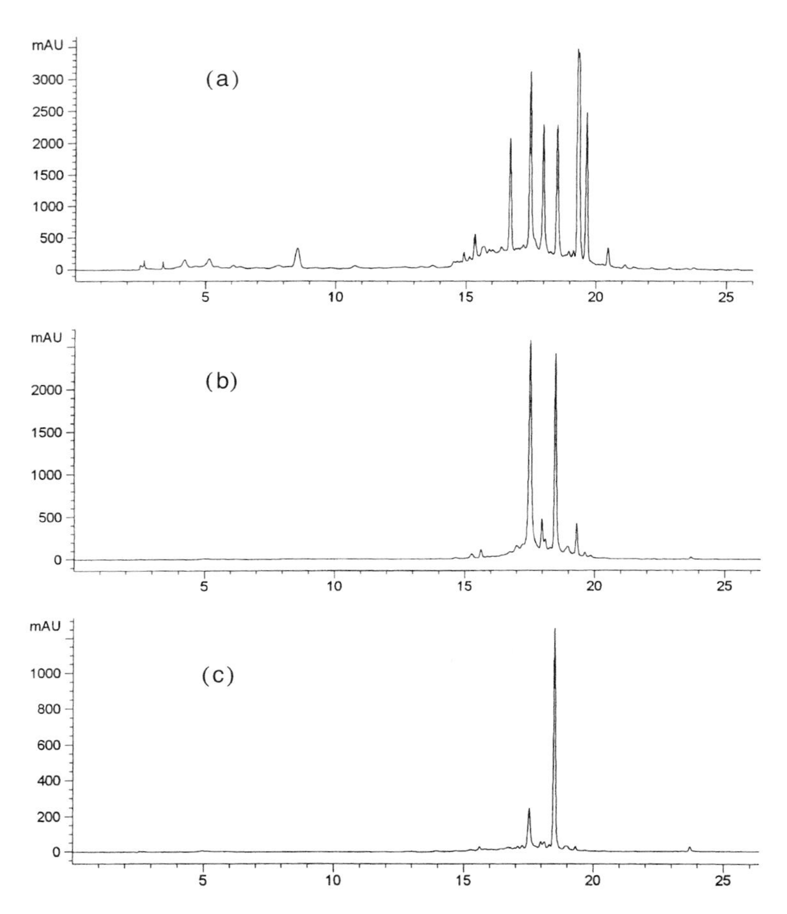

Fig. 5a–c. HPLC spectra of extracts from the fungal strain GT76111. (**a**) Manual nonfractionated extract (Amberlite XAD-16); (**b**) selected fraction of automated chromatographic SPE after the first fractionation step; (**c**) further fractionation of **b** after a second fractionation step

The analytical power of chemical screening starting from fractionated extracts (fast, inexpensive, parallel, easy-to-handle, use of UV/VIS and staining reagents) makes it possible to add physicochemical information on "hit lists" out of target-directed screening approaches. Therefore, we regard the integration of TLC analysis into secondary biological screening and hit verification as a remarkable tool in lead structure-finding strategies, e.g., for assigning the active principle, for fast and efficient de-replication, and for speed-up isolation and purification procedures. More recently, coupling techniques with mass spectrometry or even TLC-FID coupling (Vreven 1997) have been described which obviously can be integrated into our concepts.

Both the generation of natural compound libraries and access to a wanted bioactive principle from fractionated high-quality samples rely on the efficiency of isolation procedures resulting in pure compounds for further evaluation. Therefore, in cooperation with the HKI Pilot Plant for Natural Products, we addressed the transfer of our SPE fractionation process into a pilot scale chromatography in a preparative MPLC system on Amberchrom 161c. Fractionation characteristics by stepwise elution with water/methanol mixtures can be "translated" into water/methanol gradients with nearly identical separation characteristics.

8.5 Combinatorial Libraries Based on Natural Products

8.5.1 Introduction

Today, chemical libraries consisting of more than 100,000 compounds are routinely evaluated for striking biological activity. In order to generate these numbers of compounds the "classical" strategies of organic and medicinal chemistry nowadays have been surpassed by a group of technologies commonly called "combinatorial" chemistry (solid phase, solution phase, as well as split- and split-pool strategies). Recent concepts address the synthesis of medium-sized compound libraries consisting of single components.

Since its implementation several years ago, some distinct problems of combinatorial synthesis have become apparent: (a) purity of the samples often is insufficient; (b) parallel processing often yielded unexpected reaction products; (c) physicochemical analysis of the products generated is necessary; (d) reproduction and purification, especially on a larger scale for more detailed pharmacological studies, is often difficult and a logistical problem; and (e) overall, structural diversity is not as broad as expected.

With regard to the enhancement of structural diversity, there exists a need in development of combinatorial chemistry for more sophisticated synthesis concepts (e.g., multistep synthesis, larger molecules, stereochemical approaches, synthesis of more reactive compounds, making use of complex templates and building blocks). Future success in combinatorial chemistry will depend substantially on both the structural diversity and the quality of compound libraries submitted to HTS. As a logical consequence, the exploitation of natural sources for combinatorial synthesis strategies comes into focus (Bertels et al. 1999; Paululat et al. 1999). However, nature itself uses the principles of combinatorial synthesis to generate large and structurally diverse libraries by combining small biosynthetic building blocks (e.g., nucleic acids, amino acids, and other building blocks from primary metabolism such as activated C_2-, C_3-, and C_4-carboxylic acids for polyketide-type assembly).

In order to take advantage of molecular diversity from nature for combinatorial chemistry, two major strategies are addressed. On the one hand, combinatorial synthesis allows efficient and systematic structural variation of a given natural product which is used as a kind of template for synthetic "decoration" (Bertels et al. 1999). The second strategy of integrating natural products into combinatorial chemistry involves the total synthesis approach of natural products. This approach does not depend on the availability of natural products and makes it possible to generate a broader structure variation of the basic skeleton of a natural product via divergent synthesis and using various reagents and building blocks (Bertels et al. 1999).

It should be highlighted that low-molecular-mass compounds ranging from 200 to 500 daltons are in the focus of synthetic combinatorial libraries. In the same molecular range the majority of known secondary metabolites can be isolated from natural sources. An interesting approach is to generate compound libraries on the basis of natural products ranging from 500 to 900 daltons. These molecules are expected to be more suitable for targets on the basis of protein-protein, protein-DNA, and protein-RNA interactions, which play important roles, for instance, in cell regulation and differentiation processes.

Besides organic chemistry, biological methods for structural modification and subsequent compound library generation can be a supplementary tool for the derivatization of structurally complex molecules from both natural and synthetic sources. The various methods can be categorized into those employing the native biosynthetic machinery of a producing organism, those involving a manipulation of the biosynthetic pathways on the enzymatic or genetic level, and those involving the application of individual biosynthetic enzymes. The experimental demands on the application of the various methods range from "simple" feeding of biosynthetic precursors into standard cultivations to more sophisticated approaches involving genetic engineering of biosynthetic enzymes. Genetics are applied in the cell-based combination of biosynthetic genes from different strains or the in vitro reconstitution of biosynthetic pathways with over-expressed enzymes (Sattler et al. 1999). All of the various methods of biological derivatization become possible due to a relaxed substrate specificity of some of the biosynthetic enzymes, especially those of microbial secondary metabolism.

Actually, the biomimetic combinatorial synthesis of polyketides starting from simple building blocks represents a challenging target for the production of compound libraries. First attempts towards the preparation of libraries have been reported (Reggelin et al. 1998; Keinan et al. 1997). However, combinatorial chemical synthesis has to compete with combinatorial biosynthesis (Rohr 1995) for gaining access to "unnatural" natural products with a polyketide-type skeleton.

8.5.2 Combinatorial Chemistry with Natural Products

So far, a restricted number of natural products, e.g., steroids, have been used as templates for combinatorial synthesis (Bertels et al. 1999). A new approach to obtaining increased diversity in the search for new lead compounds, kombiNATURik, has been co-developed by the German enterprises AnalytiCon AG (Potsdam, Germany) and Jerini Bio Tools GmbH (Berlin, Germany). The kombiNATURik program starts with natural compounds which are further diversified by solid-support chemistry, introducing for instance peptide or carbohydrate moieties. Kombi-

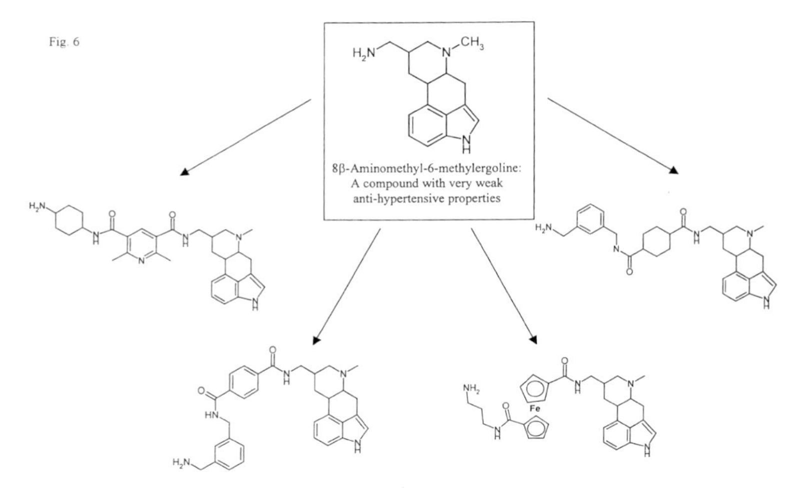

Fig. 6. β-Aminomethyl-6-methyl-ergoline derivatives exhibiting strong antihypertensive properties

NATURik libraries generally comprise several hundreds to thousands of single molecules derived from multi-parallel synthesis (Bertels et al. 1999).

In our own laboratory, we started developing methods towards automated solid-phase chemistry in parallel in order to decorate functionalized natural product skeletons on the multi-milligram scale. We started our studies by generating compound libraries with molecular masses of 500–900 daltons by modifying plant secondary metabolites e.g., alkaloids, with various building blocks such as proteinogenic and unusual amino acids, bromocarboxylic acids, dicarboxylic acids, diisocyanates, and amines. A recent highlight was the finding that various novel 8β-aminomethyl-6-methyl-ergoline derivatives with amino-substituted dicarboxylic amide residues in the side chain exhibit strong antihypertensive activities (α-adrenergic receptor binding; in vitro activity confirmed by in vivo studies on arterial blood pressure of normotensive Wistar-rats; Fig. 6; Paululat et al. 1999).

Currently, we are focusing on the modification of appropriate natural products that have been isolated within our screening programs during the past years, such as various alkaloids from microbial sources, terrecyclic acid A, naphthoquinone-type compounds, terpenoids, and O- and N-heterocycles.

Our concept for library synthesis targets single-compound strategies yielding more than 1-mg amounts of pure end products. We employ block synthesis in a format that guarantees compatibility to the 96-well microplate format for bioassaying. In general, our libraries comprise fewer than 300 single compounds. Scale-up to about 200-mg quantities for hit validation and more detailed biological studies can easily be realized.

8.6 Aspects of Application

8.6.1 Handling of Compound Libraries

Within the application of compound libraries for the HTS process, sample handling is crucial. While liquid handling in volumes >5 μl is reliable, such procedures as sample storage and retrieval, weighing, dissolving, and distributing are more sophisticated. In principle, sample storage is performed following two different strategies: in diluted form [e.g., in dimethylsulfoxide (DMSO)], or as pure samples either in tubes or in deep-well plates. Besides defined storage conditions, which should minimize stability problems, a reliable and fast retrieval of samples is required. Automated systems should be able to handle hundreds of thousands of different compounds [storage, multiplication, sample (back-) tracking, movement, etc.] with an efficient logistic. Due to the different demands, the automated storage and retrieval concepts installed in pharmaceutical companies are customer designed (Thiericke et al. 1999).

In most companies, the sample storage and retrieval facilities are centralized. Therefore, the test samples have to be transferred to the screening groups at various locations. Consequently, logistics such as bar-coding, data transfer, and delivery are crucial points. Delivery of the samples is performed usually in the 96-well format after sealing with removable films in solution (e.g., on dry ice in DMSO), or as neat-films.

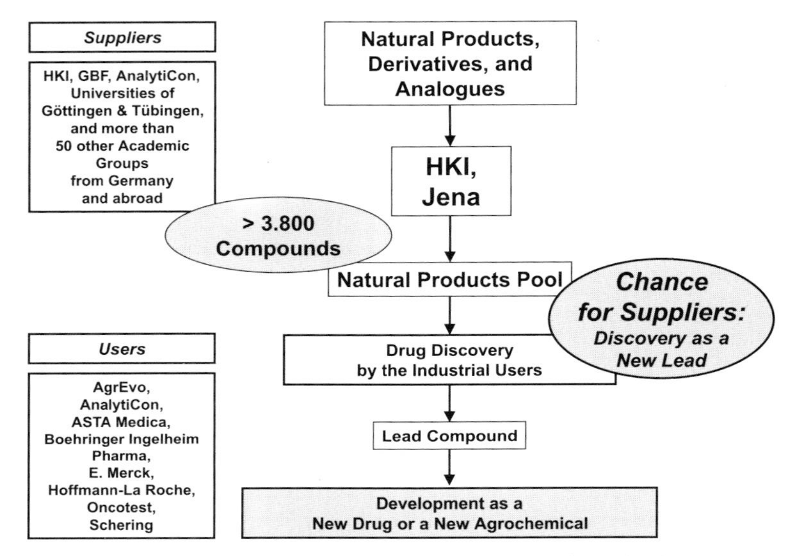

Fig. 7. The Natural Products Pool for the identification of new lead structures

8.6.2 The Natural Products Pool for Lead Discovery

As traditional natural product screening is done by testing crude extracts, followed by the crucial work of back-tracing the active compounds from the hit extracts, much experience is required to exclude both false-positive results and doublets. Considerable efforts are required to gain access to sufficient quantities of raw material for the purpose of reproduction, compound isolation, structure elucidation, and subsequent verification of biological activity. The complete process is highly time and capacity consuming. As a consequence, screening with pure compounds rather than with crude extracts has to be considered. For individual units, however, the problem is to get access to sufficient numbers of natural compounds covering substantial structural diversity.

Therefore, our concept of building up a comprehensive Natural Products Pool for industrial drug discovery purposes has gained considerable interest (Fig. 7) (Koch et al. 1997a,b, 1999). Academic research groups can supply their compounds to this pool and have them tested in target-directed bioassays of therapeutic value. However, proprietary rights of the suppliers with regard to their compounds are not affected. In addition, the suppliers receive financial incentives for each compound provided to the Natural Products Pool. In case of hit identification, the provider is informed and possibly bilateral arrangements are made in order to enable further studies. A contract regulates supply, delivery, and use of the Natural Products Pool. So far, a number of hits have been identified. Subsequent bilateral arrangements to access additional amounts for hit verification have been realized.

Within an initial 3-year period the project was supported by the BMBF (German Federal Ministry of Education, Science, Research and Technology) and German enterprises with key activities in lead discovery. Now, beyond the BMBF-supported period, the Natural Products Pool is funded exclusively by our industrial partners. In order to maintain our concept of providing about 800 compounds per year, the acquisition of natural products from abroad is being strengthened.

The Natural Products Pool aims at gaining importance in current industrial lead-discovery programs, thus strengthening the role of natural products in drug discovery. However, the HKI will also benefit from the compound collection in cooperation with compound suppliers for own lead-finding purposes, and by providing the Pool to research groups in biochemistry as well as cell and molecular biology for their investigations, e.g., bioassay systems targeting cell signaling processes.

Within the starting phase 3500 natural compounds, derivatives, and analogues have been obtained in amounts of 10–20 mg from about 40 German academic groups and AnalytiCon (Potsdam, Germany). In order to be compatible with standards of modern screening programs, the Natural Products Pool is organized in the 96-well microplate format. The Pool is delivered to each industrial partner in quantities of 1 mg/compound, and is accompanied by a database covering industrial standards. The database comprises chemical/physical data and information about known biological activities, references, and suppliers.

The range of producing organisms covers micro-organisms such as streptomycetes, rare actinomycetes, myxobacteria, Fungi Imperfecti, and basidiomycetes, mosses, a broad variety of higher plant species, some marine organisms, and a few animals. With reference to structural diversity, the Natural Products Pool makes available representatives of most biosynthetic pathways. Furthermore, nature-derived structural diversity is supplemented by synthetic analogues and derivatives of secondary metabolites. At present, the spectrum of the molecular masses of the Pool compounds centers around 300–400 daltons.

8.6.3 Signal Transduction Inhibitors from Libraries of Natural Products

Prolyl-specific enzymes such as peptidyl-prolyl *cis/trans* isomerases (PPIases) and prolyl peptidases are reported to play an important role in a number of biological processes. The first drugs on the market targeting prolyl-specific enzymes are the immunosuppressant PPIase ligands cyclosporin A, which interacts with cyclophilins, and FK506 (tacrolimus), which binds to FK506-binding proteins (FKBPs). Both drugs are used mainly to prevent organ rejection after transplantation surgery.

PPIases accelerate the slow *cis/trans* isomerization of peptidyl prolyl amide bonds in peptides and unfolded proteins (Fischer 1994). To date, three distinct PPIase families have been identified according to amino acid sequence homology and characteristics of inhibition by microbial drugs: cyclophilins, FKBPs, and parvulins (Fischer et al. 1998; Fischer 1999). Prolyl-specific enzymes are essentially involved in cell signaling processes. Therefore, it has been speculated that they represent promising new types of therapeutic targets for the treatment of diseases such as Parkinson's and Alzheimer's, depression, stroke, and disorders of the myocardium (cardiomyopathies).

So far, various FKBPs besides the cyclophilins have been proven to have regulatory functions as stable or dynamic components of hetero-oligomeric complexes containing physiologically relevant proteins, e.g., hormone and growth factor receptors and ion channels (Fischer et al. 1998; Marks 1996; Kay 1996). Recently, FK506 and structurally related FKBP12 ligands that comprise the minimal FKBP12-binding domain (see Fig. 8) and are devoid of immunosuppressive activity were shown to increase axonal regeneration in neuronal cell cultures and different animal models via a yet unknown calcineurin-independent mechanism (Hamilton and Steiner 1997). Furthermore, FKBP inhibitors were proven to exert powerful anti-neurodegenerative effects on damaged central and peripheral neurons (Hamilton and Steiner 1997).

With regard to the clinical potential of selectively acting effectors of prolyl-specific enzymes, target-directed screening on the molecular

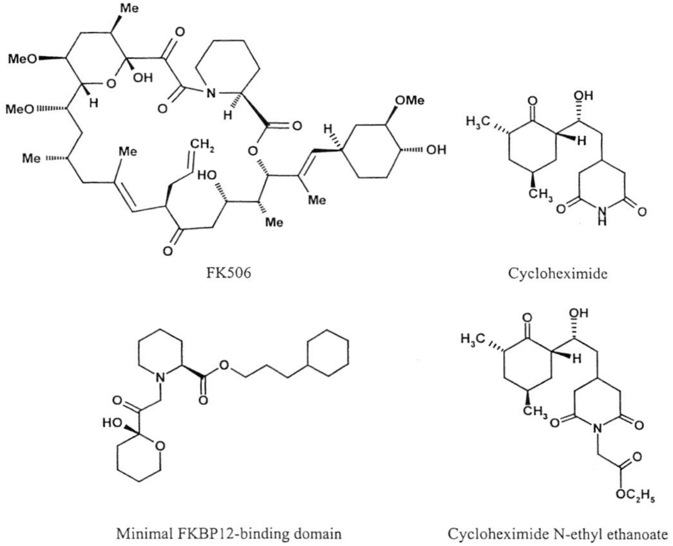

FK506

Cycloheximide

Minimal FKBP12-binding domain

Cycloheximide N-ethyl ethanoate

Fig. 8. Chemical structures of FKBP12 ligands: FK506, the minimal FKBP12-binding domain of FK506, cycloheximide, and cycloheximide N-ethyl ethanoate

level was initiated by G. Fischer, Max-Planck Research Unit (Halle, Germany). The screening was performed in close cooperation with our group, providing a collection of structurally diverse pure secondary metabolites for the identification of potential new lead structures. In order to gain information on the selectivity of enzyme inhibition, about ten PPIases representing all three enzyme families as well as prolyl peptidases were investigated in parallel (Christner 1998). The discovery of hFKBP12-affecting compounds was defined as a key screening issue because nonimmunosuppressive FKBP inhibitors are expected to enable breakthroughs in the treatment of human nerve injuries and neurodegenerative disorders, such as Parkinson's and Alzheimer's disease.

Out of a collection of about 650 natural compounds, their derivatives and analogues, surprisingly, cycloheximide (Fig. 8) (Lost et al. 1984), a potent inhibitor of eukaryotic protein synthesis, was found to specifically inactivate PPIase activity of hFKBP12. Due to its toxicity, the glutarimide antibiotic cycloheximide which is produced by several strains of *Streptomyces* is applied exclusively in biochemical research.

Structurally, cycloheximide differs completely from known FKPB12 ligands. Therefore, its binding properties were studied in detail. The reversibility of the cycloheximide-hFKBP12 interaction was demonstrated by gel filtration. Evaluation of the inhibition kinetics by the Lineweaver-Burk plot revealed a competitive mode of inhibition with a K_i value of 3.4 µ*M*. Compared with the effect on hFKBP12, cycloheximide caused a 2- to 60-fold weaker inhibition of other members of the FKBP family of PPIases. The immunosuppressive properties of FK506 are due to complex formation with FKBP12 and subsequent inhibition of the calcineurin-triggered signal transduction cascade. However, it was shown that the complex formed by the interaction of FKBP12 and cycloheximide does not inhibit the Ca^{2+}/calmodulin-dependent Ser/Thr phosphatase calcineurin.

In order to address the question of whether the well-known effects of cycloheximide on inhibition of eukaryotic translation can be discriminated from the inhibition of FKBP activity, we started to systematically modify the structure of cycloheximide. The biological profile of the derivatives was determined with respect to their inhibitory potency towards hFKBP12 and their cytotoxicity against eukaryotic cell lines (mouse L-929 fibroblasts, K-562 leukemic cells) in order to characterize their protein biosynthesis inhibition potency. As a result, several N-substituted less toxic or even nontoxic cycloheximide derivatives were identified, exhibiting IC_{50} values in the range of 22.0–4.4 µ*M* for the inhibition of hFKBP12 (Christner 1998; Christner et al. 1999). The best results with respect to both FKBP12 binding (K_i = 4.1 µ*M*) and toxicity (IC_{50} >100 µ*M*) were achieved with cycloheximide N-ethyl ethanoate (see Fig. 8), which was therefore selected for further investigations. Direct evidence for the profound deterioration of the ability of cycloheximide N-ethyl ethanoate to inhibit protein translation was obtained by analyzing the influence of cycloheximide (IC_{50} = 0.1 µ*M*) and cycloheximide N-ethyl ethanoate (IC_{50} = 115.0 µ*M*) on protein synthesis of the α-1 mating pheromone (α-factor) from *Saccharomyces cerevisiae* by means of a rabbit reticulocyte type-I translation assay. Furthermore, testing the acute toxicity of cycloheximide N-ethyl ethanoate on 15-day-old embryonized hen's eggs (HEST) indicated tolerated dosages of up to 50 mg/kg in animal models.

In order to characterize the neuroregenerative properties of cycloheximide N-ethyl ethanoate, the rat sciatic nerve crush model (Hamilton and Steiner 1997) was chosen by the group of D. Schumann, Friedrich Schiller University (Jena, Germany). The compound significantly speeded nerve regeneration at subtoxic dosages of 30 mg/kg after being administered directly to the site of sciatic nerve lesion (Christner 1998; Christner et al. 1999). Assessment of walking behavior of the animals for a period of 8 weeks following axotomy revealed an approximately 46% increase of functional recovery, whereas a less marked effect was obvious after 10 weeks.

The present results characterize the inhibition of FKBP12 by cycloheximide, thus demonstrating a new biological activity of the long-known antibiotic compound. Today, cycloheximide is utilized exclusively as a biochemical tool to inhibit eukaryotic protein synthesis. However, routinely used concentrations of cycloheximide in the range of 0.1–100 μM may also cause significant inhibition of FKBPs. This has to be taken into account for the interpretation of experimental data.

Besides cycloheximide, our studies yielded various other natural compounds that specifically inhibit prolyl-specific enzymes: (a) juglone irreversibly affects the enzymatic activity of several parvulins (K_i values in the nanomolar to micromolar range; Christner 1998; Hennig et al. 1998); (b) albocyclin and structurally related macrolides competitively inhibit prolyl endopeptidase (PEP) from human placenta (albocyclin: K_i = 14.0 μM), and its bacterial homologue from *Flavobacterium meningosepticum* (albocyclin: K_i = 106.0 μM; Christner 1998; Christner et al. 1998); (c) terrecyclic acid A was found to be a new type of PEP inhibitor, selectively and competitively inhibiting only the human placenta-derived enzyme (K_i = 2.4 μM; in vitro effects of terrecyclic acid A have been verified in in vivo animal and cell culture experiments; Christner 1998; Zerlin et al. 1996).

8.7 Perspectives

If one considers the diversity of chemical structures found in nature with the narrow spectrum of structural variation of even the largest combinatorial library it can be expected that natural products will regain their importance in drug discovery. Mainly actinomycetes, fungi, and higher plants have been proven to biosynthesize secondary metabolites of obviously unlimited structural diversity that can be further enlarged by structural modification, applying strategies of combinatorial chemistry. Probably a variety of novel concepts in natural product research is required to draw interest to incorporating natural sources into the HTS process.

Natural products libraries comprising only pure and structurally defined compounds will probably contribute to more successful completion of compounds from natural sources within the industrial drug discovery process. Only a minority of the natural products known so far have been biologically characterized in detail. Therefore, any novel target-directed screening assay may result in identification of a new lead structure, even from sample collections comprising already described compounds. Today, selected targets of interest are transferred to HTS with the aim of discovering a hit, and subsequently a lead structure within a 1- to 2-month period. After that time, the target is replaced by a new one. Thus, rapid characterization and structure elucidation of the active principles of interest from natural sources are critical, referring to competition with synthetic libraries consisting of pure and structurally defined compounds. High-quality test sample preparation and the elaboration of LC-MS and LC-NMR techniques to accelerate structure elucidation of bioactive principles from natural sources are currently underway. Combination of these techniques with databases comprising a maximum of known natural compounds will probably contribute substantially to more interest in natural products for application in drug discovery.

The fact that today only a small percentage of the organisms living in the biosphere are described implies that there is an enormous reservoir of natural compounds still undiscovered. The United Nations Convention on Biological Diversity, adopted in Rio de Janeiro in 1992, sets the basic principles for access to and exploitation of global biological sources in the future. The convention introduced national ownership of biological resources. As a result, various pharmaceutical companies fear that this will choke off the flow of genetic resources for industrial drug discovery.

With the expectation of increasing effectivity and decreasing costs for HTS technologies, the basic limiting factor in the search for new lead compounds will be the supply of structural diversity. Consequently, the relevance of natural products in drug discovery will depend greatly on the efficiency and costs of access to compounds of natural origin compared with the supply from synthetic sources (Beese 1996).

References

Attaway DH, Zaborsky OR (eds) (1993) Pharmaceutical and bioactive natural products. Plenum, New York (Marine biotechnology, vol 1)

Bach G, Breiding-Mack S, Grabley S, Hammann P, Hütter K, Thiericke R, Uhr H, Wink J, Zeeck A (1993) Liebigs Ann Chem 241

Becker EW (1994) Microalgae: biotechnology and microbiology. Cambridge University Press, Cambridge

Beese K (1996) Pharmaceutical bioprospecting and synthetic molecular diversity. Draft discussion paper, 15 May

Bertels S, Frormann S, Jas G, Bindseil KU (1999) In: Grabley S, Thiericke R (eds) Drug discovery from nature. Springer, Berlin Heidelberg New York, p 72

Bindseil KU, God R, Gumm H, Mellor F (1997) GIT Spezial Chromatogr 1:19

Blum S, Fiedler HP, Groth I, Kempter C, Stephan H, Nicholson G, Metzger J, Jung, G (1995) J Antibiot 48:619

Blum S, Groth I, Rohr J, Fiedler H-P (1996) J Basic Microbiol 36:19

Burkhardt K, Fiedler HP, Grabley S, Thiericke R, Zeeck A (1996) J Antibiot 49:432

Christner C (1998) PhD thesis, Halle-Wittenberg

Christner C, Küllertz G, Fischer G, Zerlin M, Grabley S, Thiericke R, Taddei A, Zeeck A (1998) J Antibiot 51:368

Christner C, Wyrwa R, Marsch S, Küllertz G, Thiericke R, Grabley S, Schumann D, Fischer G (1999) J Med Chem 42:3615

de Vries DJ, Beart PM (1995) TIBS 16:275

Dräger G, Kirschning A, Thiericke R, Zerlin M (1996) Nat Prod Rep 13:365

Drautz H, Reuschenbach P, Zähner H, Rohr J, Zeeck A (1985) J Antibiot 38:1292

Drautz H, Zähner H, Rohr J, Zeeck A (1986) J Antibiot 39:1657

Ebert G (1999) PhD thesis. Jena and Berlin

Ebert G, Thiericke R, Grabley S (1997) Patent application. FN 19753900.9

Fiedler H-P (1984) J Chromatogr 316:487

Fiedler H-P, Rohr J, Zeeck A (1986) J Antibiot 39:856

Fiedler HP, Kulik A, Schütz T, Volkmann C, Zeeck A (1994) J Antibiot 47:1116

Fischer G (1994) Angew Chem Int Ed 33:1415

Fischer G (1999) In: Grabley S, Thiericke R (eds) Drug discovery from nature. Springer, Berlin Heidelberg New York, p 257
Fischer G, Tradler T, Zarnt T (1998) FEBS Lett 426:17
Fuchser J, Grabley S, Noltemeyer M, Philipps S, Thiericke R, Zeeck A (1994) Liebigs Ann Chem 831
Göhrt A, Zeeck A, Hütter K, Kirsch R, Kluge H, Thiericke R (1992) J Antibiot 45:66
Göhrt A, Grabley S, Thiericke R, Zeeck A (1996) Liebigs Ann Chem 627
Goodfellow M, Williams ST, Mordarski (eds) (1988) Actinomycetes in biotechnology. Academic, London
Grabley S, Thiericke R (1999) Adv Biochem Engin/Biotechnol 64:101
Grabley S, Granzer E, Hütter K, Ludwig D, Mayer M, Thiericke R, Till G, Wink J (1992a) J Antibiot 45:56
Grabley S, Hammann P, Hütter K, Kirsch R, Kluge H, Thiericke R, Mayer M, Zeeck A (1992b) J Antibiot 45:1176
Grabley S, Kretzschmar G, Mayer M, Philipps S, Thiericke R, Wink J, Zeeck A (1993) Liebigs Ann Chem 573
Grabley S, Thiericke R, Zeeck A (1994a) Antibiotika und andere mikrobielle Wirkstoffe. In: Präve P, Faust U, Sittig W, Sukatsch DA (eds) Handbuch der Biotechnologie, 4th edn. Oldenbourgverlag, Munich, p 663
Grabley S, Thiericke R, Wink J, Henne P, Philipps S, Wessels P, Zeeck A (1994b) J Nat Prod 57:541
Grabley S, Thiericke R, Zerlin M, Göhrt A, Philipps S, Zeeck A (1996) J Antibiot 49:593
Grabley S, Thiericke R (eds) (1999a) Drug discovery from nature. Springer, Berlin Heidelberg New York
Grabley S, Thiericke R, Zeeck (1999b) In: Grabley S, Thiericke R (eds) Drug discovery from nature. Springer, Berlin Heidelberg New York, p 124
Gräfe U (1992) Biochemie der Antibiotika. Spektrum, Heidelberg
Grote R, Zeeck A, Drautz H, Zähner H (1988a) J Antibiot 41:1178
Grote R, Zeeck A, Beale Jr. JM (1988b) J Antibiot 41:1186
Gu Z-M, Zhou D, Wu J, Shi G, Zeng L, McLaughlin JL (1997) J Nat Prod 60:242
Gusovsky F, Daly JW (1990) Biochem Pharmacol 39:1633
Hamilton GS, Steiner JP (1997) Curr Pharmacol Design 3:405
Henkel T, Zeeck A (1991) Liebigs Ann Chem 367
Henkel T, Breiding-Mack S, Zeeck A, Grabley S, Hammann PE, Hütter K, Till G, Thiericke R, Wink J (1991) Liebigs Ann Chem 575
Henkel T, Brunne RM, Müller H, Reichel F (1999) Angew Chem 111:688
Henne P, Thiericke R, Grabley S, Hütter K, Wink J, Jurkiewicz E, Zeeck A (1993) Liebigs Ann Chem 565
Henne P, Grabley S, Thiericke R, Zeeck A (1997) Liebigs Ann Recueil 937

Hennig L, Christner C, Kipping M, Schelbert B, Rücknagel KP, Grabley S, Küllertz G, Fischer G (1998) Biochemistry 37:5953
Hoff H, Drautz H, Fiedler H-P, Zähner H, Schultz JE, Keller-Schierlein W, Philipps S, Ritzau M, Zeeck A (1992) J Antibiot 45:1096
Hostettmann K, Potterat O, Wolfender J-L (1997) Pharm Ind 59:339
Hutchinson CR (1999) In: Grabley S, Thiericke R (eds) Drug discovery from nature. Springer, Berlin Heidelberg New York, p 233
Jensen PR, Fenical W (1996) J Ind Microbiol 17:346
Kay JE (1996) Biochem J 314:361
Keinan E, Sinha A, Yazbak A, Sinha Santosh C, Sinha Subhash C (1997) Pure Appl Chem 69:423
Koch C, Neumann T, Thiericke R, Grabley S (1997a) Nachr Chem Tech Lab 45:16
Koch C, Neumann T, Thiericke R, Grabley S (1997b) BIOspektrum 3:43
Koch C, Neumann T, Thiericke R, Grabley S (1999) In: Grabley S, Thiericke R (eds) Drug discovery from nature. Springer, Berlin Heidelberg New York, p 51
König GM, Wright AD (1996) Planta Med 62:193
König GM, Wright AD (1999) In: Grabley S, Thiericke R (eds) Drug discovery from nature. Springer, Berlin Heidelberg New York, p 180
Kuhn W, Fiedler H-P (eds) (1995) Sekundärmetabolismus bei Mikroorganismen, Beiträge zur Forschung (engl). Attempto, Tübingen
Lost JL, Kominek LA, Hyatt GS, Wang HY (1984) Drugs Pharm Sci 22:531
Luckas B (1996) GIT 4:355
Marks AR (1996) Physiol Rev 76:631
Mayer M, Thiericke R (1993a) J Antibiot 46:1372
Mayer M, Thiericke R (1993b) J Chem Soc Perkin Trans 1:2525
Morino T, Nishimoto M, Masuda A, Fujita S, Nishikiori T, Saito S (1995) J Antibiot 48:1509
Omura S (ed) (1992) The search for bioactive compounds from microorganisms. Springer, Berlin New York New York
Omura S, Tanaka Y, Kanaya I, Shinose M, Takahashi Y (1990) J Antibiot 43:1034
Paululat T, Tang YQ, Grabley S, Thiericke R (1999) Chimica Oggi/Chemistry Today 52 May/June
Petrini O, Sieber TN, Toti L, Viret O (1992) Natural Toxins 1:185
Pfefferle C, Kempter C, Metzger J, Fiedler HP (1996) J Antibiot 49:826
Reggelin M, Brenig V, Welcker R (1998) Tetrahedron Lett 39:4801
Reichenbach H, Höfle G (1999) In: Grabley S, Thiericke R (eds) Drug discovery from nature. Springer, Berlin Heidelberg New York, p 149
Ritzau M, Philipps S, Zeeck A, Hoff H, Zähner H (1993a) J Antibiot 46:1625

Ritzau M, Keller M, Wessels P, Stetter KO, Zeeck A (1993b) Liebigs Ann Chem 871
Rodriguez S, Wolfender J-L, Hostettmann K, Odontuya G, Purev O (1996) Helv Chim Acta 79:363
Rohr J (1995) Angew Chem Int Ed Eng 34:881
Rohr J, Thiericke R (1992) Nat Prod Rep 9:103
Rohr J, Zeeck A (1987) J Antibiot 40:459
Sattler I, Grabley S, Thiericke R (1999) In: Grabley S, Thiericke R (eds) Drug discovery from nature. Springer, Berlin Heidelberg New York, p 191
Schmid I, Sattler I, Grabley S, Thiericke R (1999) J Biomol Screening 4:15
Schneider A, Späth J, Breiding-Mack S, Zeeck A, Grabley S, Thiericke R (1996) J Antibiot 49:438
Schönewolf M, Rohr J (1991) Angew Chem 103:211
Schönewolf M, Grabley S, Hütter K, Machinek R, Wink J, Zeeck A, Rohr J (1991) Liebigs Ann Chem 77
Shimizu Y (1993) In: Attaway DH, Zaborsky OR (eds) Marine biotechnology, vol 1. Plenum, New York, p 391
Strobel GA, Hess WM, Ford E, Sidhu RS, Yang X (1996) J Ind Microbiol 17:417
Tanaka Y, Kanaya I, Shiomi K, M, Tanaka H, Omura S (1993a) J Antibiot 46:1214
Tanaka Y, Kanaya I, Takahashi Y, Shinose M, Tanaka H, Omura S (1993b) J Antibiot 46:1208
Thiericke R, Zerlin M (1996) Nat Prod Lett 8:163
Thiericke R, Grabley S, Geschwill K (1999) In: Grabley S, Thiericke R (eds) Drug discovery from nature. Springer, Berlin Heidelberg New York, p 56
Vandamme EJ (ed) (1984) Biotechnology of industrial antibiotics, drugs and pharmaceutical sciences, vol 22. Dekker, New York
Volkmann C, Hartjen U, Zeeck A, Fiedler HP (1995) J Antibiot 48:522
Vreven F (1997) Labo, July, p 48
Yarbrough GG, Taylor DP, Rowlands RT, Crawford MS, Lasure LL (1993) J Antibiot 46:535
Yasumoto T, Bagnis R, Venoux JP (1976) Bull Jpn Soc Sci Fish 42:359
Zerlin M, Thiericke R (1994) J Org Chem 59:6986
Zerlin M, Christner C, Thiericke R, Hinze C, Grabley S, Küllertz G, Fischer G, Zeeck A (1996) BRD-patent application DE 196 03 510

9 Genetic Selection as a Tool in Mechanistic Enzymology and Protein Design

D. Hilvert

Protein design is a challenging problem. We do not fully understand the rules of protein folding, and our knowledge of structure-function relationships in these macromolecules is at best incomplete. It is consequently not yet possible to specify the sequence of a polypeptide with any reasonable expectation that it will adopt a unique tertiary fold, much less recognize another molecule with high selectivity or catalyze a chemical reaction.

Nature has solved the problem of protein design through the mechanism of Darwinian evolution. From primitive precursors, recursive cycles of mutation, selection, and amplification of molecules with favorable traits have given rise to all of the 100,000 or so gene products in every one of our cells. An analogous process of natural selection can be profitably exploited in the laboratory on a human time scale to create, characterize, and optimize novel protein catalysts.

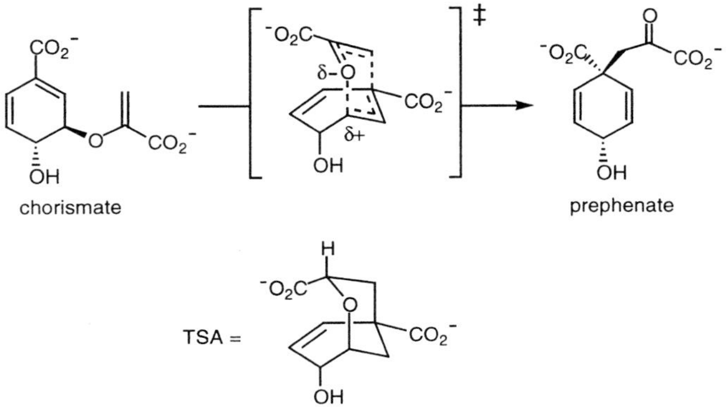

Scheme 1. The [3,3]-sigmatropic rearrangement of chorismate into prephenate. The transition state analog (TSA) shown is an effective inhibitor of chorismate mutases (K_i = 0.12–3 μM; Bartlett et al. 1988; Gray et al. 1990); it was also used to elicit the catalytic antibody 1F7 (Hilvert et al. 1988)

The basic strategy of laboratory evolution can be illustrated with proteins that catalyze a simple metabolic reaction such as the rearrangement of chorismate into prephenate (Scheme 1). This is arguably one of the simplest chemical transformations catalyzed by an enzyme. It is formally a Claisen rearrangement that proceeds via a conformationally constrained transition state in which C–O bond cleavage accompanies C–C bond formation. It is also the key step in the biosynthesis of the aromatic amino acids tyrosine and phenylalanine (Weiss and Edwards 1980).

The [3,3]-sigmatropic rearrangement of chorismate is catalyzed in vivo by enzymes called chorismate mutases. They accelerate the reaction by a factor of more than a millionfold over background. Natural chorismate mutases from different organisms exhibit similar kinetic properties but otherwise share little sequence similarity. This dissimilarity extends to their tertiary and quaternary structures, as shown by recently determined crystal structures of chorismate mutases from *Bacillus subtilis* (BsCM) (Chook et al. 1993, 1994), *Escherichia coli* (EcCM) (Lee et al. 1995), and the yeast *Saccharomyces cerevisiae* (ScCM) (Xue et al. 1994). BsCM is a symmetric homotrimer, packed as a pseudo-α/β-barrel, with three identical active sites formed at the subunit interfaces. In contrast, EcCM and ScCM are distantly related, all-α-helical homodimers. The structure of a catalytic antibody (1F7) with modest chorismate mutase activity (Hilvert et al. 1988) has also been solved (Haynes et al. 1994); it has a canonical immunoglobulin fold.

The architectural diversity nature employs to promote this relatively simple chemical transformation is truly amazing, and these different structures raise many intriguing questions. How do the natural enzymes work? Why are they 10^4 times better than the catalytic antibody? Can this insight be used to generate more effective antibody catalysts? More generally, what are the underlying chemical determinants of the individual structures? In each case, the individual domains consist of approximately 100 amino acids. Why, in one case, does the polypeptide adopt an α-helical conformation, a mixed α/β in another, and an all-β topology in a third?

Many of these questions can be investigated using random mutagenesis and genetic selection in vivo. Over the past several years, my laboratory has engineered several strains of yeast and *E. coli* that lack the enzyme chorismate mutase (Scheme 2; Bowdish et al. 1991; Tang et al. 1991; Kast et al. 1996). As a consequence, these cells are unable to grow in the absence of tyrosine and phenylalanine. Their metabolic defect can be thought of as a short circuit. This short circuit can be repaired, of course, be supplying the cells with a plasmid encoding a natural chorismate mutase. Alternatively, a mutant of the natural enzyme or a completely unrelated protein can be introduced into the cell. If the protein is able to catalyze the conversion of chorismate into prephenate, the cells will be able to produce their own tyrosine and phenylalanine and hence grow under selective conditions. If the protein is a good catalyst, the cells will grow like the wild-type strain. If the protein is a poor catalyst, the cells will grow slowly. If the protein is unable to accelerate the chorismate rearrangement, the cells will not grow at all.

This simple system thus provides an extremely powerful tool for determining whether a particular polypeptide possesses chorismate mutase activity. Because the system is limited only by the efficiency with which the chorismate mutase-deficient cells can be transformed, 10^9 different molecules can be evaluated in parallel. Because catalytic activity is such a stringent criterion for successful design – misplacement of

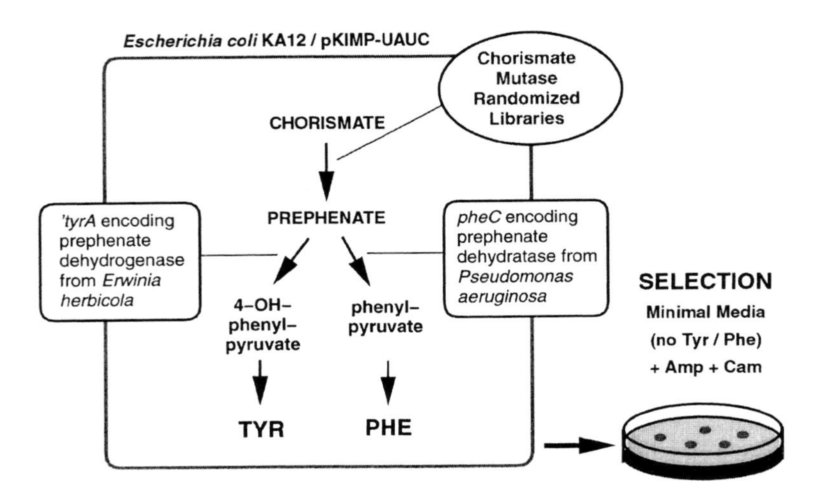

Scheme 2. An engineered *E. coli* strain lacking chorismate mutase activity (Kast et al. 1996). The genes encoding the bifunctional enzymes chorismate mutase-prephenate dehydrogenase and chorismate mutase-prephenate dehydratase were deleted, and monofunctional versions of the dehydrogenase and dehydratase were supplied on a plasmid. Potential chorismate mutases are evaluated on the basis of their ability to complement the genetic defect and allow the cells to grow in the absence of added tyrosine (Tyr) and phenylalanine (Phe)

catalytic residues by even a few tenths of an angstrom can mean the difference between full activity and no activity – this capability allows us to address the questions posed above regarding design and mechanism in a statistically meaningful way. In the sections below, some of the published applications of this system are briefly reviewed.

9.1 Mechanistic Investigations

Our understanding of the mechanism of the chorismate rearrangement derives largely from experimental investigations of the uncatalyzed reaction (Copley and Knowles 1985, 1987; Addadi et al. 1983) and from computation (Wiest and Houk 1994, 1995). These studies indicate that the reaction occurs via a concerted but asynchronous pericyclic transition state. How a protein might stabilize this species, however, has been a matter of debate. Recently, heavy atom isotope effects were used to characterize the structure of the transition state bound to BsCM (Gustin et al. 1999). A very large ^{18}O isotope effect at O(5) (ca. 5%), the site of bond breaking, and a small, normal isotope effect at C(1) (ca. 6%), the site of bond making, show that chemistry is significantly rate determining for this enzyme and confirm that the enzymatic reaction proceeds through a concerted but asymmetric transition state. However, comparison with theoretical isotope effects obtained by Becke3LYP/6–31G* calculations indicates that the transition state for the enzymatic reaction is more highly polarized than its solution counterpart, with more C–O bond cleavage and less C–C bond formation.

Examination of the residues that line the active site of BsCM (Chook et al. 1993, 1994) suggests a possible reason for the greater polarization of the bound transition state. The side chain of Arg90 places its positively charged guanidinium group within hydrogen bonding distance of the ether oxygen of bound substrate, where it would be able to stabilize additional negative charge that builds up at this site in the transition state. The contribution to catalysis of such a residue would normally be investigated by site-directed mutagenesis: Arg90 could be replaced with lysine or methionine, for example. Because a selection system is available, all 20 natural acids can be evaluated at position 90 simultaneously. Those amino acids that yield a functional enzyme can be quickly identified by their ability to complement the chorismate mutase deficiency. When this experiment was performed, the results were dramatic (Kast et al. 1996). Only arginine at position 90 yielded an active enzyme. Even the conservative replacement of arginine with lysine gave an enzyme that was dead in vivo. Kinetic characterization of representative mutants showed that removal of the arginine side chain reduces catalytic efficiency by 5–6 orders of magnitude (Kast et al. 1996; Cload et al. 1996).

If the role of Arg90 is electrostatic stabilization of the developing negative charge on the ether oxygen of chorismate in the transition state, one might wonder why complementation was not observed when arginine was replaced with another positively charged residue such as lysine. One possibility is that the Arg90Lys variant is poorly expressed in *E. coli*. The growth phenotype depends on the total amount of chorismate mutase activity present in the cell, which is determined by the specific activity of the catalyst and its concentration. If the catalyst is very active but present only at extremely low concentrations, no growth will be observed. In general, BsCM mutants appear to be produced at comparable levels within the selection system (Kast et al. 1996), but exceptions are possible.

An alternative explanation is that the active site of the enzyme places intrinsic structural constraints on possible substitutions. The lysine side chain is shorter than that of arginine and it is conceivable that it would not be able to reach far enough into the active site to place its ε-ammonium group within hydrogen bonding distance of the critical ether oxygen. If this is true, introduction of additional mutations at other positions within the active site might be necessary to accommodate a substitution at position 90.

To test this idea, we simultaneously mutated two residues – Arg90 and Cys88 – within the binding pocket of BsCM (Kast et al. 1996). The side chain of Cys88 is about 7 Å distant from bound ligand, but it nestles up against the side chain of Arg90 and might influence the conformation of the latter within the active site. Combinatorial mutagenesis of these two residues yields 400 (20×20) possibilities, which again can be rapidly evaluated in the chorismate mutase-deficient selection strain. The results of this experiment proved quite informative (Kast et al. 1996). They showed first that Cys88 is not essential for catalysis. It can be replaced by large, medium-sized, and small amino acids, and as long as an arginine is at position 90, an active enzyme is obtained. If a small residue (glycine or alanine, for example) replaced Cys88, active double mutants with an additional Arg90Lys mutation were found, showing that the guanidinium group is not crucial for catalysis. However, the most interesting result to emerge from this study is that a cation at position 90 can be dispensed with, provided that Cys88 is replaced with a lysine.

Kinetic studies on the Cys88Lys/Arg90Ser double mutant show that lysine at position 88 restores three of the five orders of magnitude in catalytic efficiency lost upon removal of the arginine at position 90. Crystallographic studies of this double mutant (Y. Xue, unpublished) show that the side chain of Lys88 extends into the active site and places its ammonium group on top of the guanidinium group of Arg90 in the wild-type enzyme, within hydrogen bonding distance of the ether oxygen of bound ligand. Thus, combinatorial mutagenesis and selection

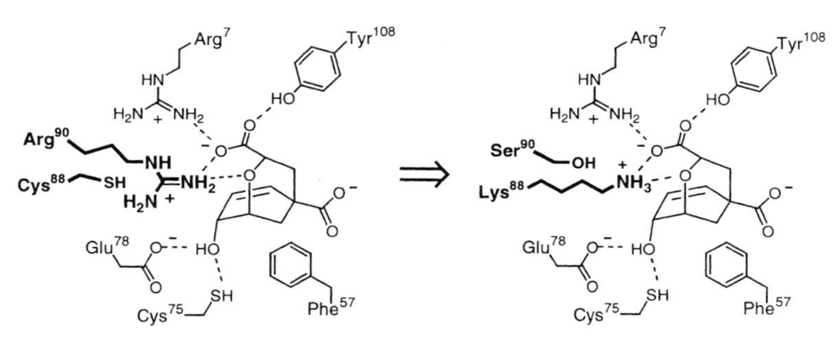

Scheme 3. Redesign of the BsCM active site. Combinatorial mutagenesis and selection experiments with wild-type BsCM (*left*) yielded an active double mutant (Cys88Lys/Arg90Ser, *right*) in which an essential cationic interaction with the ether oxygen of bound ligand is moved to a different location within the active site (Kast et al. 1996). Interactions between the protein and a TSA inhibitor are shown (Chook et al. 1993)

resulted in a redesigned active site, providing a novel solution to the chemical problem of catalyzing the chorismate mutase reaction (Scheme 3).

The results of these selection experiments are mechanistically significant, insofar as they support a critical role for a cation in the mechanism of chorismate mutase. No active catalysts were found that did not have such a cation in the vicinity of the substrate's ether oxygen. These experiments allow us to conclude that a cation is crucial for high chorismate mutase activity with much greater certainty than would be possible in a conventional mutagenesis experiment, in which only single substitutions are considered. Although other, equally effective arrangements of catalytic groups lacking this functionality might be found in the future through more extensive mutagenesis, it is interesting to note that both the *E. coli* (Lee et al. 1995) and yeast chorismate mutases (Xue and Lipscomb 1995; Sträter et al. 1997) have a cationic lysine in an equivalent position in their respective active sites despite their otherwise unrelated tertiary structures.

Interesting, too, is the fact that the 10^4-fold less efficient catalytic antibody 1F7 (Haynes et al. 1994) lacks such a feature. This is not surprising, given the structure of the transition state analog used to generate this catalyst (see Scheme 1). It does, however, suggest possible strategies for producing antibodies with much higher activity. For example, additional negative charges might be designed intentionally into the hapten to elicit the catalytically essential cation. Such a cation might also be directly engineered into the 1F7 active site using site-directed mutagenesis. A selection approach may ultimately be the most effective strategy for optimizing antibody activity, however.

The Fab fragment of 1F7 has already been shown to function in the cytoplasm of a chorismate mutase-deficient yeast strain (Tang et al. 1991). When expressed at a sufficiently high level, the catalytic antibody is able to replace the missing enzyme and complement the metabolic defect. Conceivably, therefore, it can be placed under selection pressure to identify variants that have higher catalytic efficiency. Such experiments are currently under way, and preliminary results appear quite promising (Tang 1996).

9.2 Topological Redesign

Selection methods also lend themselves to problems of design. Proteins are highly overdetermined structures and their construction is basically a combinatorial problem, requiring efficient methods for simultaneous evaluation of many different alternatives. Even though a population of 10^9 molecules is small compared with the 20^{100} ($=10^{130}$) possibilities for a 100-amino-acid polypeptide, libraries of this size can provide us with statistically valuable insights into intrinsic secondary structural preferences in proteins, constraints on segments that link secondary structural elements, and optimal packing arrangements of residues in the interior of a protein (Beasley and Hecht 1997; Sauer 1996; Reidhaar-Olson and Sauer 1988).

The conversion of a dimeric chorismate mutase into a monomer illustrates the potential of selection methods for redesigning enzyme scaffolds (Scheme 4; MacBeath et al. 1998d). All known chorismate mutases are multimeric proteins. EcCM, for instance, is an intricately entwined all-α-helical homodimer with two identical active sites constructed from residues contributed by both polypeptide chains (Lee et al. 1995). It seemed conceivable that this molecule might be converted into a catalytically active monomer by inserting a flexible loop into the long H1 helix that spans the dimer. This loop would allow the N-terminal half

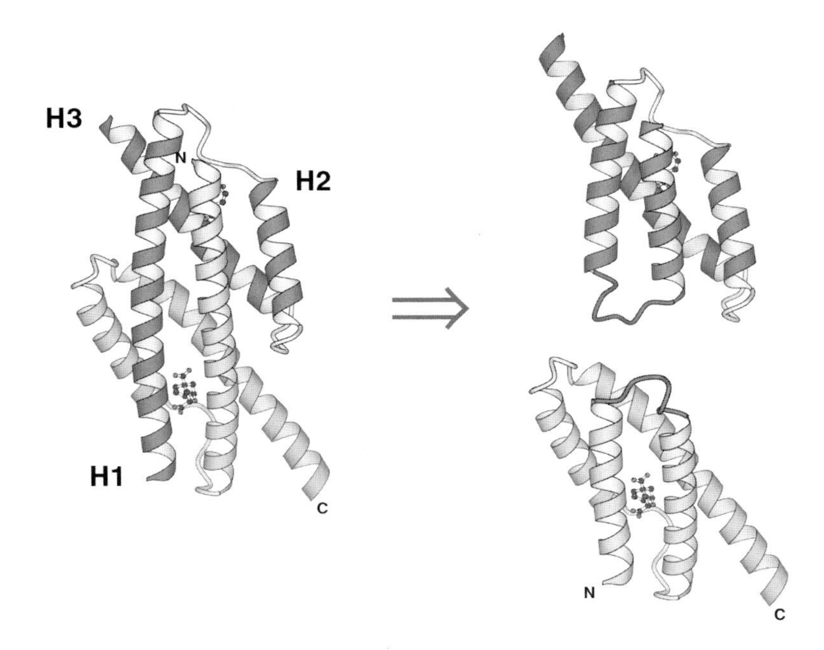

Scheme 4. Conversion of dimeric EcCM (*left*, with bound TSA inhibitor) into a monomer (MacBeath et al. 1998d). Topological redesign of the dimer involved insertion of a flexible loop into the dimer-spanning H1 helix, followed by selection for active catalysts and screening for well-behaved monomer enzymes

of H1 to bend back on itself, displacing the second polypeptide, to yield a monomeric 4-helix bundle.

The chemically interesting questions here center on the optimal length and sequence of the insert. Conceivably, any insert could work. Studies on several helix bundle proteins have shown that there are relatively few constraints on interhelical segments with respect to length and composition. For example, >90% of all possible three-residue interhelical turns were found to yield correctly folded cytochrome b-562 (Brunet et al. 1993), and studies on the 4-helix-bundle protein ROP have shown that the length of the interhelical loop is not critical (Predki and Regan 1995; Nagi and Regan 1997). Similar results were obtained in selection experiments on the solvent-exposed residues in the interhelical segment connecting the H2 and H3 helices of EcCM, although it was found that long-range tertiary interactions can impose strong constraints on allowable substitutions of buried residues (MacBeath et al. 1998a).

Although helix-helix interactions, rather than the interhelical turns, appear to dictate the overall structure of helix bundle proteins, initial attempts to convert dimeric EcCM into a monomeric chorismate mutase with a 4-helix bundle topology were surprisingly unsuccessful. There are several reasons why this might be so. First, biophysical studies on the wild-type dimer suggest that its stability derives largely from a cluster of hydrophobic residues located at the center of the dimer (MacBeath et al. 1998b). When the dimer is "cut in half", these stabilizing interactions are lost. In addition, previously buried apolar groups are exposed to solvent, which could either lead to aggregation or result in alternative conformations being preferred. A second problem may be the limited intrinsic stability of the putative monomeric 4-helix bundle. In contrast to a typical helical bundle protein, the hypothetical monomer has a very polar core. This is where the active site is located. This pocket is rich in cationic residues for binding the chorismate dianion and stabilizing additional charge that develops in the transition state. Thus, the hydrophobic interactions that provide the driving force for folding a typical protein are largely absent.

Use of a more thermostable chorismate mutase as the starting point for monomer design was considered as a possible solution to these problems. A large number of mutases related to EcCM, some from thermophilic organisms, are known (MacBeath et al. 1998b). For example, the enzyme from the archaebacterium *Methanococcus jannaschii* (MjCM), an organism that normally grows at 85°C, shares 21% identity with EcCM, and all six residues that line the active site are completely conserved. It is also 25°C more stable than EcCM (MacBeath et al. 1998b). Since the hydrophobic cores of MjCM and EcCM are very similar, interactions distant from the dimer interface must be responsible for the additional stability. These same interactions could conceivably stabilize the desired monomer.

The design of the flexible segment to be inserted into the middle of the H1 helix was based on a comparison with a helix-turn-helix motif in seryl-tRNA synthetase (MacBeath et al. 1998d). Homology modeling suggested duplicating two residues in helix H1 (Leu^{20}-Lys^{21}) and introducing a six-residue turn between the repeated units (that is, -Leu^{20}-Lys^{21}-Xaa-Xaa-Xaa-Xaa-Xaa-Xaa-Leu^{20a}-Lys^{21a}-). Two point mutations (Leu^{20a} to Glu and Ile^{77} to Arg) were also introduced as an element of negative design to destabilize the dimer and minimize unproductive aggregation. This design yields a library containing 20^6 ($=6\times10^7$) individual members. The library was constructed at the genetic level, transformed into the chorismate mutase-deficient *E. coli* strain ($>10^8$ members), and placed under selection pressure by excluding tyrosine and phenylalanine from the growth medium.

Only those cells that have a functional chorismate mutase are able to complement the chorismate mutase deficiency. In the original population, only 0.7% fulfilled this criterion (MacBeath et al. 1998d). Since those clones possessing an active enzyme have a significant advantage over those that do not under selective conditions, they can be amplified. Indeed, after 3 days of selection, >80% of the clones in the library were able to complement the chorismate mutase deficiency. This represents an enrichment of >100-fold in variants exhibiting high enzymatic activity. In essence, the selection process narrows the search for a monomeric catalyst to a more manageable subset of molecules.

Because function rather than topology serves as the basis for selection, a subsequent screen of representative clones by size-exclusion chromatography was necessary to identify a monomeric chorismate mutase. Interestingly, gel filtration of wild-type MjCM revealed that 75% of the protein is produced as a misfolded and catalytically inactive aggregate; only 25% elutes as an active dimer (MacBeath et al. 1998b,d). In contrast, none of the 26 proteins that emerged from the selection showed any tendency to form higher-order aggregates, although most of these proteins were dimeric or mixtures of dimers and monomers. Only one of the proteins exhibited an elution profile expected for a monomeric enzyme. The six-residue insert in this molecule (mMjCM, for monomeric MjCM) had the sequence Ala-Arg-Trp-Pro-Trp-Ala (MacBeath et al. 1998d). The presence of a proline residue is compatible with a turn structure, while the large number of apolar groups might interact with the exposed hydrophobic face created by disruption of the subunit interface.

The combination of selection (>100-fold enrichment) and screening (one in 26) shows that fewer than 0.05% of the possible turn sequences are capable of yielding well-behaved, monomeric proteins. This result contradicts the simple expectation that most interhelical turn sequences are functionally equivalent. It also underscores in dramatic fashion the advantage of the selection approach. Individual characterization of 1000–10,000 proteins would represent a daunting experimental undertaking.

The monomeric protein identified by selection was characterized by a range of techniques (MacBeath et al. 1998d). The best evidence that the molecule is monomeric was provided by analytical ultracentrifugation. Sedimentation equilibrium data gave an average molecular mass of 14,700±850 daltons for mMjCM, which can be compared with the value of 13,174 daltons expected for the monomer. Under the same conditions, the wild-type protein gave 23,000±1100, in good agreement with the value of 24,212 for a dimer. Like the parent enzyme, mMjCM is α-helical and undergoes reversible chemical denatuaration. The cooperative transition from folded to unfolded states is independent of protein concentration but relatively broad, consistent with a protein with a relatively small hydrophobic score. The free energy of unfolding [$\Delta G_U(H_2O)$ = 2.7 kcal/mol] is fairly low for a typical protein of this size, which is also in agreement with this conclusion.

Perhaps most impressively, mMjCM is highly active as a catalyst. It exhibits saturation kinetics for the chorismate mutase reaction with a k_{cat} of 3.2 s^{-1} and a K_m of 170 μM. These values compare remarkably well with those for wild-type MjCM (k_{cat} = 3.2 s^{-1} and a K_m of 50 μM) and for EcCM (k_{cat} = 9.0 s^{-1} and a K_m of 300 μM). These results demonstrate how effectively design, random mutation, and selection can be combined to achieve topological redesign of an enzyme.

Of course, the monomeric chorismate mutase is also an effective catalyst in vivo. After all, that is how it was originally identified. Interestingly, however, cells harboring the monomer grow much more effectively than cells containing the wild-type MjCM dimer in the absence of tyrosine and phenylalanine. Although the precise reasons for this difference are unknown, it is likely a consequence of the fact that the wild-type protein aggregates extensively, whereas the monomer shows no tendency to do so. This observation illustrates another important advantage of genetic selection. Not only is catalytic activity maximized in such an approach, but problems associated with aggregation or toxicity during production are simultaneously minimized.

Analogous experiments in which the length of the interhelical segment was varied have yielded a variety of other monomeric chorismate mutases in addition to a hexamer (MacBeath et al. 1998c). Proteins with redesigned topology may serve as model systems for studying how structure, stability, and function interrelate. Such experiments may also shed light on the evolutionary origins of multimeric proteins.

9.3 Perspectives

Over the last two decades, a variety of strategies for designing enzymes have emerged. While much progress has been made in the area of de novo design (DeGrado et al. 1999), much remains to be learned. In general, efforts to redesign existing proteins (Bell and Hilvert 1994) and exploitation of the diversity and selectivity of the mammalian immune system (Hilvert 2000; Lerner et al. 1991) have been much more successful in providing access to *functional* macromolecules. Even so, the first-generation catalysts are typically orders of magnitude less active than corresponding enzymes, where available. General strategies for optimizing these molecules would obviously be of great value.

Approaches based on genetic selection are like to be especially powerful in this regard. In principle, selection schemes can be imagined for virtually any metabolic process, as well as for transformations involving the creation or destruction of a vital nutrient or toxin. As shown here for chorismate mutase enzymes, these approaches can be applied with great success to problems of chemical mechanism and to design. They can also be used to increase stability, enhance production (Martineau et al. 1998), tailor substrate specificity (Yano et al. 1998), or even confer completely new chemical activities to a given active site (Altaminrano et al. 2000). Selection methods have even also been applied with great success to the production of catalysts based on oligonucleotide rather than polypeptide backbones (Wilson and Szostak 1999). Future applications of evolutionary approaches to the study of macromolecules are expected to contribute to our knowledge of structure-function relationships in these systems and to provide access to novel catalysts for a wide range of practical applications.

References

Addadi L, Jaffe EK, Knowles JR (1983) Secondary tritium isotope effects as probes of the enzymic and nonenzymic conversion of chorismate to prephenate. Biochemistry 22:4494–4501

Altaminrano MM, Blackburn JM, Aguayo C, Fersht AR (2000) Directed evolution of new catalytic activity using the α/β-barrel scaffold. Nature 403:617–622

Bartlett PA, Nakagawa Y, Johnson CR, Reich SH, Luis A (1988) Chorismate mutase inhibitors: synthesis and evaluation of some potential transition-state analogues. J Org Chem 53:3195–3210

Beasley JR, Hecht MH (1997) Protein design: The choice of de novo sequences. J Biol Chem 272:2031–2034

Bell IM, Hilvert D (1994) New biocatalysts via chemical modification. In: Behr J-P (ed) The lock-and-key principle. Wiley, Sussex, pp 73–88

Bowdish K, Tang Y, Hicks JB, Hilvert D (1991) Yeast expression of a catalytic antibody with chorismate mutase activity. J Biol Chem 266:11910–11908

Brunet AP, Huang ES, Huffine ME, Loeb JE, Weltman RJ, Hecht MH (1993) The role of turns in the structure of an α-helical protein. Nature 364:355–358

Chook YM, Ke H, Lipscomb WN (1993) Crystal structures of the monofunctional chorismate mutase from Bacillus subtilis and its complex with a transition state analog. Proc Natl Acad Sci U S A 90:8600–8603

Chook YM, Gray JV, Ke H, Lipscomb WN (1994) The monofunctional chorismate mutase from Bacillus subtilis. Structure determination of chorismate mutase and its complexes with a transition state analog and prephenate, and implications for the mechanism of the enzymatic reaction. J Mol Biol 240:476–500

Cload ST, Liu DR, Pastor RM, Schultz PG (1996) Mutagenesis study of active site residues in chorismate mutase from Bacillus subtilis. J Am Chem Soc 118:1787–1788

Copley SD, Knowles JR (1985) The uncatalyzed Claisen rearrangement of chorismate to prephenate prefers a transition state of chairlike geometry. J Am Chem Soc 107:5306–5308

Copley SD, Knowles JR (1987) The conformational equilibrium of chorismate in solution: implications for the mechanism of the non-enzymic and the enzyme-catalyzed rearrangement of chorismate to prephenate. J Am Chem Soc 109:5008–5013

DeGrado WF, Summa CM, Pavone V, Nastri F, Lombardi A (1999) De novo design and structural characterization of proteins and metalloproteins. Annu Rev Biochem 68:779–819

Gray JV, Eren D, Knowles JR (1990) Monofunctional chorismate mutase from Bacillus subtilis: kinetic and ^{13}C NMR studies on the interactions of the enzyme with its ligands. Biochemistry 29:8872–8878

Gustin DJ, Mattei P, Kast P, Wiest O, Lee L, Cleland WW, Hilvert D (1999) Heavy atom isotope effects reveal a highly polarized transition state for chorismate mutase. J Am Chem Soc 121:1756–1757

Haynes MR, Stura EA, Hilvert D, Wilson IA (1994) Routes to catalysis: structure of a catalytic antibody and comparison with its natural counterpart. Science 263:646–652

Hilvert D (2000) Critical analysis of antibody catalysis. Annu Rev Biochem 69 (in press)

Hilvert D, Carpenter SH, Nared KD, Auditor M-TM (1988) Catalysis of concerted reactions by antibodies: the Claisen rearrangement. Proc Natl Acad Sci U S A 85:4953–4955

Kast P, Asif-Ullah M, Jiang N, Hilvert D (1996) Exploring the active site of chorismate mutase by combinatorial mutagenesis and selection: the importance of electrostatic catalysis. Proc Natl Acad Sci U S A 93:5043–5048

Lee AY, Karplus AP, Ganem B, Clardy J (1995) Atomic structure of the buried catalytic pocket of Escherichia coli chorismate mutase. J Am Chem Soc 117:3627–3628

Lerner RA, Benkovic SJ, Schultz PG (1991) At the crossroads of chemistry and immunology: catalytic antibodies. Science 252:659–667

MacBeath G, Kast P, Hilvert D (1998a) Exploring sequence constraints on an interhelical turn using in vivo selection for catalytic activity. Protein Sci 7:325–335

MacBeath G, Kast P, Hilvert D (1998b) A small thermostable and monofunctional chorismate mutase from the archeon Methanococcus jannaschii. Biochemistry 37:10062–10073

MacBeath G, Kast P, Hilvert D (1998c) Probing enzyme quaternary structure by mutagenesis and selection. Protein Sci 7:1757–1767

MacBeath G, Kast P, Hilvert D (1998d) Redesigning enzyme topology by directed evolution. Science 279:1958–1961

Martineau P, Jones P, Winter G (1998) Expression of an antibody fragment at high levels in the bacterial cytoplasm. J Mol Biol 280:117–127

Nagi AD, Regan L (1997) An inverse correlation between loop length and stability in a four-helix-bundle protein. Fold Des 2:67–75

Predki PF, Regan L (1995) Redesigning the topology of a four-helix-bundle protein: monomeric Rop. Biochemistry 34:9834–9839

Reidhaar-Olson JF, Sauer RT (1988) Combinatorial cassette mutagenesis as a probe of the informational content of protein sequences. Science 241:53–57

Sauer RT (1996) Protein folding from a combinatorial perspective. Fold Des 1:R27–R30

Sträter N, Schnappauf G, Braus G, Lipscomb WN (1997) Mechanisms of catalysis and allosteric regulation of yeast chorismate mutase from crystal structures. Structure 5:1437–1452

Tang Y (1996) Evolutionary studies with a catalytic antibody. PhD, Scripps Research Institute

Tang Y, Hicks JB, Hilvert D (1991) In vivo catalysis of a metabolically essential reaction by an antibody. Proc Natl Acad Sci U S A 88:8784–8786

Weiss U, Edwards JM (1980) The biosynthesis of aromatic amino compounds. Wiley, New York

Wiest O, Houk KN (1994) On the transition state of the chorismate-prephenate rearrangement. J Org Chem 59:7582–7584

Wiest O, Houk KN (1995) Stabilization of the transition state of the chorismate-prephenate rearrangement: an ab initio study of enzyme and antibody catalysis. J Am Chem Soc 117:11628–11639

Wilson DS, Szostak JW (1999) In vitro selection of functional nucleic acids. Annu Rev Biochem 68:611–647

Xue Y, Lipscomb WN (1995) Location of the active site of allosteric chorismate mutase from Saccharomyces cerevisiae, and comments on the catalytic and regulatory mechanisms. Proc Natl Acad Sci U S A 92:10595–10598

Xue Y, Lipscomb WN, Graf R, Schnappauf G, Braus G (1994) The crystal structure of allosteric chorismate mutase at 2.2 Å resolution. Proc Natl Acad Sci U S A 91:10814–10818

Yano T, Oue S, Kagamiyama H (1998) Directed evolution of an aspartate aminotransferase with new substrate specificities. Proc Natl Acad Sci U S A 95: 5511–5515

10 The Biotechnological Exploitation of Medicinal Plants

T.M. Kutchan

10.1 Introduction

During the second half of this century, the world population has increased from 2.5 billion in 1950 to approximately 6 billion in the late 1990s. The global population growth peaked at 2.1% per year by the late 1960s. This annual rate has decreased by approximately one third to 1.4%. Even with this reduced growth rate, however, a global population of 8–10 billion people by the year 2050 is plausible (Ruttan 1999; Federoff and Cohen 1999)]. Cereal yield almost doubled between 1960 and 1990 but has not kept pace with population growth during this same period (Kishore and Shewmaker 1999). The next half-century will most certainly see another 2–4 billion inhabitants on our planet, accompanied by fewer nonrenewable resources with which to provide the additional pharmaceuticals, fuel, and fiber that will be necessary to maintain at least the current standard of living. Estimates are that we have depleted approximately 50% of our oil reserves in the past century, indicating that complete depletion may occur in the current one.

Advances in plant molecular biology and genetic engineering (plant biotechnology) in the past 15 years form the basis for development of sustainable agriculture and renewable resources over the next 50 years. Since 1995, plant biotechnology has produced crops with improved agronomic traits. These are mostly single-gene traits such as herbicide or pest resistance that improve grower productivity. Improved food-processing single-gene traits have also been introduced, for example with processing tomatoes.

Plant biotechnological expectations for the new millennium will be the use of plants as factories for the production of pharmaceuticals and specialty chemicals. We look to a shift away from fossil fuel towards plants for a more readily renewable source of carbon for chemical and pharmaceutical industry. Plants are nature's best manufacturing system. They were the sole source of food, fuel, and fiber to mankind for many centuries until the use of fossil fuel began. Using plants as factories would enable us to harness solar energy to meet future global fuel, fiber, and specialty chemical needs. In addition, plants as raw material can be domestically produced in many countries and would reduce pollution due to the petrochemical industry.

The chemical diversity provided by plant secondary metabolites far surpasses that of the highly reduced carbon of fossil fuel. The structures of over 200,000 plant natural products have already been identified, thereby making the biotechnological potential of plant secondary metabolism enormous. The classes of compounds that are already particularly amenable to exploitation by food and the pharmaceutical industry are the flavonoids, isoflavonoids, glucosinolates, terpenoids, and alkaloids. Advances have been made in our understanding of how plants biosynthesize many secondary metabolites, and we are beginning to accumulate some of the biosynthetic genes that are necessary to any genetic manipulation.

According to current thought, there are several ways in which plants can be biotechnologically exploited for the production of pharmaceuticals. Selected examples of the results to date aimed at this exploitation are reviewed herein.

10.2 Innovations in Pharmaceutical Production

10.2.1 Antibodies Expressed in Plants

A modern combination of molecular genetics and agriculture is providing the basis for novel methods of production of a variety of pharmaceutically important peptides and proteins. The production of recombinant proteins in plants is called "molecular farming". The first experimental plant of choice was tobacco, due to optimized transformation and regeneration protocols that resulted from many years of research worldwide. For human peptide hormones, such as insulin for example, integrated *Agrobacterium*-mediated expression would be suitable. Expression could be achieved in plant cell suspension culture or in field-grown plants. The plantation growth characteristics and harvesting of tobacco are well understood from at least a century of experience within the tobacco industry. National guidelines appropriate to agriculture-based manufacturing are being developed in the USA and are expected to be less stringent than those for the isolation of recombinant proteins from bacteria.

A more challenging problem for green biotechnology is, however, the production of antibodies in plants for the development of new immunotherapeutic and immunodiagnostic reagents. Functional full-size recombinant antibodies were first expressed in transgenic plants in 1989 (Hiatt et al. 1989). In many applications of recombinant antibodies, the Fc (complement-binding) region of the immunoglobulin is not necessary. Focus has therefore been on the expression of single-chain variable regions, so-called scFv fragments. Anti-phytochrome scFv antibody was expressed in up to 0.5% of total soluble protein in transgenic tobacco leaf (Firek et al. 1993). In cell suspension cultures derived from the transgenic plants, functional antibody was secreted into the culture medium. Human antibodies have been expressed in transgenic corn, and humanized antibodies have been expressed in transgenic soybean. If gene expression can be directed to the seed, then the recombinant gene product can be stored in grain over long periods of time at low cost. In the case of expression of tumor-specific antibodies, however, speed becomes essential, and the months of work that are involved in the production of stable tobacco transformants are too long. The development of viral transfection vectors in the past decade provides a method for accelerated gene transfer to cultivated crop plants (Turpen 1999). Recombinant vectors based on tobacco mosaic virus (TMV) are now used to transfect field-grown tobacco plants in quantities that have commercial significance.

The following example of how this technology could be exploited was published this year. B-cell malignancies are typically incurable and are characterized by variability in treatment and prognosis. They share, however, a common tumor-specific marker, a unique cell-surface immunoglobulin. Tumor surface immunoglobulin has been used to treat patients in chemotherapy-induced remission. Extended application of this therapy requires rapid and reliable production of therapeutic quantities of antibody. An idiotype-specific scFv fragment of the immunoglobulin from a 38C13 mouse B-cell lymphoma was expressed in tobacco plants using TMV transfection within 4 weeks after molecular cloning (McCormick et al. 1999). The plant-derived protein generated an antibody response in injected animals that was relevant to the tumor idiotype. This rapid production of tumor-specific proteins may provide a viable strategy for the treatment of non-Hodgkin's lymphoma. This suggests that a TMV-based transient expression system could be a feasible means of commercially producing patient-specific antibodies for therapeutic use. Clinical trials should begin within a year (Fischer et al. 1999).

10.2.2 Vaccines Expressed in Plants

An alternative to eukaryotic cell culture systems for the production of safe vaccines at a lower cost may be provided by plants. Rapid, cost-effective production can be achieved with recombinant TMV transfection. To investigate the possibilities for a model disease, malarial epitopes were genetically engineered on the surface of TMV. Tobacco plants infected with recombinant virus produced viral particles containing coat protein and malarial epitope fusion protein (Turpen et al. 1995). TMV coat protein is a highly immunogenic carrier, which shows the potential for its use in the production of stable subunit vaccines.

Genetically engineered plant systems are currently being developed for the production of safe, stable, and cost-effective edible vaccines. To investigate the possibilities for vaccine production in a model system, the potato has been used to express Norwalk virus capsid protein (Mason et al. 1996). Norwalk virus causes epidemic acute gastroenteritis in humans. When transgenic potato tubers expressing recombinant Norwalk virus-like particle were fed to mice, the animals developed serum IgG specific for the virus-like particle. An additional example of plant-produced edible vaccines is the expression of *Escherichia coli* heat-labile enterotoxin B subunit in the potato. Mice were fed transgenic potato tubers and challenged with heat-labile enterotoxin. The mice were partially protected against the effects of the toxin, suggesting that an edible vaccine against enterotoxigenic *E. coli* is feasible (Mason et al. 1998). Although the raw transgenic potato tubers used in these studies would be found unpalatable by most human beings, the development of an oral vaccine in a plant normally eaten in uncooked form would offer the hope of convenient, practical, and less expensive vaccination programs in developing countries.

10.3 Natural Product Combinatorial Biosynthesis

10.3.1 Microbial Polyketides

Micro-organisms synthesize a plethora of secondary metabolites that have a role in chemical defense, offense, and communication. The compounds can also possess physiological activities that make them useful pharmaceuticals. Microbial-produced antibiotics have been extremely valuable drugs over the past 60 years, but their use in the future is threatened by ever-increasing multidrug resistances. The targeted screening programs of the pharmaceutical industry yield fewer and fewer new drugs with novel mechanisms of action, which would be essential to overcoming bacterial drug resistance.

A commercially important class of microbial-derived metabolites are the polyketide antibiotics. The polyketides are polymerized simple fatty acids. Thousands of polyketides have been discovered thus far in nature. The genetics of antibiotic production by *Streptomyces* is now well understood and formed the basis for a new approach to the production of novel antibiotics. All micro-organisms that biosynthesize polyketides possess a core set of genes called minimal polyketide synthase, or PKS genes. This is comprised of a ketosynthase, an acyltransferase, and an

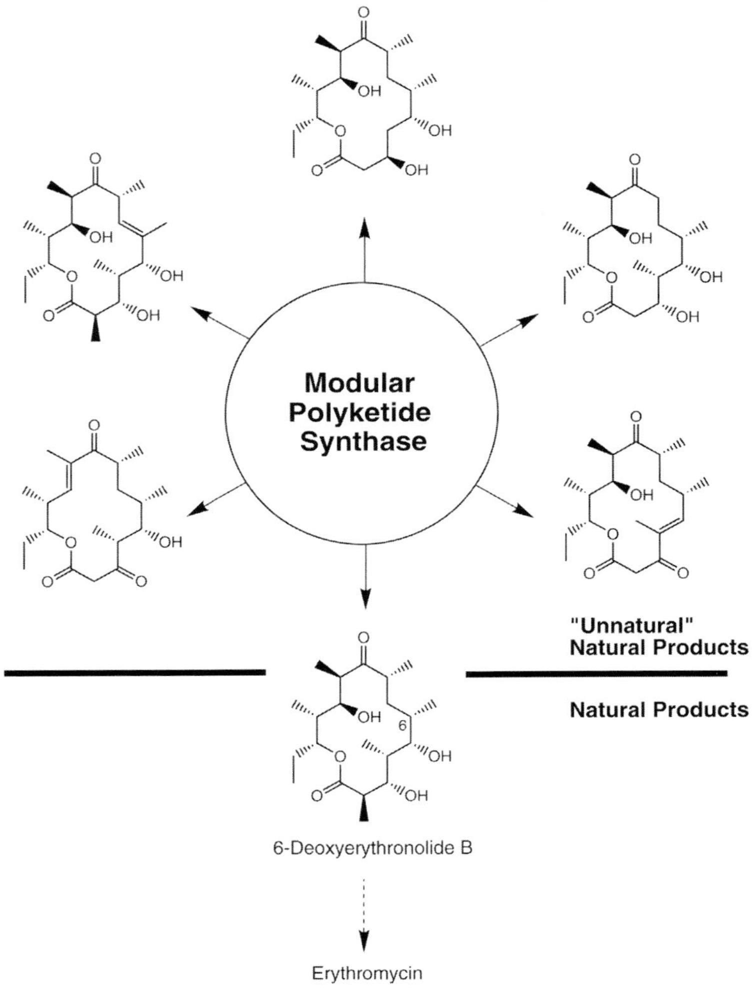

Modular Polyketide Synthase

"Unnatural" Natural Products

Natural Products

6-Deoxyerythronolide B

Erythromycin

Fig. 1. Examples of novel polyketides that have been produced by combinatorial biosynthesis. Up to two modules of the 6-deoxyerythronolide B polyketide synthase gene cluster were exchanged with modules involved in other polyketide biosynthetic pathways to result in the synthesis of up to 50 new polyketides. Examples of five of these new molecules are shown here. These structures have not yet been discovered in nature and are referred to as "unnatural" natural products acyl carrier protein. The PKS gene products catalyze the formation of the polyketide backbone structure (Fig. 1). The myriad of structures that are produced in a species-specific manner are formed by additional unique gene products, such as ketoreductases, dehydratases, and enoyl reductases, that are characteristic to particular bacterial species. The modular functional domain structure of these biosynthetic enzymes led several laboratories to attempt module swapping between various polyketide-producing organisms. The idea was to arrange modules through genetic engineering into an unnatural constellation in a heterologous host in order to produce polyketide structures not yet discovered in nature. The recombination of related, but not identical, gene products to produce novel metabolites is called combinatorial biosynthesis and has been tremendously successful in the polyketide field (Hutchinson 1999). Well over 100 new "unnatural" natural products have been thus far produced through combinatorial biosynthesis, demonstrating the potential for synthesis of new chemotypes by this methodology (McDaniel et al. 1999).

10.3.2 Plant Alkaloids

The biosynthesis of complex natural products in plants involves enzymes that, due to their high substrate specificity, appear to have evolved for secondary metabolism. In some cases, transformations are catalyzed that are difficult to reproduce chemically in equivalent yield or enantiomeric excess. A wealth of knowledge has been gained on the enzymology of selected classes of monoterpenoid indole- and isoquinoline-alkaloids (Kutchan 1998). In more recent years, this knowledge has been complemented by results of molecular genetic investigations. We understand very well which types of enzymes are catalyzing many of the multiple steps of alkaloid formation (Chou and Kutchan 1998). Oxidoreductases frequently catalyze reactions that result in the formation of the species-specific parent ring systems of the various classes of alkaloids. These have been shown to be either cytochrome P-450-dependent oxidases or flavin-dependent oxidases. The parent ring systems are decorated by functional groups such as hydroxyl moieties. Mono-oxygenation of alkaloids is achieved by the action of either cytochrome P-450-dependent hydroxylases or 2-oxo-glutarate-dependent dioxy-

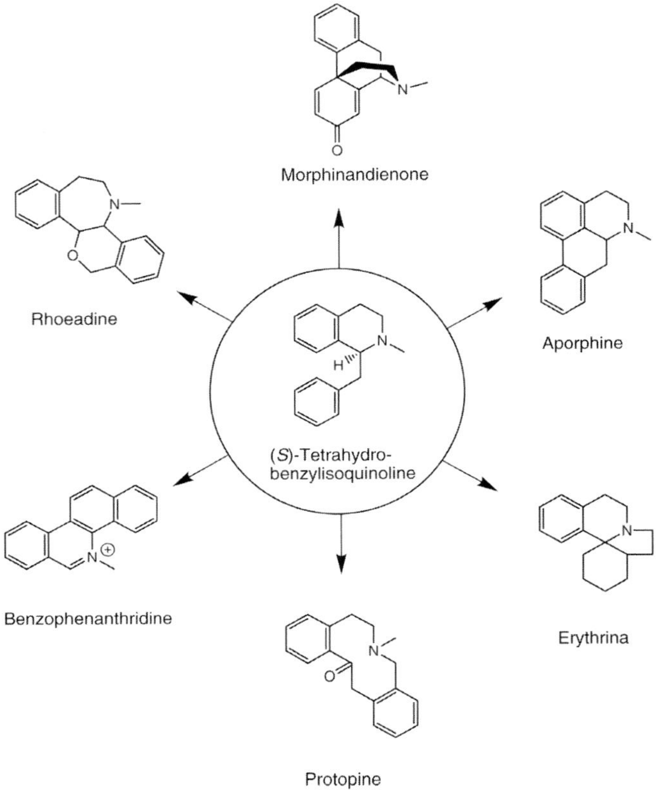

Fig. 2. Examples of selected carbon skeletons that are biosynthetically derived from the simple benzyltetrahydroisoquinoline nucleus in plants. As for polyketides, isoquinoline alkaloid structures may be amenable to combinatorial biosynthesis for the production of novel compounds when more biosynthetic genes have been identified

genases. *O*- and *N*-methylation is carried out by S-adenosylmethionine-dependent methyltransferases. As the nucleotide sequences of genes encoding various enzymes of alkaloid biosynthesis become known, apparent conserved regions become evident. Knowledge of these conserved regions facilitates the generation, by the polymerase chain reaction, of cDNAs encoding alkaloid biosynthetic genes without the need to first purify the native protein. This can greatly accelerate the speed at which at we isolate new genes and reduces the amount of fresh plant material necessary for the isolation.

Analogous to microbial polyketide formation, alkaloid classes also have central pathways that involve a subset of biosynthetic enzymes and encoding genes that are present in all plant species that form that particular alkaloid class (Fig. 2). In a species-specific manner, additional genes are present, encoding enzymes that determine exactly which alkaloids a particular plant species will produce. It is conceivable that the species-specific plant alkaloid genes from a variety of unrelated plant species can be combined in a way in a heterologous system which has not yet been found in nature to produce novel alkaloid structures. The limiting factor to this approach is the number of alkaloid genes that are currently available. It is clear that we need even faster systems with which to identify plant gene function.

10.4 Metabolic Engineering of Medicinal Plants

10.4.1 Deadly Nightshade, *Atropa belladonna*

The current commercial source of the anticholinergic alkaloid scopolamine, which is used in the prevention and treatment of motion sickness, is the plant species *Duboisia*, originally cultivated in Australia. Certain tropane alkaloid-producing plant species, such as deadly nightshade, *Atropa belladonna*, accumulate the scopolamine biosynthetic precursor hyoscyamine. (In racemic form, hyoscyamine is called atropine, previously used to dilate the pupil of the eye.) To address whether the expression of a transgene in a medicinal plant could alter the alkaloid pattern of the plant, a cDNA encoding the enzyme hyoscyamine 6β-hydroxylase that converts hyoscyamine to scopolamine was introduced via *Agrobacterium*-mediated transformation into *A. belladonna* (Fig. 3).

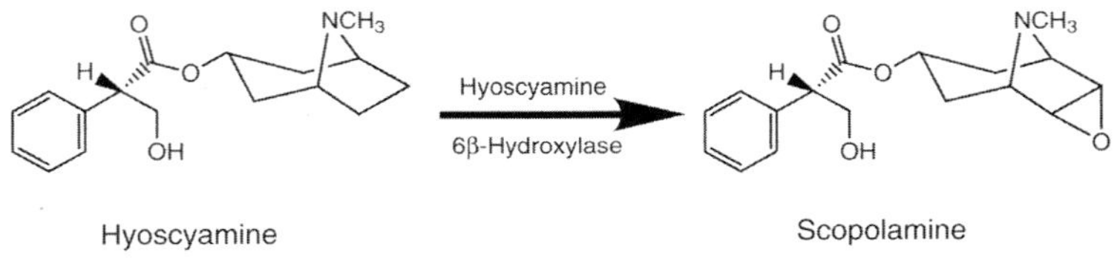

Hyoscyamine

Hyoscyamine 6β-Hydroxylase

Scopolamine

Fig. 3. The ultimate step in the biosynthesis of the tropane alkaloid scopolamine in plants. The enzyme hyoscyamine 6β-hydroxylase is a 2-oxoglutarate-dependent diooxygenase that first hydroxylates hyoscyamine, then catalyzes epoxide formation

The resultant transgenic plants contained elevated levels of scopolamine (Yun et al. 1992). This use of *A. belladonna* is clearly as a model system to demonstrate metabolic engineering in medicinal plants. The more commercially significant experiments will come when transformation and regeneration protocols for *Duboisia* have been developed.

10.4.2 Opium Poppy, *Papaver somniferum*

The opium poppy, *Papaver somniferum,* is one of our most important renewable resources for the production of pharmaceutical alkaloids. The narcotic analgesic morphine, the antitussive and narcotic analgesic codeine, the antitussive and apoptosis-inducer noscapine, and the vasodilator papaverine are the most important physiologically active alkaloids from *P. somniferum*. Of these four alkaloids, only papaverine is chemically synthesized for commercial use. Per annum, 90%–95% of the 160 tons of morphine that are legally purified from *P. somniferum* is chemically converted to codeine. Codeine is then either used directly or further chemically modified to a variety of antitussives and analgesics. Codeine is the direct biosynthetic precursor of morphine. The final step of morphine biosynthesis is the demethylation of codeine, but codeine does not accumulate to any significant degree in the plant. *P. somniferum* produces more than 100 different alkaloids that are biosynthesized, as is morphine, from the aromatic amino acid L-tyrosine. Given the complex alkaloid pattern and the commercial importance of *P. somniferum*, it is a prime target for metabolic engineering of secondary metabolism.

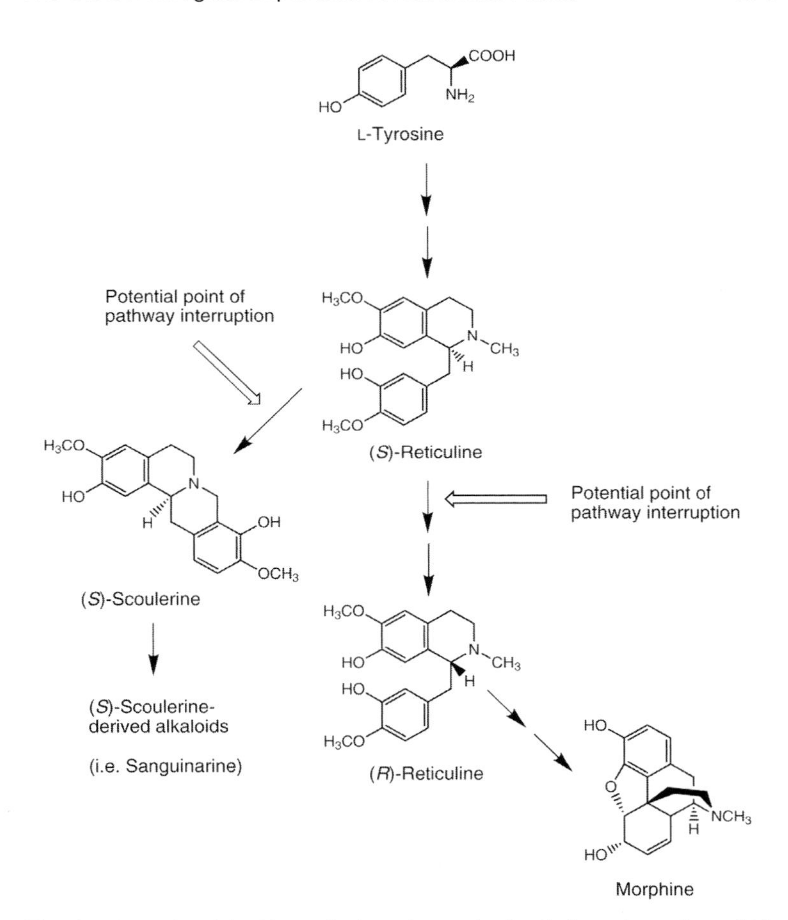

Fig. 4. Schematic of the biosynthetic pathway that leads from the amino acid l-tyrosine to the more than 100 isoquinoline alkaloids in the opium poppy *Papaver somniferum*. Two potential points of genetic interruption to effect metabolic engineering of the highly branched alkaloid biosynthetic pathway are indicated

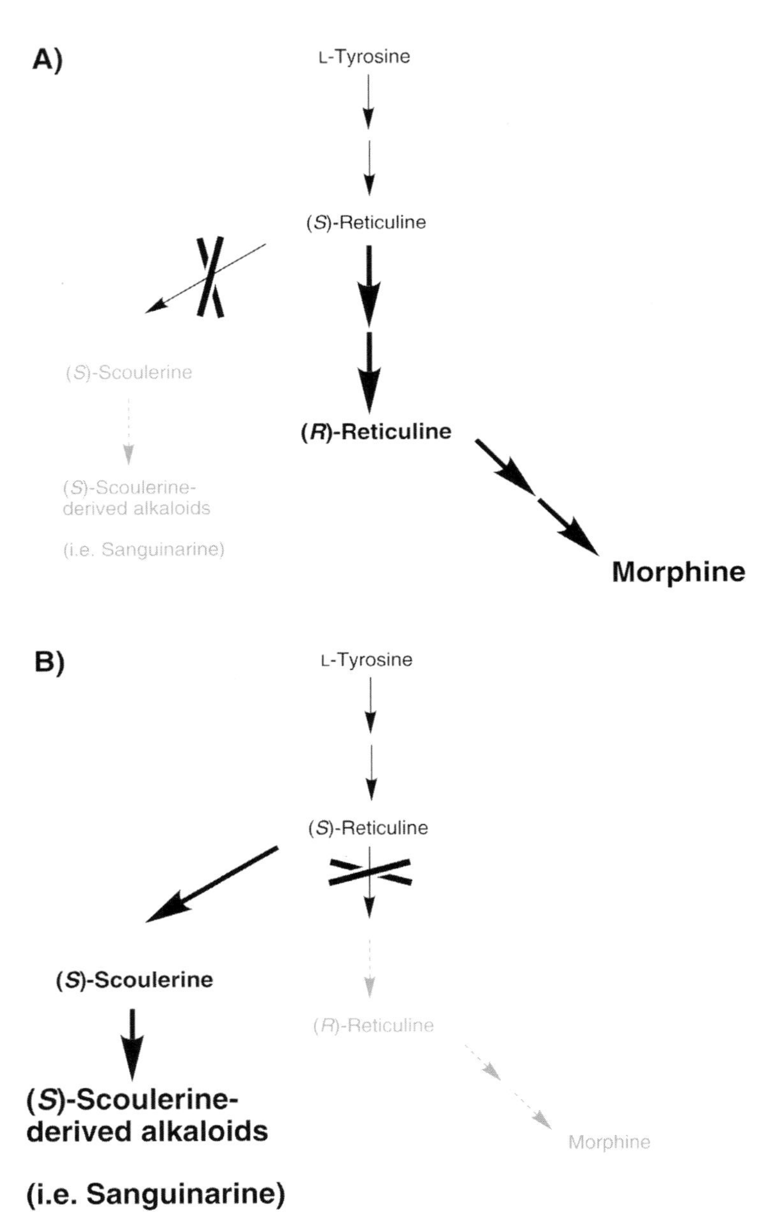
A)
L-Tyrosine
(S)-Reticuline
(S)-Scoulerine
(R)-Reticuline
(S)-Scoulerine-
derived alkaloids
(i.e. Sanguinarine)
Morphine
B)
L-Tyrosine
(S)-Reticuline
(S)-Scoulerine
(R)-Reticuline
(S)-Scoulerine-
derived alkaloids
(i.e. Sanguinarine)
Morphine

Along the biosynthetic pathway that leads from L-tyrosine to the vast array of benzylisoquinoline alkaloids of *P. somniferum*, a bifurcation exists at the position of (*S*)-reticuline (Fig. 4). (*S*)-Reticuline can be enzymatically oxidized at C1 to form the dihydroreticulinium ion, which can then be stereospecifically reduced by dihydroreticuline reductase to the correct epimer for morphine biosynthesis, (*R*)-reticuline. Alternatively, (*S*)-reticuline can be oxidized at the *N*-methyl group by the berberine bridge enzyme to the intermediary methylene iminium ion, which is then cyclized to (S)-scoulerine. Using gene antisense or co-suppression technologies, the biosynthetic pathway could be interrupted in transgenic plants at the position of the berberine bridge enzyme, which may result in an increased flow of carbon into the morphine biosynthetic pathway (Fig. 5a). Likewise, the morphine biosynthetic pathway could be interrupted to produce transgenic *P. somniferum* plants that produce no morphine and would be suitable for seed production for the food industry (Fig. 5b). The idea of tailored *P. somniferum* alkaloid profiles would also have application at later stages in the morphine biosynthetic pathway. Transgenic plants that produce either thebaine or codeine as the major alkaloid are also conceivable, given the availability of the proper biosynthetic genes and high-efficiency transformation and regeneration protocols. We are beginning to accumulate a number of the genes of alkaloid biosynthesis in *P. somniferum*, so that metabolically engineered opium poppy is close to being a reality (Dittrich and Kutchan 1991; Rosco et al. 1997; Pauli and Kutchan 1998; Unterlinner et al. 1999; Frick and Kutchan 1999).

◄

Fig. 5a,b. Proposed schemes of metabolic engineering of *P. somniferum*. **a** Using antisense or co-suppression technologies, the branch point in the morphine biosynthetic pathway that leads to (*S*)-scoulerine could be eliminated, possibly leading to poppies that accumulate more morphine. **b** Likewise, the pathway that specifically leads to morphine could be suppressed, thereby channeling more carbon into (*S*)-scoulerine-derived alkaloids such as the benzophenanthridine sanguinarine. Morphine-free poppies could find use in the food and oils industry

10.5 Future Innovations in Plant Natural Product Gene Isolation and Identification

We are currently at the early stages of several very exciting technologies in plant science. Molecular farming is soon to become a commercial reality, creating new "medicinal plants" out of traditional crop species such as tobacco, corn, and soybean. The green biotechnological production of pharmaceutically important recombinant peptides and proteins will certainly be commercially feasible within the next several years. The synthesis of novel microbial secondary metabolites belonging to the polyketide class through combinatorial biosynthesis has been amply demonstrated. The upper limit of novel structures that are theoretically possible to biosynthesize lies between 10^7–10^{14}. This new methodology could potentially serve as a very rich source of new, biologically active low-molecular-weight compounds.

The metabolic engineering of plant natural product biosynthetic pathways has the potential to yield novel structures or to develop plants with improved natural product profiles. A rate-limiting step in the exploitation of natural product pathways is the speed with which we can isolate the biosynthetic genes. Unlike antibiotic biosynthetic genes in microbes, plant natural product biosynthetic genes do not appear to be tightly clustered. It has not been possible to isolate "cassettes" of plant genes. Although the rate of plant natural product biosynthetic gene discovery is greater now than 10 years ago, new methodologies are needed.

The most promising new method of gene discovery lies in plant genomics. Genome-sequencing programs alone will not, however, lead to the immediate identification of the species-specific genes of natural product biosynthesis. For this, novel methods of functional genomics are necessary. One very promising technology is the use of plant viral vectors for high-throughput gene function discovery (Pogue et al. 1998). Normalized cDNA libraries can potentially be singly introduced into plants to produce gain-of-function/loss-of-function phenotypes. That an approach of this type may be feasible has been demonstrated with carotenoid biosynthesis in *Nicotiana benthamiana*. A gain-of-function phenotype was achieved by functional integration of non-native capsanthin-capsorubin synthase into *Nicotiana benthamiana* (Kumagai et al. 1998). A loss-of-function phenotype was attained by inhibition of carotenoid biosynthesis with virus-derived RNA (Kumagai et al. 1995). The challenge to this technology is to broaden the range of plant species that can be subjected to gain-of-function/loss-of-function analysis with viral vectors by increasing the vector palate.

International efforts in plant biotechnology have begun in recent years to include the exploitation of plant secondary pathways. This rich source of chemical structures will most certainly serve as raw materials and final products for the chemical and pharmaceutical industries. Molecular farming will also most certainly supplement tissue culture and microbial heterologous hosts for production of medical and diagnostic peptides and proteins and, through metabolic engineering, will become a source of readily renewable chemical resources.

References

Chou W-M, Kutchan TM (1998) Enzymatic oxidations in the biosynthesis of complex alkaloids. Plant J 15:289–300

Dittrich H, Kutchan TM (1991) Molecular cloning, expression and induction of berberine bridge enzyme, an enzyme essential to the formation of benzophenanthridine alkaloids in the response of plants to pathogenic attack. Proc Natl Acad Sci U S A 88:9969–9973

Federoff NV, Cohen JE (1999) Plants and population: is there time? Proc Natl Acad Sci U S A 96:5903–5907

Firek S, Draper J, Owen MR, Gandecha A, Cockburn B, Whitelam GC (1993) Secretion of a functional single-chain Fv protein in transgenic tobacco plants and cell suspension cultures. Plant Mol Biol 23:861–870

Fischer R, Liao Y-C, Hoffmann K, Schillberg S, Emans N (1999) Molecular farming of recombinant antibodies in plants. Biol Chem 380:825–839

Frick S, Kutchan TM (1999) Molecular cloning and functional expression of *O*-methyltransferases common to isoquinoline alkaloid and phenylpropanoid biosynthesis. Plant J 17:329–339

Hiatt A, Cafferkey R, Bowdish K (1989) Production of antibodies in transgenic plants. Nature 342:76–78

Hutchinson CR (1999) Microbial polyketide synthases: more and more prolific. Proc Natl Acad Sci U S A 96:3336–3338

Kishore GM, Shewmaker C (1999) Biotechnology: enhancing human nutrition in developing and developed worlds. Proc Natl Acad Sci U S A 96:5968–5972

Kumagai MH, Donson J, della-Cioppa G, Harvey D, Hanley K, Grill LK (1995) Cytoplasmic inhibition of carotenoid biosynthesis with virus-derived RNA. Proc Natl Acad Sci U S A 92:1679–1683

Kumagai MH, Keller Y, Bouvier F, Clary D, Camara B (1998) Functional integration of non-native carotenoids into chloroplasts b viral-derived expression of capsanthin-capsorubin synthase in *Nicotiana benthamiana*. Plant J 14:305–315

Kutchan TM (1998) Molecular genetics of plant alkaloid biosynthesis. In: Cordell, G (ed) The alkaloids, vol 50. Academic, San Diego, pp 257–316

Mason HS, Ball JM, Shi JJ, Jiang X, Estes MK, Arntzen CJ (1996) Expression of Norwalk virus capsid protein in transgenic tobacco and potato and its oral immunogenicity in mice. Proc Natl Acad Sci U S A 28:5335–5340

Mason HS, Haq TA, Clements JD, Arntzen CJ (1998) Edible vaccine protects mice against *Escherichia coli* heat-labile enterotoxin (LT): potatoes expressing a synthetic LT-B gene. Vaccine 16:1336–1343

McCormick AA, Kumagai MH, Hanley K, Turpen TH, Hakim I, Grill LK, Tuse D, Levy S, Levy R (1999) Rapid production of specific vaccines for lymphoma by expression of the tumor-derived single-chain Fv epitopes in tobacco plants. Proc Natl Acad Sci U S A 96:703–708

McDaniel R, Thamchaipenet A, Gustafsson C, Fu H, Betlach M, Betlach M, Ashley G (1999) Multiple genetic modifications of the erythromycin polyketide synthase to produce a library of novel "unnatural" natural products. Proc Natl Acad Sci USA 96:1846–1851

Pauli HH, Kutchan TM (1998) Molecular cloning and functional heterologous expression of two alleles encoding (*S*)-N-methylcoclaurine 3'-hydroxylase (CYP80B1), a new methyl jasmonate-inducible cytochrome P-450 dependent monooxygenase of benzylisoquinoline alkaloid biosynthesis. Plant J 13:793–801

Pogue GP, Lindbo JA, Dawson WO, Turpen TH (1998) Tobamovirus transient expression vectors: tools for plant biology and high-level expression of foreign proteins in plants. In: Gelvin SB, Schilperoort RA (eds) Plant molecular biology manual, 2nd edn. Kluwer Academic, Dordrecht

Rosco A, Pauli HH, Priesner W, Kutchan TM (1997) Cloning and heterologous expression of cytochrome P450 reductases from the Papaveraceae. Arch Biochem Biophys 348:369–377

Ruttan VW (1999) The transition to agricultural sustainability. Proc Natl Acad Sci U S A 96:5960–5967

Turpen TH (1999) Tobacco mosaic virus and the virescence of biotechnology. Philos Trans R Soc Lond B Biol Sci 354:665–673

Turpen TH, Reinl SJ, Charoenvit Y, Hoffman SL, Fallarme V, Grill LK (1995) Malarial epitopes expressed on the surface of recombinant tobacco mosaic virus. Biotechnology 13:53–57

Unterlinner B, Lenz R, Kutchan TM (1999) Molecular cloning and functional expression of codeinone reductase – the penultimate enzyme in morphine biosynthesis in the opium poppy *Papaver somniferum*. Plant J 18:465–475

Yun D-J, Hashimoto T, Yamada Y (1992) Metabolic engineering of medicinal plants: transgenic *Atropa belladonna* with an improved alkaloid composition. Proc Natl Acad Sci U S A 89:11799–11803

11 Multifunctional Asymmetric Catalysis

M. Shibasaki

11.1 Introduction

The development of catalytic asymmetric reactions is one of the major areas of research in the field of organic chemistry. So far, a number of chiral catalysts have been reported, and some of them have exhibited a much higher catalytic efficiency than enzymes, which are natural catalysts (Herrmann and Cornils 1996; Noyori 1994; Ojima 1994; Bosnich 1986; Morrison 1985). Most of the synthetic asymmetric catalysts, however, show limited activity in terms of either enantioselectivity or chemical yields. The major difference between synthetic asymmetric catalysts and enzymes is that the former activate only one side of the substrate in an intermolecular reaction, whereas the latter not only activate both sides of the substrate but can also control the orientation of the substrate. If this kind of synergistic cooperation can be realized in synthetic asymmetric catalysis, the concept will open up a new field in asymmetric synthesis, and a wide range of applications may well ensue. This minireview covers two types of asymmetric two-center catalyses promoted by complexes showing Lewis acidity and Brønsted basicity and/or Lewis acidity and Lewis basicity (Steinhagen and Helmchen 1996; Shibasaki et al. 1997).

11.2 Heterobimetallic Asymmetric Catalysis

Our preliminary attempts to obtain a basic chiral rare earth complex have led us to create several new chiral heterobimetallic complexes which catalyze various types of asymmetric reactions. The rare earth-alkali metal-tris(1,1'-bi-2-naphthoxide) complexes (LnMB, where Ln=rare earth, M=alkali metal, and B=1,1'-bi-2-naphthoxide) have been efficiently synthesized from the corresponding metal chloride and/or alkoxide (Sasai et al. 1992, 1993a,b, 1997b), and the structures of the LnMB complexes have been unequivocally determined by a combination of X-ray crystallography and LDI-TOF-mass spectroscopy as shown in Fig. 1 (Sasai et al. 1993d, 1995a; Takaoka et al. 1997).

For example, the effective procedure for the synthesis of LLB (where L=lanthanum and lithium, respectively) is the treatment of $LaCl_3 \bullet 7H_2O$ with 2.7 mol equiv of BINOL dilithium salt, and NaO-*t*-Bu (0.3 mol equiv) in THF at 50°C for 50 h. Alternatively, we established another

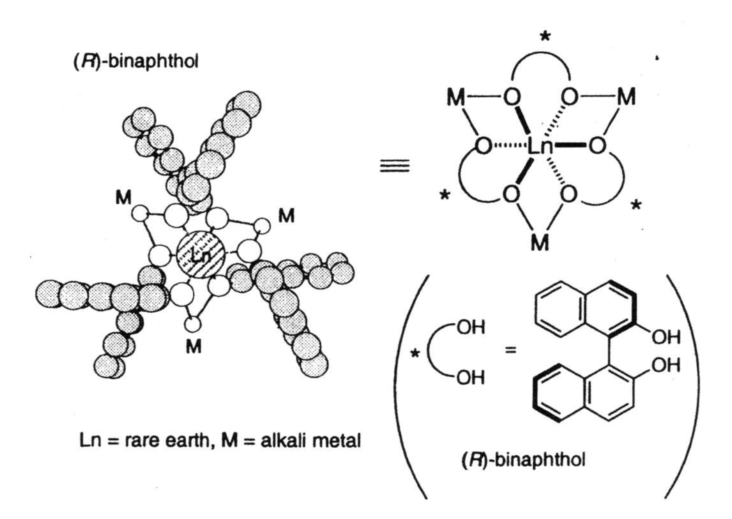

Fig. 1. The structure of rare earth-alkali metal binaphthoxide complexes

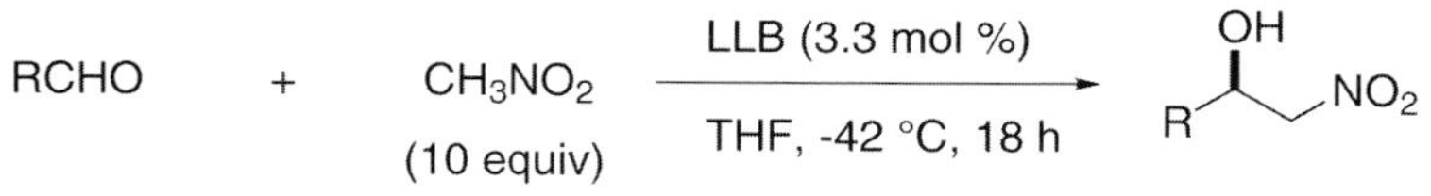

$$\text{RCHO} + \underset{(10\ \text{equiv})}{CH_3NO_2} \xrightarrow[\text{THF, -42 °C, 18 h}]{\text{LLB (3.3 mol \%)}} \text{R-CH(OH)-CH}_2\text{NO}_2$$

1: R = $PhCH_2CH_2$ **2:** 79% (73% ee), R = $PhCH_2CH_2$
3: R = *i*-Pr **4:** 80% (85% ee), R = *i*-Pr
5: R = cyclohexyl **6:** 91% (90% ee), R = cyclohexyl

Scheme 1. Catalytic asymmetric nitroaldol reactions promoted by LLB

efficient procedure for the preparation of LLB, this time starting from La(O-*i*-Pr)3 (purchased from Kojundo Chemical Laboratory Co, Saitama, Japan), the exposure of which to 3 mol equiv of BINOL in THF is followed by the addition of butyllithium (3 mol equiv) at 0°C. It is noteworthy that these heterobimetallic asymmetric complexes including LLB are stable in organic solvents such as THF, CH_2Cl_2, and toluene, which contain small amounts of water, and are also insensitive to oxygen. These heterobimetallic complexes can promote a variety of efficient asymmetric reactions such as nitroaldol, aldol, Michael, hydrophosphonylation, hydrophosphination, protonation, and Diels-Alder reactions by choosing suitable rare earth metals or alkali metals. A catalytic asymmetric nitroaldol reaction and a direct catalytic asymmetric aldol reaction are discussed here in detail.

The nitroaldol (Henry) reaction has been recognized as a powerful synthetic tool and has also been utilized in the construction of numerous natural products and other useful compounds. As shown in Scheme 1, we succeeded in realizing the first example of a catalytic asymmetric nitroaldol reaction using a catalytic amount of LLB.

The rare earth metals are generally regarded as a group of 17 elements with similar properties, especially with respect to their chemical reactivity. However, in the case of the above-mentioned catalytic asymmetric nitroaldol reaction, we observed pronounced differences in both the reactivity and the enantioselectivity of various rare earth metals used. For example, when benzaldehyde and nitromethane were used as starting materials, the corresponding Eu complex gave the nitroaldol in 72% ee (91%), in contrast to 37% ee (81%) in the case of the LLB (–40°C, 40 h). These results suggest that small changes in the structure of the catalyst (ca. 0.1 Å in the ionic radius of the rare earth cation)

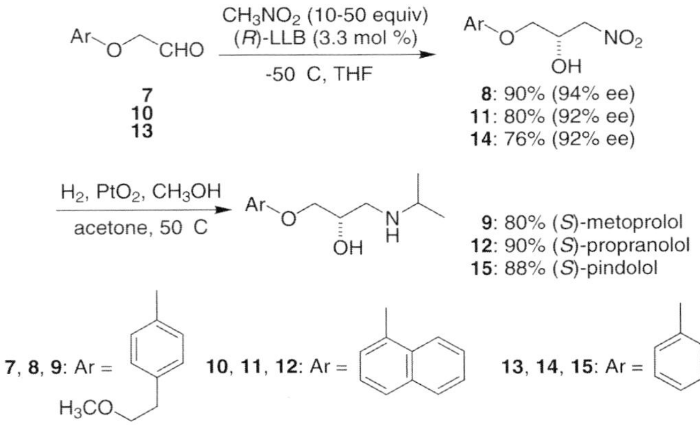

Ar–O–CHO (CH_3NO_2 (10-50 equiv), (*R*)-LLB (3.3 mol %), -50 C, THF)

7, **10**, **13**

8: 90% (94% ee)
11: 80% (92% ee)
14: 76% (92% ee)

H_2, PtO_2, CH_3OH, acetone, 50 C

9: 80% (*S*)-metoprolol
12: 90% (*S*)-propranolol
15: 88% (*S*)-pindolol

7, **8**, **9**: Ar = ; **10**, **11**, **12**: Ar = ; **13**, **14**, **15**: Ar =

Scheme 2. Catalytic asymmetric synthesis of β-blockers using (*R*)-LLB as a catalyst

cause a drastic change in the optical purity of the nitroaldols produced. Although in general, nitroaldol reactions are regarded as equilibrium processes, no detectable retronitroaldol reactions were observed in the Ln-BINOL complex-catalyzed asymmetric nitroaldol reactions. Having succeeded in obtaining the first results from a catalytic asymmetric nitroaldol reaction, we then attempted to apply the method to catalytic asymmetric synthesis of biologically important compounds. The nitroaldol products were readily converted into β-amino alcohols and/or α-hydroxy carbonyl compounds. Thus, convenient syntheses of three kinds of optically active β-blockers are presented in Scheme 2 (Sasai et al. 1993c, 1994a, 1995b). Interestingly, the nitroaldol products **8**, **11**, and **14** were found to have (*S*)-absolute configuration when (*R*)-LLB was used. The nitronates thus appear to react preferentially with the *si* face of the aldehydes, in contrast to the enantiofacial selectivity which might have been expected on the basis of previous results (cf. Scheme 1). These results suggest that the presence of an oxygen atom at the β-position greatly influences the enantiofacial selectivity. LLB-type catalysts were also able to promote diastereoselective and enantioselective nitroaldol reactions starting from prochiral materials. However, limited enantioselectivities (<78% ee) and diastereoselectivities (ca. 2:1–3:1) were obtained using LLB. In order to obtain both high enantio- and

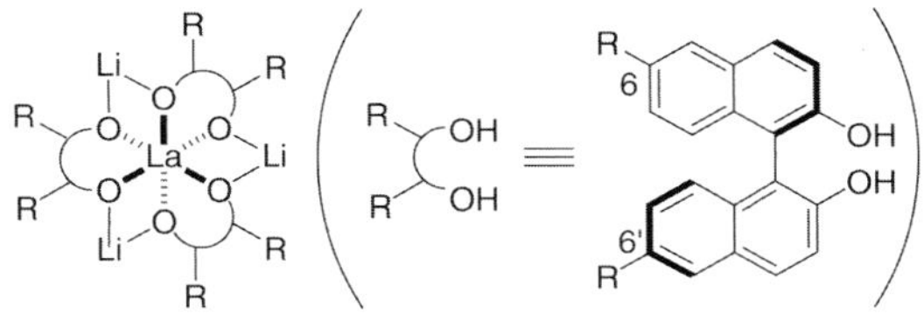

LLB: R = H
16: R = Br
17: R = CH_3
18: R = C N
19: R = C CH

20: R = C CPh
21: R = C $CSi(CH_3)_3$
22: R = C $CSiEt_3$
23: R = C CTBS
24: R = C $CSi(CH_3)_2Ph$

Fig. 2. Structural modification of LLB

Table 1. *syn*-Selective catalytic asymmetric nitroaldol reaction

RCHO + $R'CH_2NO_2$ → catalyst (3.3 mol %), THF → *syn* + *anti*

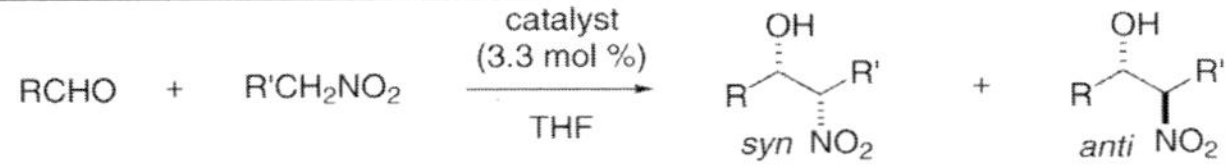

1: R = $PhCH_2CH_2$
28: R = $CH_3(CH_2)_4$

25: R' = CH_3
29: R' = Et
32: R' = CH_2OH

26 (*syn*), **27** (*anti*): R = $PhCH_2CH_2$, R' = CH_3
30 (*syn*), **31** (*anti*): R = $PhCH_2CH_2$, R' = Et
33 (*syn*), **34** (*anti*): R = $PhCH_2CH_2$, R' = CH_2OH
35 (*syn*), **36** (*anti*): R = $CH_3(CH_2)_4$, R' = CH_2OH

Entry	Aldehyde	Nitroalkane	Catalyst	Time (h)	Temp (C)	Nitro-aldols	Yield (%)	*syn*/*anti*	ee of *syn* (%)
1	**1**	**25**	LLB	75	-20	**26** + **27**	79	74:26	66
2	**1**	**25**	**21**	75	-20	**26** + **27**	72	85:15	92
3	**1**	**25**	**22**	75	-20	**26** + **27**	70	89:11	93
4	**1**	**29**	LLB	138	-40	**30** + **31**	89	85:15	87
5	**1**	**29**	**22**	138	-40	**30** + **31**	85	93:7	95
6	**1**	**32**	LLB	111	-40	**33** + **34**	62	84:16	66
7	**1**	**32**	**22**	111	-40	**33** + **34**	97	92:8	97
8	**28**	**32**	LLB	93	-40	**35** + **36**	79	87:13	78
9	**28**	**32**	**22**	93	-40	**35** + **36**	96	92:8	95

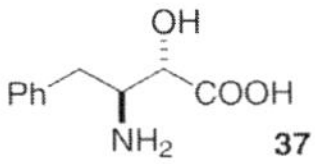

37

diastereoselectivity, we focused our attention on the preparation of a novel asymmetric catalyst. Among many catalysts prepared, catalysts **16–24** were first found to give higher enantioselectivity in the catalytic asymmetric nitroaldol reaction of hydrocinnamaldehyde **1** using nitromethane. With more effective asymmetric catalysts in hand, we next applied the most efficient catalysts **21** and/or **22** to diastereoselective nitroaldol reactions. We were very pleased to find that, in all cases, high

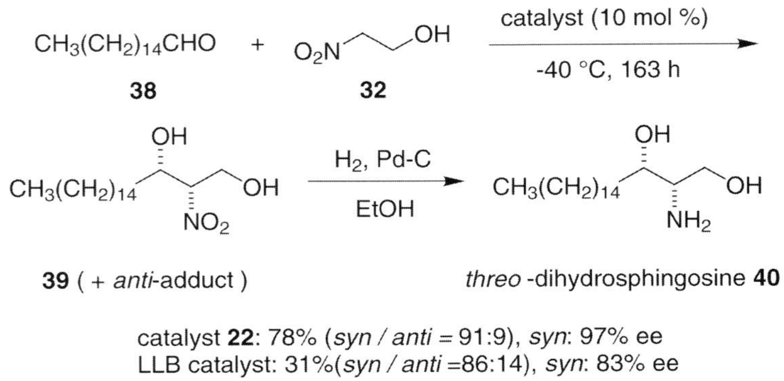

$CH_3(CH_2)_{14}CHO$ + O_2N~~OH — catalyst (10 mol %), -40 °C, 163 h →

38 **32**

$CH_3(CH_2)_{14}$ (OH, NO_2) OH — H_2, Pd-C, EtOH → $CH_3(CH_2)_{14}$ (OH, NH_2) OH

39 (+ *anti*-adduct) *threo* -dihydrosphingosine **40**

catalyst **22**: 78% (*syn* / *anti* = 91:9), *syn*: 97% ee
LLB catalyst: 31%(*syn* / *anti* =86:14), *syn*: 83% ee

Scheme 3. Catalytic asymmetric synthesis of *threo*-dihydrosphingosine

syn selectivity and enantioselectivity were obtained using 3.3 mol% of the catalysts (Sasai et al. 1995c). Representative results are shown in Table 1. It appears that the *syn* selectivity in the nitroaldol reaction can best be explained as arising from steric hindrance in the bicyclic transition state, and higher stereoselectivities obtained by using catalysts **21** and **22** seem to be ascribed to the increase of catalyst stability in the presence of excess nitroalkanes. The *syn*-selective asymmetric nitroaldol reaction was successfully applied to the catalytic asymmetric synthesis of *threo*-dihydrosphingosine **40**, which elicits a variety of cellular responses by inhibiting protein kinase C. Moreover, an efficient synthesis of *erythro*-AHPA **37** from l-phenylalanine was achieved using LLB (Sasai et al. 1994b). Catalytic asymmetric nitroaldol reactions promoted by LLB (Scheme 3) or its derivatives require at least 3.3 mol% of asymmetric catalysts for efficient conversion. Moreover, even in the case of 3.3 mol% of catalysts, the reactions are rather slow. A consideration of the possible mechanism for catalytic asymmetric nitroaldol reactions is clearly a necessary prerequisite to the formulation of an effective strategy for enhancing the activity of the catalyst. One possible mechanism for the catalytic asymmetric nitroaldol reactions is shown at the top of Scheme 4. We strove to detect the postulated intermediate **I** using various methods. These attempts proved to be unsuccessful, however, probably owing to the low concentrations of the intermediate, which we

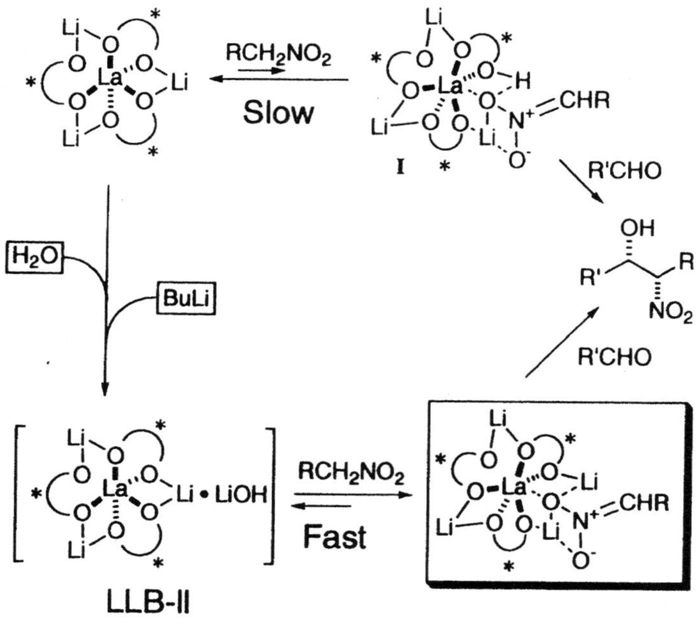

Scheme 4. Proposed mechanism for the catalytic asymmetric nitroaldol reaction promoted by LLB and/or LLB-II

thought might be ascribable to the presence of an acidic OH group in close proximity.

In order to remove a proton from **I**, we added almost 1 equiv of base to the LLB catalyst. After many attempts, we were finally pleased to find that 1 mol% of second-generation LLB (LLB-II), prepared from LLB, 1 mol equiv of H_2O, and 0.9 mol equiv of butyllithium, efficiently promoted the catalytic asymmetric nitroaldol reactions. Moreover, we found that the use of LLB-II (3.3 mol%) accelerated these reactions. The use of other bases such as NaO-*t*-Bu, KO-*t*-Bu and $Ca(O\text{-}i\text{-}Pr)_2$ gave less satisfactory results. The results are shown in Table 2. The structure of LLB-II has not yet been unequivocally determined. We suggest here, however, that it is a complex of LLB and LiOH: a proposed reaction course for its use in an improved catalytic asymmetric nitroaldol reaction is shown at the bottom of Scheme 4 (Arai et al. 1996c). Industrial application of a catalytic asymmetric nitroaldol reaction is being examined.

Having developed an efficient catalytic asymmetric nitroaldol reaction, we next directed our attention to a direct catalytic asymmetric aldol reaction. The aldol reaction is generally regarded as one of the most powerful of the carbon-carbon bond-forming reactions. The development of a range of catalytic asymmetric aldol-type reactions has proven to be a valuable contribution to asymmetric synthesis. (For recent examples of catalytic asymmetric Mukayama-aldol reactions, see Evans et al. 1998, Krüger and Carreira 1998, Yanagisawa et al. 1997, Denmark et al. 1997 and references cited therein.) In all of these catalytic asymmetric aldol-type reactions, however, preconversion of the ketone moiety to a more reactive species such as an enol silyl ether, enol methyl ether, or ketene silyl acetal is an unavoidable necessity (Scheme 5). Development of a direct catalytic asymmetric aldol reaction, starting from aldehydes

Table 2. Comparisons of catalytic activity between either LLB and second-generation LLB (LLB-II) or **22** and **22**-II

RCHO + $R'CH_2NO_2$ —catalyst→ $R-CH(OH)-CH(NO_2)-R'$

5: R = C_6H_{11}
1: R = $PhCH_2CH_2$

41 : R' = H
25 : R' = CH_3
29 : R' = Et
32 : R' = CH_2OH

6 : R = C_6H_{11}, R' = H
26 : R = $PhCH_2CH_2$, R' = CH_3
30 : R = $PhCH_2CH_2$, R' = Et
33 : R = $PhCH_2CH_2$, R' = CH_2OH

Entry	Substrate	Catalyst[a] (mol %)	Time (h)	Temp (°C)	Product	Yield (%) (*syn/anti*)	ee (%) of *syn*
1	**5** + **41**	LLB (1)	24	-50	**6**	5.6	88
2	**5** + **41**	LLB-II (1)	24	-50	**6**	73	89
3	**5** + **41**	LLB-II (3.3)	4	-50	**6**	70	90
4	**1** + **25**	**22** (1)	113	-30	**26**	25 (70/30)	62
5	**1** + **25**	**22**-II (1)	113	-30	**26**	83 (89/11)	94
6	**1** + **29**	**22** (1)	166	-40	**30**	trace	–
7	**1** + **29**	**22**-II (1)	166	-40	**30**	84 (95/5)	95
8	**1** + **32**	**22** (1)	154	-50	**33**	trace	–
9	**1** + **32**	**22**-II (1)	154	-50	**33**	76 (94/6)	96

[a]LLB-II : LLB + H_2O (1 mol equiv) + BuLi (0.9 mol equiv); **22**-II : **22** + H_2O (1 mol equiv) + BuLi (0.9 mol equiv)

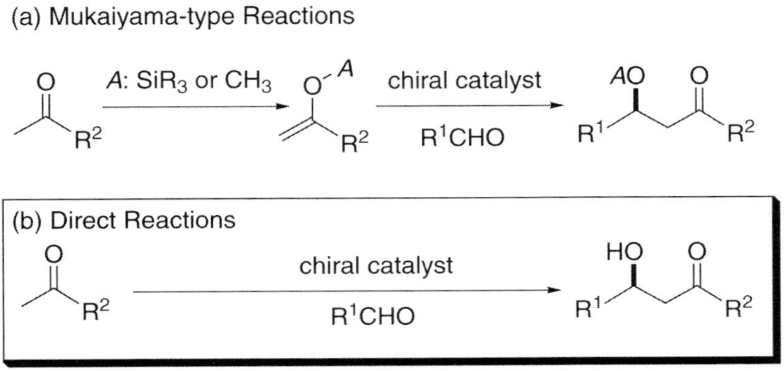

(a) Mukaiyama-type Reactions

A: SiR_3 or CH_3 — chiral catalyst, R^1CHO

(b) Direct Reactions

chiral catalyst, R^1CHO

Scheme 5. Several types of catalytic asymmetric aldol reactions

and unmodified ketones, is thus a noteworthy endeavor. Such reactions are known in enzyme chemistry (Fessner et al. 1996), the fructose-1,6-bisphosphate and/or DHAP aldolases being characteristic examples. The mechanism of these enzyme-catalyzed aldol reactions is thought to involve co-catalysis by a Zn^{2+} cation and a basic functional group in the enzyme's active site, with the latter abstracting a proton from a carbonyl compound while the former functions as a Lewis acid to activate the other carbonyl component.

We speculated that it might be possible to develop a direct catalytic asymmetric aldol reaction of aldehydes and unmodified ketones by employing heterobimetallic catalysts. However, our initial concerns were dominated by the possibility that our heterobimetallic asymmetric catalysts would be ineffective at promoting aldol reactions due to their rather low Brønsted basicity. We were thus pleased to find that, first of all, aldol reactions of the desired type using tertiary aldehydes proceeded smoothly in the presence of LLB as catalyst. It is noteworthy that **56** can be obtained in 94% ee. The achievement of developing an efficient catalytic asymmetric aldol reaction using aldehydes with α-hydrogens clearly represents a much greater challenge than for cases such as those above, since self-aldol products can easily be formed. However, we found that, for example, the reaction of cyclohexanecarboxaldehyde **5** with acetophenone **43** proceeded smoothly without significant formation of the self-aldol product of **5**, giving **51** in 44% ee and in 72% yield.

On the other hand, the reaction between hydrocinnamaldehyde **1**, which possesses two α-hydrogens, and **43** proved to be more difficult. Although **53** was obtained in 52% ee, the yield was low (28%) due to the formation of self-condensation by-products (–20°C). The results are summarized in Table 3 (Yamada et al. 1997). Thus, we have achieved success in carrying out direct catalytic asymmetric aldol reactions of aldehydes with unmodified ketones for the first time. However, in order to attain a synthetically useful level for this methodology, the challenge remains to reduce the amounts of ketones and catalysts used, shorten

Table 3. Direct catalytic asymmetric aldol reactions promoted by (*R*)-LLB (20 mol%)

R^1CHO + $CH_3C(O)R^2$ —(*R*)-LLB (20 mol %), THF, -20 °C→ $R^1CH(OH)CH_2C(O)R^2$

42: R^1 = *t*-Bu
45: R^1 = $PhCH_2C(CH_3)_2$
5: R^1 = cyclohexyl
3: R^1 = *i*-Pr
1: R^1 = $Ph(CH_2)_2$

43: R^2 = Ph
46: R^2 = 1-naphthyl
48: R^2 = CH_3
50: R^2 = Et

44: R^1 = *t*-Bu, R^2 = Ph
47: R^1 = *t*-Bu, R^2 = 1-naphthyl
49: R^1 = $PhCH_2C(CH_3)_2$, R^2 = Ph
51: R^1 = cyclohexyl, R^2 = Ph
52: R^1 = *i*-Pr, R^2 = Ph
53: R^1 = $Ph(CH_2)_2$, R^2 = Ph
54: R^1 = $PhCH_2C(CH_3)_2$, R^2 = CH_3
55: R^1 = *t*-Bu, R^2 = CH_3
56: R^1 = $PhCH_2C(CH_3)_2$, R^2 = Et

Entry	Aldehyde	Ketone (equiv)	Product	Time (h)	Yield (%)	ee (%)
1[a]	**42**	**43** (5)	**44**	88	43	89
2	**42**	**43** (5)	**44**	88	76	88
3	**42**	**43** (1.5)	**44**	135	43	87
4	**42**	**43** (10)	**44**	91	81	91
5	**42**	**46** (8)	**47**	253	55	76
6	**45**	**43** (7.4)	**49**	87	90	69
7	**5**	**43** (8)	**51**	169	72	44
8[b]	**3**	**43** (8)	**52**	277	59	54
9	**1**	**43** (10)	**53**	72	28	52
10	**45**	**48** (10)	**54**	185	82	74
11	**42**	**48** (10)	**55**	100	53	73
12	**45**	**50** (50)	**56**	185	71	94

[a] (*R*)-LLB and addition of 1 equiv of H_2O to LLB.
[b] The reaction was carried out at -30 °C.

reaction times, and increase enantioselectivities. As mentioned above, we observed for an asymmetric nitroaldol reaction that the LLB•LiOH tight complex enhanced the catalytic activity of LLB. Encouraged by this result, development of a new strategy to activate LLB for the direct catalytic asymmetric aldol reaction was attempted. As a result the catalyst generated from LLB, KHMDS (0.9 equiv to LLB) and H_2O (1 equiv to LLB), which presumably forms a heteropolymetallic complex (LLB-II'), was found to be a superior catalyst for the direct catalytic asymmetric aldol reaction, giving **49** in 89% yield and 79% ee (using 8 mol% of LLB). We employed this method to generate KOH in situ because of its insolubility in THF. The use of KO-*t*-Bu instead of KHMDS gave a similar result, indicating that HMDS dose not play a key role. Interestingly, further addition of H_2O (1 equiv with respect to LLB) resulted in the formation of **49** in 83% yield and higher ee. The powder obtained from the catalyst solution by evaporation of the solvent showed a similar result. This powder can be easily handled without the need of an inert atmosphere. In addition, we were pleased to find that as little as 3 mol% of the catalyst promoted the reaction efficiently, to give **49** in 71% yield and 85% ee. In contrast to catalytic asymmetric nitroaldol reactions, the generation of LiOH or other bases was found to give less satisfactory results. The results are summarized in Table 4.

Table 4. Direct catalytic asymmetric aldol reactions of **45** with **43** under various conditions

45 + **43** (5 eq) → (*R*)-LLB (8 mol %), base (7.2 mol %), THF, H_2O, -20 °C → **49**

Entry	Base	H_2O (mol %)	Time (h)	Yield (%)	ee (%)
1	- (LLB itself)	-	18	trace	-
2	KHMDS	0	18	83	58
3	KHMDS	8	18	89	79
4	KHMDS	16	18	83	85
5[a]	KHMDS	16	33	71	85
6	KHMDS	32	18	67	89
7	LHMDS	16	5	22	80
8	NHMDS	16	5	28	86
9	KHMDS	16	5	74	84

[a] 3 mol % of catalyst was used.

This newly developed heteropolymetallic catalyst system (LLB-II') was applied to a variety of direct catalytic asymmetric aldol reactions, giving aldol products **44–67** in modest to good ees, as shown in Table 5. It is noteworthy that even **65** can be produced from hexanal **59** in 55% yield and 42% ee without the formation of the corresponding self-aldol product (–50°C). This result can be understood by considering that, in general, aldehyde enolates are not generated by the catalyst at low temperature. In fact, this assumption was confirmed by several experimental results. It is also noteworthy that the direct catalytic asymmetric aldol reaction between **45** and cyclopentanone **61** also proceeded smoothly to afford **67** in 95% yield (*syn*/*anti* = 93/7, *syn* = 76% ee, *anti* = 88% ee). Several of the aldol products obtained were readily converted to their corresponding esters by Baeyer-Villiger oxidation. The results are summarized in Table 5. Ester **69** was further transformed into key epothilone A intermediate **72** and also a key synthetic intermediate for bryostatin 7 **73**. What is the mechanism of the present direct catalytic asymmetric aldol reactions using LLB-II'? It is obvious that the self-assembly of LLB and KOH takes place, because of the formation of a variety of aldol products in high ees and yields. In addition, the ^{13}C NMR spectrum of LLB•KOH and also the LDI-TOF(+)MS spectrum show that there is a rapid exchange between Li^+ and K^+. We have already found that LPB[La_3K_3-tris(binaphthoxide)] itself is not a useful catalyst for aldol reactions, and that the complexes LPB•KOH or LPB•LiOH give rise to much less satisfactory results.

Consequently we believe that the BINOL core of the active complex is essentially LLB. Therefore, the heteropolymetallic complex of LLB and KOH, with KOH axially coordinated to La, among other possible complexes, would be the most effective catalyst for the present reaction. To clarify the reaction mechanism, we carried out kinetic studies. Significant isotope effects ($k_H/k_D \sim 5$) were observed, and the reaction rate was found to be independent of the concentration of the aldehyde. Both of these results indicate that the rate-determining step is the deprotonation of the ketone, and they also suggest that the catalyst readily forms a relatively tight complex with the aldehyde, thus activating it. This coordination of an aldehyde was supported by the ^1H NMR spectrum. Although the precise role of H_2O is not clear at present, we have suggested a working model of the catalytic cycle and a possible mechanism which allows us to explain the observed absolute configurations of the products (Scheme 6).

Table 5. Direct catalytic asymmetric aldol reactions promoted by heteropolymetallic asymmetric catalyst and following Baeyer-Villiger oxidations

(*R*)-LLB (8 mol %), KHMDS (7.2 mol %), H_2O (16 mol %), THF, -20 °C; **1-59** + **43-61** → **44-67**; [O] → **68-70**; [O] → **71**

42: R^1 = *t*-Bu
45: R^1 = $PhCH_2C(CH_3)_2$
3: R^1 = *i*-Pr
1: R^1 = $PhCH_2CH_2$
57: R^1 = $BnOCH_2C(CH_3)_2$
58: R^1 = Et_2CH
59: R^1 = *n*-C_5H_{11}

43: R^2 = Ph
48: R^2 = CH_3
50: R^2 = Et
60: R^2 = 3-NO_2-C_6H_4
61: R^2 = -$(CH_2)_3$-

Entry	Aldehyde (R^1)	Ketone[a] (R^2) (eq)	Aldol	Time (h)	Yield (%)	ee (%)	Yield of ester[b]
1	**42**	**43** (5)	**44**	15	75	88	
2	**45**	**43** (5)	**49**	28	85	89	**68**: 80%[c]
3	**45**	**48** (10)	**54**	20	62	76	
4[d]	**45**	**50** (15)	**56**	95	72	88	
5	**57**	**43** (5)	**62**	36	91	90	
6[e]	**57**	**43** (5)	**62**	24	70	93	**69**: 73%[f]
7[g]	**3**	**43** (5)	**52**	15	90	33	
8[h]	**3**	**60** (3)	**63**	70	68	70	**70**: 80%[i]
9[j]	**58**	**60** (3)	**64**	96	60	80	
10[h,k]	**59**	**60** (5)	**65**	96	55	42	
11[l]	**1**	**60** (3)	**66**	31	50	30	
12	**45**	**61** (5)	**67**	99	95 (*syn*/*anti* = 93/7)	76/88 (*syn*/*anti*)	**71**: 85%[c]

[a] Excess of ketone was recovered after reaction. [b] The yield from aldol product. [c] Conditions: $SnCl_4$ (cat.), $(TMSO)_2$, *trans*-*N*,*N*′-bis(*p*-toluene-sulfonyl)cyclohexane-1,2-diamine (cat.), MS 4A, CH_2Cl_2. [d] 8 mol % of H_2O was used. [e] The reaction was carried out in 5.7 mmol (**57**) scale. [f] Conditions: *m*CPBA, NaH_2PO_4, DCE. [g] The reaction was carried out at -30 °C. [h] The reaction was carried out at -50 °C. [i] Conditions: i) PtO_2, H_2, MeOH; ii) ZCl, Na_2CO_3, MeOH-H_2O; iii) $SnCl_4$ (cat.), $(TMSO)_2$, *trans*-*N*,*N*′-bis(*p*-toluene-sulfonyl)cyclohexane-1,2-diamine (cat.), MS 4A, CH_2Cl_2. R^2 (**70**) = 3-ZNH-C_6H_4. [j] Conditions: (*R*)-LLB (15 mol %), KHMDS (13.5 mol %), H_2O (30 mol %), -45 °C. [k] Conditions: (*R*)-LLB (30 mol %), KHMDS (27 mol %), H_2O (60 mol %). [l] The reaction was carried out at -40 °C.

72 **73**

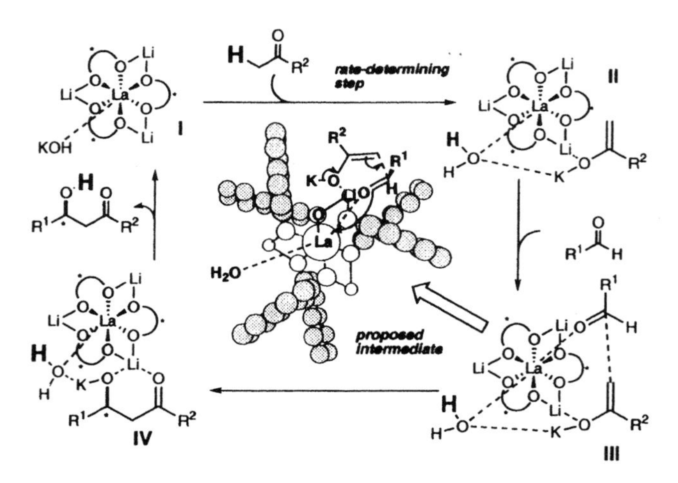

Scheme 6. Working model for direct catalytic asymmetric aldol reactions promoted by the heteropolymetallic asymmetric catalyst

The stereoselectivities appear to be kinetically controlled. In fact, the ee of the aldol product was constant during the course of the reaction. Thus, we have succeeded in carrying out the first catalytic asymmetric aldol reaction between aldehydes and unmodified ketones by using LLB or LLB-II'. Several reactions are already synthetically useful, especially in the case of tertiary aldehyde, leading to the catalytic asymmetric synthesis of key intermediates en route to natural products (Yoshikawa et al. 2000). Further studies are currently underway. Moreover, these rare earth-containing heterobimetallic complexes can be utilized for a variety of efficient catalytic asymmetric reactions, as shown in Scheme 7.

Next we began with the development of an amphoteric asymmetric catalyst assembled from aluminum and an alkali metal (Arai et al. 1996b). The new asymmetric catalyst was prepared efficiently from $LiAlH_4$ and 2 mol equiv of (*R*)-BINOL, and the structure was unequivocally determined by X-ray crystallographic analysis (Scheme 8). This aluminum-lithium-BINOL complex (ALB) was highly effective in the Michael reaction of cyclohexenone **75** with dibenzyl malonate **77**, giving **82** with 99% ee and 88% yield at room temperature. Although LLB

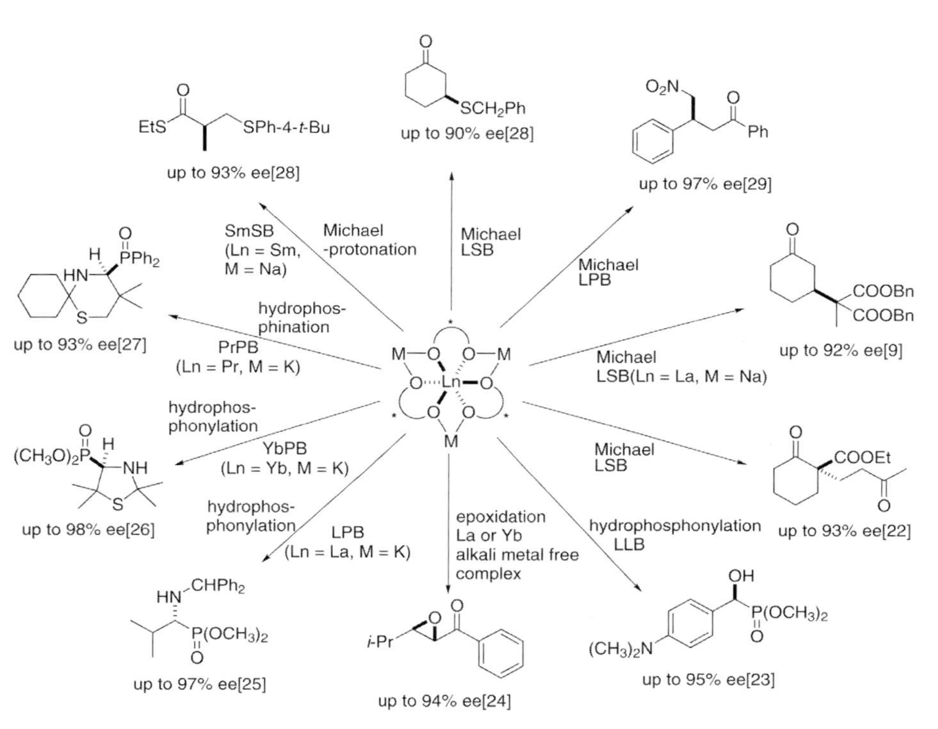

Scheme 7. Other efficient and general catalytic asymmetric reactions promoted by heterobimetallic complexes. (9=Sasai et al. 1995a; 22=Sasai et al. 1996; 23=Sasai et al. 1997a; 24=Bougauchi et al. 1997, Watanabe et al. 1998a,b; 25=Sasai et al. 1995d; 26=Gröger et al. 1996, 1998; 27=Yamakoshi et al. 1999; 28=Emori et al. 1998; 29=Funabashi et al. 1998)

and LSB complement each other in their ability to catalyze asymmetric nitroaldol and Michael reactions, aluminum-M-(*R*)-BINOL complexes (M=Li, Na, K, and Ba) are useful for the catalytic asymmetric Michael reactions. Moreover, we have developed a strategy for the activation of ALB: the addition of nearly 1 equiv of base, such as BuLi or KO-*t*-Bu, to ALB can accelerate a catalytic asymmetric Michael addition without lowering the high enantiomeric excess. However, 3–5 mol% of the catalyst is still required to obtain the product in excellent yield and high

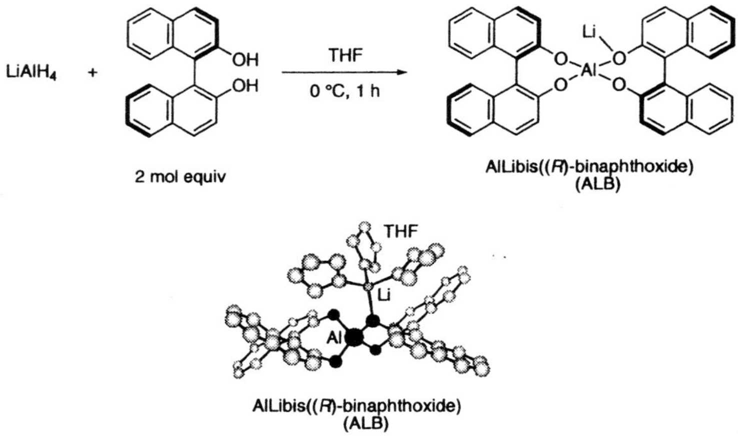

Scheme 8. Preparation of ALB and its X-ray structure

enantiomeric excess. We intended to improve the catalytic asymmetric Michael addition to a practically useful level. We were pleased to find that addition of MS 4A to the reaction medium greatly improved the catalytic asymmetric Michael addition. Actually, as shown in Table 6, the use of ALB (0.3 mol%), KO-*t*-Bu (0.27 mol%), and MS 4A gave **83** in 99% ee and 94% yield even at room temperature. Furthermore, we successfully carried out this reaction on a 100-g scale. Using the Michael adduct **83**, catalytic asymmetric syntheses of tubifolidine **85** (Shimizu et al. 1998) and coronafacic acid **86** (Table 7) have been achieved (Nara et al. 1997).

The mechanistic considerations of a catalytic asymmetric Michael reaction suggest that the reaction of an alkali metal enolate derived from a malonate derivative with an enone should lead to an intermediary aluminum enolate. Is it possible that such an Al enolate could then be trapped by an electrophile such as an aldehyde? As was anticipated, the reaction of **74**, diethyl methylmalonate **76**, and hydrocinnamaldehyde **1** in the presence of 10 mol% of ALB gave the three-component coupling product **87** as a single isomer in 91% ee (64% yield) (Table 8). This cascade Michael-aldol reaction has been successfully applied to a catalytic asymmetric synthesis of 11-deoxy-PGF$_{1}\alpha$ **88** (Yamada et al. 1998).

Table 6. Catalytic asymmetric Michael reactions promoted by the AlMbis[(*R*)-binaphthoxide] complex (AMB)

AMB (10 mol %), THF, rt

74: n = 1
75: n = 2

76: R^1 = Et, R^2 = CH_3
77: R^1 = Bn, R^2 = H
78: R^1 = CH_3, R^2 = H
79: R^1 = Et, R^2 = H

80: n = 1, R^1 = Et, R^2 = CH_3
81: n = 1, R^1 = Bn, R^2 = H
82: n = 2, R^1 = Bn, R^2 = H
83: n = 2, R^1 = CH_3, R^2 = H
84: n = 2, R^1 = Et, R^2 = H

Entry	Enone	Michael Donor	Product	M	Time (h)	Yield (%)	ee (%)
1	**74**	**76**	**80**	Li	72	84	91
2	**74**	**77**	**81**	Li	60	93	91
3	**75**	**77**	**82**	Li	72	88	99
4	**75**	**77**	**82**	Na	72	50	98
5	**75**	**77**	**82**	K	72	43	87
6	**75**	**77**	**82**	Ba	6	100	84
7	**75**	**78**	**83**	Li	72	90	93
8	**75**	**79**	**84**	Li	72	87	95

Moreover, ALB was found to be useful for the hydrophosphonylation of aldehydes (Arai et al. 1996a). ALB and LLB can thus be used in a complementary manner for the hydrophosphonylation of aldehydes.

The enantioselective ring opening of epoxides is an attractive and quite powerful method for asymmetric synthesis. Although various types of stoichiometric or catalytic asymmetric epoxide ring openings have been described, only a few practical methods have been reported so far (Tokunaga et al. 1997). We became very interested in the development of catalytic asymmetric epoxide ring openings using nucleophiles such as RSH, ROH, HCN, and HN_3. First of all, we examined a catalytic asymmetric ring opening of symmetrical epoxides with thiols using the heterobimetallic complexes. We envisioned that these complexes would

Table 7. A greatly improved catalytic asymmetric Michael addition of **78** to **75**

75 + 78 —[(*R*)-ALB (x mol %)[a], KO-t-Bu (0.9 eq to ALB) / MS 4A, THF, rt]→ 83

Entry	ALB (x mol %)	KO-t-Bu	MS 4A	Time (h)	Yield (%)	ee (%)
1[b]	10	-	-	72	90	93
2[c]	5	+	-	48	97	98
3[c]	0.3	+	-	120	74	88
4[c]	0.3	+	+[e]	120	94	99
5[d]	1.0	+	+[f]	72	96	99

[a] (*R*)-AlLibis(binaphthoxide). [b] 200 mg scale reaction. [c] 400 mg scale reaction.
[d] 100 g scale reaction. [e] MS 4A (8.3 g) was used for ALB (1 mmol).
[f] MS 4A (2.0 g) was used for ALB (1 mmol).

85 **86**

prove to be useful for the catalytic asymmetric ring openings of **90** with a nucleophile such as $PhCH_2SH$. However, LaM_3tris(binaphthoxide) (M=Li or Na) or ALB showed only low catalytic activity, giving 2-(benzylthio) cyclohexanol in 1%–10% yields, although modest to high ees (27–86% ee) were observed. We then examined the new heterobimetallic asymmetric complexes with group 13 elements (B, Ga, In) other than Al. Of these, the GaLibis[(*R*)-binaphthoxide] complex [(*R*)-GaLB], which was readily prepared from $GaCl_3$, (*R*)-binaphthol (2 mol equiv to

Table 8. Tandem Michael-aldol reactions

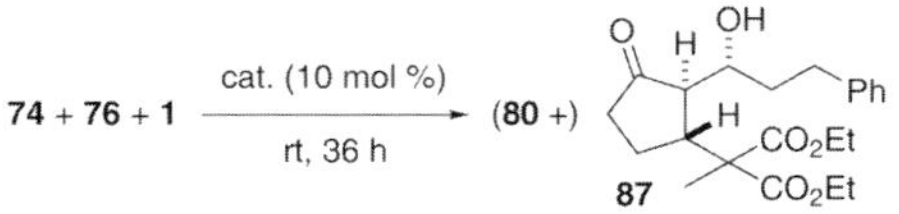

74 + 76 + 1 —cat. (10 mol %), rt, 36 h→ (80 +) 87

Catalyst	80 Yield (%)	80 ee (%)	87 Yield (%)	87 ee (%)
(*R*)-ALB	7	90	64	91
(*R*)-LLB	46	3	30[a]	–
(*R*)-LSB	73	86	trace	–
Li-free-La-(*R*)-BINOL	57	83	trace	–

[a] Inseparable mixture.

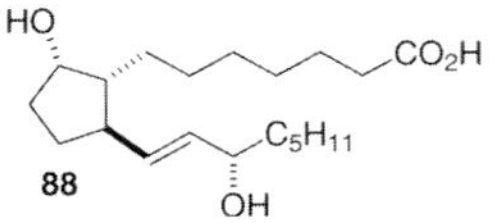

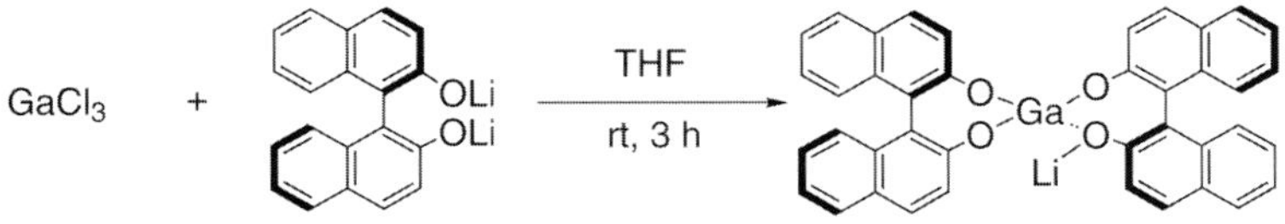

88

$GaCl_3$ + (2 mol equiv) —THF, rt, 3 h→ (*R*)-GaLB

Scheme 9. Preparation of (*R*)-GaLibis(binaphthoxide) [(*R*)-GaLB]

$GaCl_3$), and BuLi (4 mol equiv to $GaCl_3$) in THF showed a high catalytic activity for the present reaction (Scheme 9; Iida et al. 1997).

Following many attempts, GaLB was found to be quite useful for the asymmetric ring opening of symmetrical epoxides with *t*-BuSH, as shown in Table 9.

The almost optically pure **97** has been utilized for the preparation of the attractive chiral ligand by Prof. D.A. Evans (personal communication).

Moreover, as shown in Table 10, GaLB was found to be suitable to the asymmetric ring opening of symmetrical epoxides with 4-methoxyphenol (Iida et al. 1998). In addition, a direct catalytic asymmetric Mannich-type reaction has quite recently been achieved by the

Table 9. Catalytic asymmetric ring openings of symmetrical epoxides with t-BuSH (**89**) catalyzed by (*R*)-GaLB with MS 4A

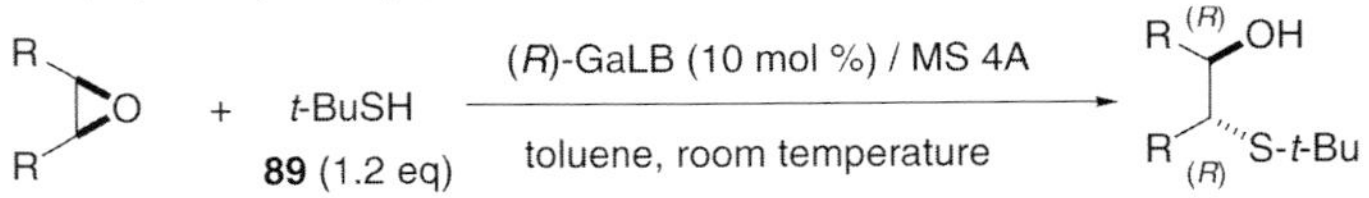

R,R-epoxide + *t*-BuSH **89** (1.2 eq) → [(*R*)-GaLB (10 mol %) / MS 4A; toluene, room temperature] → R(*R*)–OH, R(*R*)–S-*t*-Bu

Entry	Epoxide		MS 4A[a] (g)	Time (h)	Product	Yield (%)	ee (%)
1	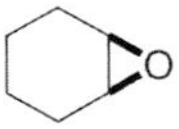 cyclohexene oxide	**90**	none	65	**97**	35	98
2	 cyclohexene oxide	**90**	0.2	9	**98**	80	97
3	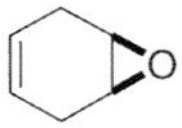 cyclohexadiene monoxide	**91**	0.2	36	**99**	74	95
4[b]	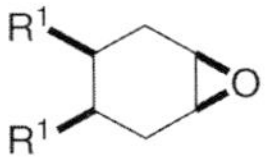 R¹, R¹ (cis) cyclohexene oxide	**92**	0.2	12	**100**	83	86
5[b]	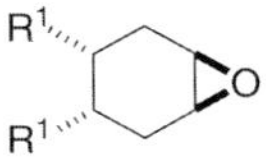 R¹, R¹ (trans) cyclohexene oxide	**93**	0.2	137	**101**	64	91
6	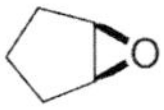 cyclopentene oxide	**94**	0.2	24	**102**	89	91
7[c]	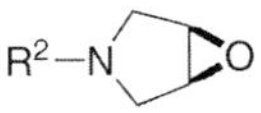 R^2–N pyrrolidine epoxide	**95**	0.2	72	**103**	89	89
8	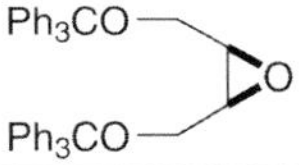 Ph_3CO, Ph_3CO epoxide	**96**	2.0	48	**104**	89	82

[a] Weight per 0.1 mmol of GaLB. [b] R^1 = CH_2OSiPh_2t-Bu. [c] Carried out at 50 °C in the presence of 30 mol % of GaLB. R^2 = 2,4,6-Trimethylbenzenesulfonyl.

cooperative catalysis of ALB and $La(OTf)_3 \bullet nH_2O$ (Yamasaki et al. 1999).

Table 10. Catalytic enantioselective epoxide ring opening with 4-methoxyphenol **105** promoted by gallium heterobimetallic complexes in the presence of MS 4A

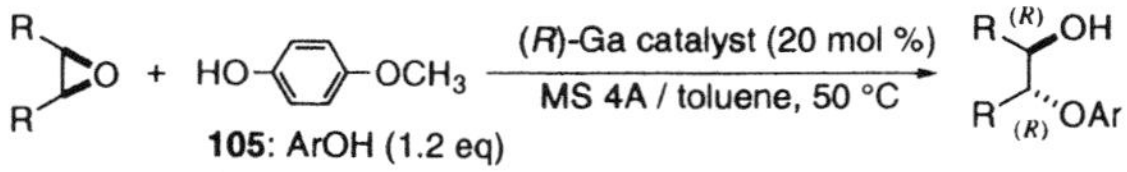

(R)-Ga catalyst (20 mol %)
MS 4A / toluene, 50 °C
105: ArOH (1.2 eq)

Entry	Epoxide	Product	GaLB Time (h)	GaLB Yield (%)	GaLB ee (%)	GaSO Time (h)	GaSO Yield (%)	GaSO ee (%)
1	 **94**	**107**	72 (72)[a]	75 (73)[a]	86 (89)[a]	4	77	54
2	**90**	**108**	72 (72)[a]	48 (60)[a]	93 (94)[a]	4 (4)[b]	73 (61)[b]	56 (51)[b]
3	 **106**	**109**	72	31	67	4	67	58
4	 **91**	**110**	72 (72)[a]	70 (69)[a]	87 (92)[a]	24	90	55
5[c]	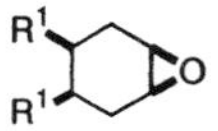 **92**	**111**	96	34	80	48	83	43
6[d]	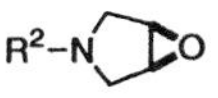 **95**	**112**	160	51	90	19	44	34
7	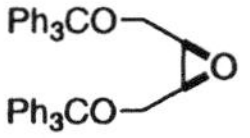 **96**	**113**	72	e	e	7	75	50
8[f]	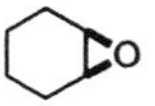 **90**	**114**	–	–	–	4	75	61

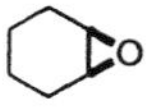

[a] Values in parentheses show the results of the GaLB* catalyzed reaction (B* = 6,6'-bis((triethylsilylethynyl) binaphthol)). [b] 5 mol % GaSO was used. [c] R^1 =CH_2OSiPh_2tBu. [d] R^2 = 2,4,6-Trimethylbenzenesulfonyl ; 30 mol % GaLB was used. [e] No reaction. [f] 4-Methoxy-1-naphthol was used instead of **105**.

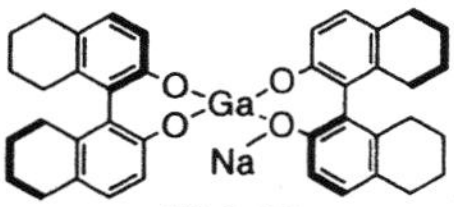

(R)-GaSO

11.3 Bifunctional Asymmetric Catalysis Promoted by Chiral Lewis Acid–Lewis Base Complexes

Based on the achievements described above, it seemed rational to design a new bifunctional asymmetric catalyst consisting of Lewis acid and Lewis base moieties, which activate both electrophiles and nucleophiles at defined positions simultaneously. This type of asymmetric catalysis is seen in only a few examples (Corey et al. 1987; Noyori and Kitamura 1991; Kobayashi et al. 1991). We designed the chiral Lewis acid–Lewis base catalyst **115**. We assumed that the aluminum would work as a Lewis acid to activate the carbonyl group, and that the oxygen atom of the phosphine oxide would work as a Lewis base to activate the silylated nucleophiles. Catalyst **115** has been found to be a highly efficient catalyst for the cyanosilylation of aldehydes (for other catalytic asymmetric cyanosilylation of aldehydes, see Hwang et al. 1998) with broad generality, affording products in excellent chemical yields and excellent enantioselectivities. One of the key issues for designing a Lewis acid–Lewis base catalyst is how to prevent the internal complexation of these moieties.

Molecular modeling studies suggested that **115** would avoid such a problem, because the coordination of the Lewis base to the internal aluminum seemed to be torsionally unfavorable. Considering **116**, however, which has an ethylene linker, the internal coordination seemed to be quite stable without strain. In accordance with this expectation, the reaction of TMSCN with benzaldehyde **123**, catalyzed by **116** (9 mol%), proceeded slowly at –40°C (37 h) and gave the cyanohydrin **133** in only 4% yield after hydrolysis. However, a solution of **115** (9 mol%), **123**, and TMSCN at –40°C (37 h) afforded **133** in 91% yield and in 87% ee. Encouraged by the result with benzaldehyde, we next investigated the reaction of aliphatic aldehydes. Surprisingly, aliphatic aldehydes afforded very low ee values. We anticipated that there would be competition between two reaction pathways in the case of the more reactive aliphatic aldehydes. The desired pathway involves the dual activation between the Lewis acid and the aldehyde and between the Lewis base and TMSCN, whereas the undesired pathway involves monoactivation by the Lewis acid. We assumed that these two pathways could differ more significantly if the Lewis acidity of the catalyst was decreased, and so we investigated the effect of additives which coordi-

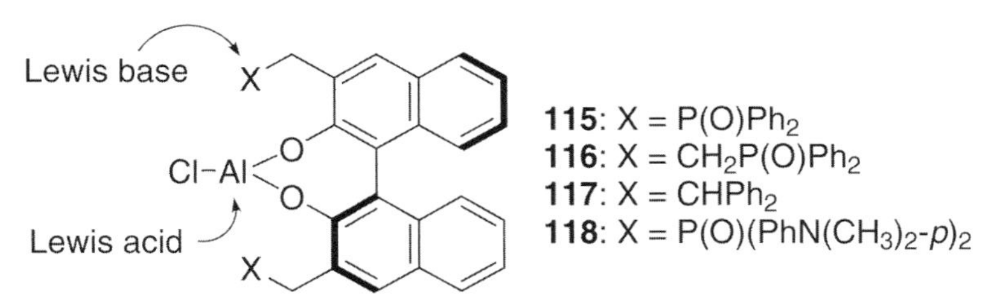

Fig. 3. Chiral Lewis acid–Lewis base catalysts

nate to the aluminum to reduce its Lewis acidity. Moreover, the additive could change the geometry of aluminum from tetrahedral to trigonal bipyramidal, which should allow the phosphine oxide to exist in a more favorable position relative to the aldehyde. We found that electron-donating phosphine oxides had a beneficial effect on ee. (For a beneficial effect on the addition of phosphine oxide, see Daikai et al. 1998.) In the case of **1**, the ee values of **126** significantly increased from 9% to 41% and 56% by the addition of 36 mol% of $CH_3P(O)Ph_2$ and $Bu_3P(O)$, respectively. Further improvement of ee (up to 97%) was achieved by the slow addition of TMSCN (10 h) via syringe pump in the presence of $Bu_3P(O)$. In the case of **123**, however, addition of $Bu_3P(O)$ resulted in a very sluggish reaction, affording only a trace amount of the product. However, the reaction proceeded in 98% yield and in 96% ee in the presence of $CH_3P(O)Ph_2$. Therefore, we used $Bu_3P(O)$ as the additive for aliphatic and α,β-unsaturated aldehydes and $CH_3P(O)Ph_2$ as the additive for aromatic aldehydes. This catalyst is practical and has a broad generality with respect to the variety of aldehydes that can be used (Table 11). To the best of our knowledge, this is the most efficient and the most general catalytic asymmetric cyanosilylation of aldehydes. Preliminary kinetic studies, using catalyst **118**, which contains a more electron-rich phosphine oxide, seem to support the dual Lewis acid–Lewis base activation pathway. The initial reaction rate with **118** (10 mol%) is 1.2 times faster than that with **115** (10 mol%) (k**118**/k**115**=1.2), reflecting the higher Lewis basicity of the phosphine oxide in the reaction of **1** in the presence of $Bu_3P(O)$. Thus, the enantioselectivity of the reaction catalyzed by **115** may be explained by the working model **136**, with the external phosphine oxide coordinating to the aluminum, thus giving a pentavalent aluminum (Hamashima et al.

Table 11. Asymmetric cyanosilylation of aldehydes catalyzed by **115**[a]

RCHO + $(CH_3)_3SiCN$ → 1) **115** (9 mol %), additive (36 mol %), CH_2Cl_2, -40 °C; 2) H^+ → (S)-RCH(OH)CN

Entry	R	Aldehyde	Product	Additive	Time (h)	Yield (%)[b]	ee (%)[c]	S/R
1	$Ph(CH_2)_2$	**1**	**126**	$Bu_3P(O)$	37	97	97	S
2	$CH_3(CH_2)_5$	**119**	**127**	$Bu_3P(O)$	58	100	98	S
3	$(CH_3)_2CH$	**3**	**128**	$Bu_3P(O)$	45	96	90	S
4	$(CH_3CH_2)_2CH$	**58**	**129**	$Bu_3P(O)$	60	98	83	S
5	*trans*-$CH_3(CH_2)_3CH{=}CH_2$	**120**	**130**	$Bu_3P(O)$	58	94	97	–[g]
6	PhCH=CH	**121**	**131**	$Bu_3P(O)$	40	99	98	S
7[d]	2-methylthiazol-4-yl-CH=C(CH_3)–	**122**	**132**	$Bu_3P(O)$	74	97	99	–[g]
8[e]	Ph	**123**	**133**	$CH_3P(O)Ph_2$	96	98	96	S
9	p-$CH_3C_6H_4$	**124**	**134**	$CH_3P(O)Ph_2$	79	87	90	S
10[f]	2-furyl	**125**	**135**	$CH_3P(O)Ph_2$	70	86	95	S

[a] TMSCN (1.8 equiv) was added over 10 h via syringe pump unless otherwise mentioned. [b] Isolated yield. [c] Determined by HPLC analysis. Configuration assigned by comparison to literature values of optical rotation. [d] 20 mol % of **115** and 80 mol % of the addtive were used. 1.2 equiv of TMSCN was used. [e] TMSCN (1.2 equiv) was added dropwise over 1 min. [f] 18 mol % of **115** and 72 mol % of the additive waer used. [g] The absolute configuration was not determined.

136

1999). This concept could provide a guide for designing new asymmetric catalysts for the reaction of a variety of nucleophiles, including silylated ones, with carbonyl compounds.

11.4 Conclusion

We believe that the successful development of the multifunctional concept has opened up a new field in asymmetric catalysis. We hope that the findings discussed herein will prove to be a significant landmark in the development of the field of catalytic asymmetric synthesis.

References

Arai T, Bougauchi M, Sasai H, Shibasaki M (1996a) Catalytic asymmetric synthesis of α-hydroxy phosphonates using the Al-Li-BINOL complex. J Org Chem 61:2926–2927

Arai T, Sasai H, Aoe K, Okamura K, Date T, Shibasaki M (1996b) A new multifunctional heterobimetallic asymmetric catalyst for Michael additions and tandem Michael-aldol reactions. Angew Chem Int Ed Engl 35:104–106

Arai T, Yamada YMA, Yamamoto N, Sasai H, Shibasaki M (1996c) Self-assembly of heterobimetallic complexes and reactive nucleophiles: a general strategy for the activation of asymmetric reactions promoted by heterobimetallic catalysis. Chem Eur J 2:1368–1372

Bosnich B (1986) Asymmetric catalysis. Nijhoff, Dordrecht

Bougauchi M, Watanabe S, Arai T, Sasai H, Shibasaki M (1997) Catalytic asymmetric epoxidation of α,β-unsaturated ketones promoted by lanthanoid complexes. J Am Chem Soc 119:2329–2330

Corey EJ, Bakshi RK, Shibata S (1987) Highly enantioselective borane reduction of ketones catalyzed by chiral oxazaborolidines. Mechanism and synthetic implications. J Am Chem Soc 109:5551–5553

Daikai K, Kamaura M, Inanaga J (1998) Remarkable ligand effect on the enantioselectivity of the chiral lanthanum complex-catalyzed asymmetric epoxidation of enones. Tetrahedron Lett 39:7321–7322

Denmark SE, Wong K-T, Stavenger RA (1997) Highly selective asymmetric aldol additions of ketone enolates. J Am Chem Soc 119:2333–2334

Emori E, Arai T, Sasai H, Shibasaki M (1998) A catalytic Michael addition of thiols to α,β-unsaturated carbonyl compounds: asymmetric protonations. J Am Chem Soc 120:4043–4044

Evans DA, Burgey CS, Paras NA, Vojkovsky T, Tregay SW (1998) C2-symmetric copper(II) complexes as chiral Lewis acids. Enantioselective catalysis of the glyoxylate-ene reaction. J Am Chem Soc 120:5824–5825

Fessner W-D, Schneider A, Held H, Sinerius G, Walter C, Hixon M, Schloss JV (1996) The mechanism of class II, metal-dependent aldolases. Angew Chem Int Ed Engl 35:2219–2221

Funabashi K, Saida Y, Kanai M, Arai T. Sasai, H, Shibasaki M (1998) Catalytic asymmetric Michael addition of nitromethane to enones controlled by (R)-LPB. Tetrahedron Lett 39: 7557–7558

Gröger H, Saida Y, Arai S, Martens J, Sasai H, Shibasaki M (1996) First catalytic asymmetric hydrophosphonylation of cyclic imines: highly efficient enantioselective approach to a 4-thiazolidinylphosphonate via chiral titanium and lanthanoid catalysis. Tetrahedron Lett 37:9291–9292

Gröger H, Saida Y, Sasai H, Yamaguchi K, Martens J, Shibasaki M (1998) A new and highly efficient asymmetric route to cyclic α-amino phosphonates: the first catalytic enantioselective hydrophosphonylation of cyclic imines catalyzed by chiral heterobimetallic lanthanoid complexes. J Am Chem Soc 120:3089–3103

Hamashima Y, Sawada D, Kanai M, Shibasaki M (1999) A new bifunctional asymmetric catalysis: an efficient catalytic asymmetric cyanosilylation of aldehydes. J Am Chem Soc 121:2641–2642

Herrmann WA, Cornils B (1996) Applied homogeneous catalysis with organometallic compounds. VCH, Weinheim

Hwang C-D, Hwang D-R, Uang B-J (1998) Enantioselective addition of trimethylsilyl cyanide to aldehydes induced by a new chiral Ti (IV) complex. J Org Chem 63:6762–6763

Iida T, Yamamoto N, Sasai H, Shibasaki M (1997) New asymmetric reactions using a gallium complex: a highly enantioselective ring opening of epoxides with thiols catalyzed by a gallium-lithium-bis(binaphthoxide) complex. J Am Chem Soc 119:4783–4784

Iida T, Yamamoto N, Woo H-G, Shibasaki M (1998) Enantioselective ring opening of epoxides with 4-methoxyphenol catalyzed by gallium heterobimetallic complexes: an efficient method for the synthesis of optically active 1,2-diol monoethers. Angew Chem Int Ed Engl 37:2223–2226

Kobayashi S, Tsuchiya Y, Mukaiyama T (1991) Enantioselective addition reaction of trimethylsilyl cyanide with aldehydes using a chiral tin (II) Lewis acid. Chem Lett: 541–544

Krüger J, Carreira EM (1998) Apparent catalytic generation of chiral metal enolates; enantioselective dienolate additions to aldehydes mediated by Tol-BINAP Cu(II) fluoride complexes. J Am Chem Soc 120:837–838

Morrison JD (1985) Asymmetric synthesis, vol 5. Academic, Orlando

Nara S, Toshima H, Ichihara A (1997) Asymmetric total syntheses of (+)-coronafacic acid and (+)-coronatine, phytotoxins isolated from Pseudomonas syringae pathovars. Tetrahedron 53:9509–9524

Noyori R (1994) Asymmetric catalysis in organic synthesis. Wiley, New York

Noyori R, Kitamura M (1991) Enantioselective addition of organometallic reagents to carbonyl compounds: chirality transfer, multiplication, and amplification. Angew Chem Int Ed Engl 30:49–69

Ojima I (1994) Catalytic asymmetric synthesis. VCH, New York

Sasai H, Suzuki T, Arai S, Arai T, Shibasaki M (1992) Basic character of rare earth metal alkoxides. Utilization in catalytic C-C bond-forming reactions and catalytic asymmetric nitroaldol reactions. J Am Chem Soc 114:4418–4420

Sasai H, Suzuki T, Itoh N, Shibasaki M (1993a) Catalytic nitroaldol reactions. A new practical method for the preparation of the optically active lanthanum complex. Tetrahedron Lett 34:851–854

Sasai H, Suzuki T, Itoh N, Arai S, Shibasaki M (1993b) Effect of rare earth metals on the catalytic asymmetric nitroaldol reaction. Tetrahedron Lett 34:2657–2660

Sasai H, Itoh N, Suzuki T, Shibasaki M (1993c) Catalytic asymmetric nitroaldol reaction: an efficient synthesis of (S) propranolol using the lanthanum binaphthol complex. Tetrahedron Lett 34:855–858

Sasai H, Suzuki T, Itoh N, Tanaka K, Date T, Okamura K, Shibasaki M (1993d) Catalytic asymmetric nitroaldol reaction using optically active rare earth BINOL complexes: investigation of the catalyst structure. J Am Chem Soc 115:10372–10373

Sasai H, Yamada YMA, Suzuki T, Shibasaki M (1994a) Synthesis of (S)-(-)-pindolol and [3'-^{13}C]-(R)-(-)-pindolol utilizing a lanthanum-lithium-(R)-BINOL ((R)-LLB) catalyzed nitroaldol reaction. Tetrahedron 50:12313–12318

Sasai H, Kim W-S, Suzuki T, Shibasaki M, Mitsuda M, Hasegawa J, Ohashi J (1994b) Diastereoselective catalytic asymmetric nitroaldol reaction utilizing rare earth-Li-(R)-BINOL complex. A highly efficient synthesis of norstatin. Tetrahedron Lett 35:6123–6126

Sasai H, Arai T, Satow Y, Houk KN, Shibasaki M (1995a) The first heterobimetallic multifunctional asymmetric catalyst. J Am Chem Soc 117:6194–6198

Sasai H, Suzuki T, Itoh N, Shibasaki M (1995b) Catalytic asymmetric synthesis of propranolol and metoprolol using La-Li-BINOL complex. Appl Organomet Chem 9:421–426

Sasai H, Tokunaga T, Watanabe S, Suzuki T, Itoh N, Shibasaki M, (1995c) Efficient diastereoselective and enantioselective nitroaldol reactions from prochiral starting materials. Utilization of La-Li-6,6'-disubstituted BINOL complexes as asymmetric catalysts. J Org Chem 60:7388–7389

Sasai H, Arai S, Tahara Y, Shibasaki M (1995d) Catalytic asymmetric synthesis of α-amino phosphonates using lanthanoid-potassium-BINOL complexes. J Org Chem 60:6656–6657

Sasai H, Emori E, Arai T, Shibasaki M (1996) Catalytic asymmetric Michael reactions promoted by the La-Na-BINOL complex (LSB). Enantioface selection on Michael donors. Tetrahedron Lett 37:5561–5564

Sasai H, Bougauchi B, Arai T, Shibaski M (1997a) Enantioselective synthesis of α-hydroxy phosphonates using the La-Li_3-tris(binaphthoxide) catalyst (LLB), prepared by an improved method. Tetrahedron Lett 38:2717–2720

Sasai H, Watanabe S, Shibasaki M (1997b) A new practical preparation method for lanthanum-lithium-binaphthol catalysts (LLBs) for use in asymmetric nitroaldol reactions. Enantiomer 2:267–271

Shibasaki M, Sasai H, Arai T (1997) Asymmetric catalysis with heterobimetallic compounds. Angew Chem Int Ed Engl 36:1236–1256

Shimizu S, Ohori K, Arai T, Sasai H, Shibasaki M (1998) A catalytic asymmetric synthesis of tubifolidine. J Org Chem 63:7547–7551

Steinhagen H, Helmchen G (1996) Asymmetric two-center catalysis: learning from nature. Angew Chem Int Ed Engl 35:2339–2342

Takaoka E, Yoshikawa N, Yamada YMA, Sasai H, Shibasaki M (1997) Catalytic asymmetric synthesis of arbutamine. Heterocycles 46:157–163

Tokunaga M. Larrow, JF, Kakiuchi F, Jacobsen EN (1997) Asymmetric catalysis with water: efficient kinetic resolution of terminal epoxides by means of catalytic hydrolysis. Science 277:936–938

Watanabe S, Arai T, Sasai H, Bougauchi M, Shibasaki M (1998a) The first catalytic enantioselective synthesis of *cis*-epoxyketones from *cis*-enones. J Org Chem 63:8090–8091

Watanabe S, Kobayashi Y, Arai T, Sasai H, Bougauchi M, Shibasaki M (1998b) Water vs. desiccant. Improvement of Yb-BINOL complex-catalyzed enantioselective epoxidation of enones. Tetrahedron Lett 39:7353–7356

Yamada K, Arai T, Sasai H, Shibasaki M (1998) A catalytic asymmetric synthesis of 11-deoxy-PGF1a using ALB, a heterobimetallic multifunctional asymmetric complex. J Org Chem 63:3666–3672

Yamada YMA, Yoshikawa N, Sasai H, Shibasaki M (1997) Direct catalytic asymmetric aldol reactions of aldehydes and unmodified ketones. Angew Chem Int Ed Engl 36:1871–1873

Yamakoshi K, Harwood S, Kanai M, Shibasaki M (1999) Catalytic asymmetric addition of diphenylphosphine oxide to cyclic imines. Tetrahedron Lett 40:2565–2568

Yamasaki S, Iida T, Shibasaki M (1999) Direct catalytic asymmetric Mannich-type reaction of unmodified ketones utilizing the coopoeration of an AlLi-bis(binaphthoxide) complex and La(OTf)3 H_20. Tetrahedron Lett 40:307–310

Yanagisawa A, Matsumoto Y, Nakashima H, Asakawa K, Yamamoto H (1997) Enantioselective aldol reaction of tin enolates with aldehydes catalyzed by BINAP silver(I) complex. Am Chem Soc 119:9319–9320

Yoshikawa N, Yamada YMA, Das J, Sasai H, Shibasaki M (2000) Direct catalytic asymmetric aldol reaction. J Am Chem Soc 121:4168–4177

12 New Directions in Immunopharmacotherapy

K.D. Janda

The application of antibodies for immunotherapy was documented as early as 1890 (von Behring and Kitasato 1890). In the past, polyclonal antibodies found in antiserum against specific antigens were used in passive immunization strategies to combat infectious diseases such as hepatitis B and HIV (Sawyer et al. 1998; Jacobson 1998). While this approach has merit, it is hampered by several unwanted side effects, including the risk of viral contamination, serum sickness, anaphylaxis (nonhuman serum) and, importantly, the difficulty of making standardized batches of antibody to insure reproducibility and proper dosing. A milestone was reached when the production of homogeneous monoclonal antibodies (mAbs) with defined specificity and high affinity became a reality through hybridoma technology, developed by Köhler and Milstein in 1975. As a result, thousands of mAbs against different antigens have been generated and used in a wide variety of biological studies. Additionally, hybridoma technology ushered in the concept of replacing polyclonal antibodies with mAbs as therapeutic drugs (Coons 1995; Scott and Welt 1997; Vaswani and Hamilton 1998). Indeed, with well-defined specificity and better reproducibility, mAbs have been touted as "magic bullets" in the fight against disease. Our research efforts have been aimed at utilizing antibodies, both in passive administration and as generated in vivo by an immune response, for cancer treatment and in the treatment of cocaine abuse. We describe the details and directions of our immunopharmacological studies in these areas.

12.1 Cancer Therapy

Attempts to develop mAbs as therapeutic agents for cancer patients have intensified with the ever-increasing identification of cancer-related antigens (LoBuglio and Saleh 1992a,b; Riethmüller et al. 1993; Dickman 1998). Many mAbs elicited from these antigens have demonstrated an ability not only to specifically recognize tumor cells (Scott and Welt 1997; LoBuglio and Saleh 1992a,b; Riethmüller et al. 1993) but also to induce complement-dependent or antibody-dependent cellular cytotoxicity (CDC or ADCC) elucidated from in vitro studies (Dillmann et al. 1984, 1986; Foon et al. 1984; Dyer et al. 1989; Brown et al. 1989; Miller et al. 1983; Ritz et al. 1981; Grossbard et al. 1992). Such antibodies can also be applied in immunoradiotherapy and radioimmunolocalization (Goldenberg et al. 1991; DeNardo et al. 1988; Press et al. 1989; Rosen et al. 1987), as well as for toxin and chemotherapeutic agent delivery (Oldham 1991; Grossbard et al. 1992; Chaudhary et al. 1989). Despite these promising results, only two mAbs have been approved by the FDA for application in cancer therapy (Table 1). Rituxan (Rituximab), approved in November 1997, is utilized for patients with non-Hodgkin's lymphoma (Marwick 1997). Here, the antibody acts by binding to the B-cell-specific antigen CD-20 and mediating both CDC and ADCC to arrest proliferation of malignant B cells (Marloney et al. 1997). Herceptin (Trastuzumab) was approved for the treatment of metastatic breast cancer in September 1998 (Robertson 1998). Mechanistically, the antibody targets an excess amount of the receptor HER2 on the tumor surface and boosts the efficiency of chemotherapy (Pegram et al. 1998).

Currently, the greatest problem associated with immunotherapy is that the mAbs, such as those above, originate from mice. The drawbacks are significant and include: (a) a circulating half-life of only 16–48 h (Khazaeli et al. 1988) compared with a 21-day half-life of all human

IgGs, except IgG_3 (Waldmann and Strober 1969); (b) an effective plasma concentration that can be obtained only with repeated administration of antibody; (c) a human anti-mouse antibody (HAMA) response with undesirable side effects unless patients receive repeated immunosuppressant treatment (Dillmanne t al. 1986). To counteract these drawbacks, numerous attempts have been made to re-engineer murine mAbs. Two endeavors include the construction of a chimeric antibody (Morrison 1985) (Rituxan) and "humanized" antibody (Jones et al. 1986) (Herceptin). Antibodies engineered in this way can have a longer circulating half-life than their murine counterparts (LoBuglio et al. 1989). The engineering strategies for developing humanized antibodies have included complementarity determining region (CDR) grafting (Jones et al. 1986), reshaping (Verjpeyem et al. 1988; Riechmann et al. 1988), hyperchimerization (Co et al. 1991, 1992, 1996), and veneering (Padlan 1991). Yet antibodies are obtained that usually retain immunogenicity and often have reduced affinity. Most clinical results indicated that HAMA responses were generally lower, and side effects reduced; however, the risk of kidney failure and other long-term manifestations could not be predicted.

A second problem has been finding appropriate cell-surface markers that allow the specific targeting of tumor cells. However, over the past three decades, substantial evidence has accumulated that now concretely associates major changes in glycosphingolipid (GSL) and glycoprotein expression and composition with oncogenic transformation (Hounsell et al. 1997; Kim et al. 1996; Bhavanandan 1991; Hakomori 1986, 1989). Significantly, these changes were found to occur in a wide range of carcinomas and to be correlated with tumor progression, metastasis, and patient survival rates (Hakomori 1996; Muramatsu 1993). Early studies indicated that mAbs could be derived from tumor-cell immunizations of syngeneic mice, and that these mAbs showed binding of cancer cells bearing GSL markers (Ito et al. 1984). Indeed, antibodies that bind GSL and mucin glycoprotein antigens are detectable in the serum of patients with cancers. Even though these markers are often present on both normal cells and carcinomas, they are immunogenic only in the cancerous state (Lloyd 1993). Therefore, gangliosides such as GD2, GD3, GM2, and GM3 and glycoprotein epitopes that include the T, Tn, sTn, Le^x, and Le^y antigens are attractive targets for cancer immunotherapy (Hakomori and Zhang 1997; Apostolopoulos and McKenzie 1994;

Oettgen and Old 1991). By targeting GD2 on tumor cells, two mAbs (3F8 and ch14.18) have been used to treat malignant melanoma in clinical trials (Cheung et al. 1987, 1998a,b; Yu et al. 1998). A recent phase-I trial also showed promising results when children with refractory neuroblastoma were treated with a murine anti-GD2 mAb (14.G2a) and interleukin-2 (Frost et al. 1997).

Passive immunotherapy could be fully exploited if it were possible to have human mAbs against any desired tumor carbohydrate antigen. Interestingly, in melanoma patients found to have serum antibodies that reacted with their own cell-surface antigens, human mAbs with GD2, GD3, GM2, and GM3 specificity were isolated using Epstein-Barr virus transformation of peripheral lymphocytes (Lloyd 1993; Yamaguchi et al. 1990; Furukawa et al. 1989; Irie et al. 1989; Tai et al. 1983). However, these were IgM antibodies undesirable for clinical applications, of low affinity, and difficult to obtain in significant quantities. In general, human mAb technology has lagged far behind that of the well-established murine-based technique. Human hybridomas are difficult to prepare, often unstable, and secrete antibody at low levels (James 1994). If immunopharmacological reagents are to be used for clinical evaluation and the treatment of cancer, they must be available in customized specificities, with high affinity, and in large quantities, and must be compatible with the human immune system. Two significant recent developments offer great promise for enhancing the applicability and efficacy of passive immunotherapeutic protocols for cancer treatment.

First, the development of antibody phage-display technology in our laboratory has allowed the selection of high-affinity human mAbs. In contrast to traditional hybridoma technology, phage display utilizes recombinant heavy and light chain DNA fragments within a phagemid so as to create diverse antibody libraries (Fab or scFv; Kang et al. 1991; Barbas et al. 1991, 1992; Barbas 1994; Burton and Barbas 1994; Lerner et al. 1992). Each antibody is then displayed as a unique fusion protein on the surface of filamentous phage that also contains the genetic information encoding the antibody (Fig. 1). In practice, these antibody phage-display libraries are exposed to an antigen coupled to a solid support to select an antigen-specific antibody. Hence, we now have access in vitro to human immunological repertoires. We have successfully selected mAbs with specific catalytic activity (Janda et al. 1994, 1997; Gao et al. 1997a) as well as the first human Fab specific against

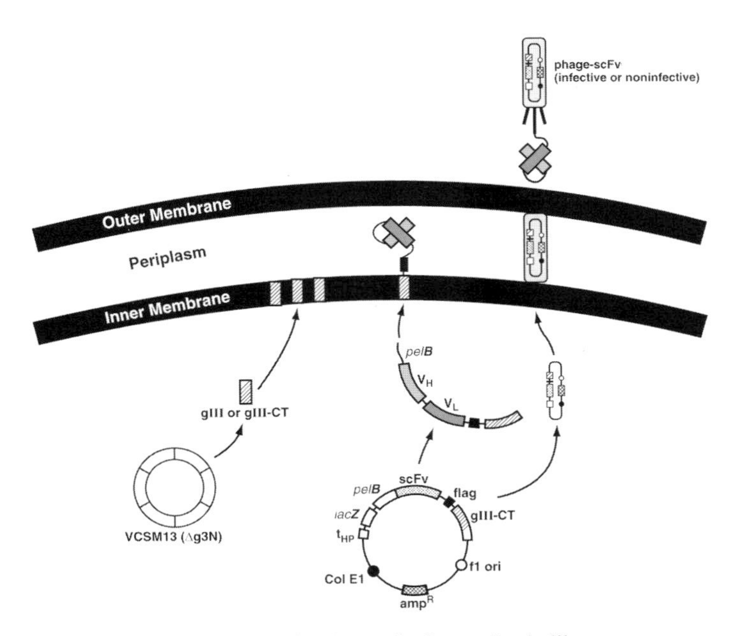

Fig. 1. Pathway for assembly of a phage-display antibody library

cholera toxin B (Mao et al. 2000). In addition, the ability to isolate human mAbs from cDNA libraries of seropositive individuals has furnished mAbs for prophylactic measures (Barbas 1994; Burton and Barbas 1994; Barbas et al. 1992; Lerner et al. 1992). The latter method proved applicable to HIV-1, respiratory syncytial virus, hepatitis B virus, and even in a case of a very low serum titer of the measles virus (Kang et al. 1991; Bender et al. 1994; Burton and Barbas 1993a,b; Williamson et al. 1993; Burton et al. 1991). Current methods should allow the isolation and production of human mAbs that satisfy the demands of specificity and affinity necessary for cancer immunotherapy.

Second, the advances made in complex oligosaccharide syntheses here at The Scripps Research Institute (Takayama and Wong 1997; Gijsen et al. 1996; Wong et al. 1995a,b) and by others (Seeberger et al. 1997; Sames et al. 1997; Castro-Palomino et al. 1997) have revolutionized the study of glycolipids and glycoproteins and their interactions with binding proteins. Now, through synthesis, we will have access to pure, fully characterized carbohydrates that are appropriately modified. These can be readily conjugated to carrier proteins for use as in vitro selection reagents against human antibody libraries. Synthetic oligosaccharides are indeed representative of the epitopes found on the cell surface, as demonstrated by an in vivo murine polyclonal response then found to recognize cancer cells with specificity (Kudryashov et al. 1998; Ragupathi et al. 1997).

The direct selection of antibodies from large libraries provides solutions to virtually all drawbacks associated with anti-carbohydrate murine mAbs obtained from immunization. By imposing constraints upon the selection process, scFvs of high affinity can be obtained for analysis. However, unlike somatic mutation as found in the natural immune system (Tonegawa 1983), antibodies procured from phage-display selection do not go through affinity maturation. Thus, only existing clones can be selected and amplified from the library. With the advent of new technologies such as error-prone PCR and DNA shuffling (Crameri et al. 1998; Stemmer 1994), random mutations can be generated in a chosen antibody. These approaches serve to introduce new amino acids into the antibody combining site and to optimize antibody-antigen interactions. They may also effect an improvement in the stability of the clone through mutation within the framework region. The combination of these techniques within the phage-display approach allows for a more refined clone to be selected (Marks and Marks 1996; Yang et al. 1995). The availability of tight-binding human antibodies derived from in vitro methods would greatly enhance the efficacy of cancer immunotherapy. The ease of manipulation of scFv genes can afford whole IgG of a desired subclass, modified scFv antibodies for drug, radionuclide, or enzyme conjugation, or recombinant scFv fusion proteins. Whether immunotherapeutic methods might someday stand alone in treatment is perhaps not the central issue, but rather their routine and effective application as part of a treatment program. To this end, the adjunct use of immunotherapy could be of the greatest immediate value (Schneider-Gädicke and Riethmüller 1995; Riethmüller et al. 1994). We anticipate that our strategy will significantly advance the utility of immunotherapy in the treatment of cancer.

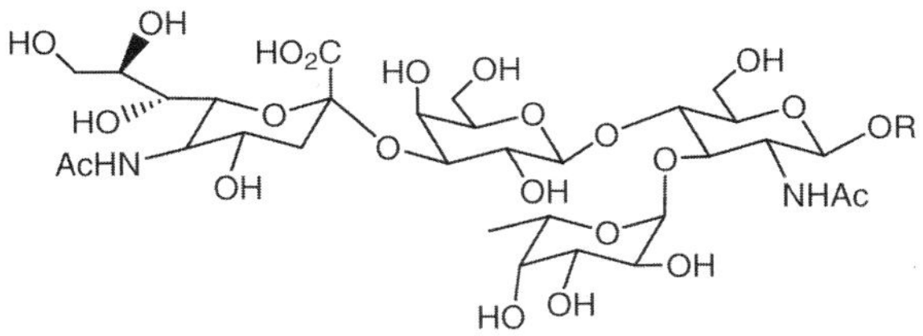

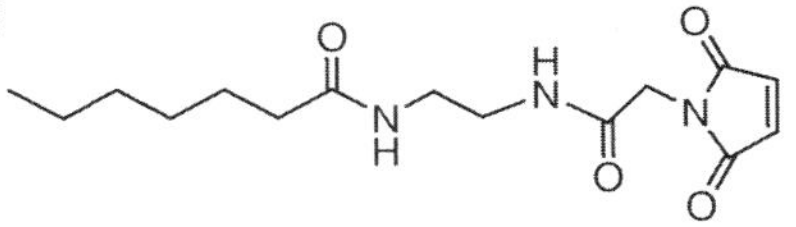

Fig. 2. Structures of sialyl Lewis X and a linker-modified sialyl Lewis X for conjugation to proteins

The carbohydrate sialyl Lewis X (sLe^{X}) **1** was characterized as a tumor-associated antigen for all types of highly malignant cancer (Hakomori 1989; Hakomori 1996; Fukushima et al. 1984). Early studies indicated that mAbs generated against sLe^{X} reacted with leukemic myeloblasts as well as human neutrophils and monocytes but not with normal hematopoietic progenitor cells (Muroi et al. 1992). Also significant, sLe^{X} was identified as a ligand for the cell adhesion molecule ELAM-1 (E-selectin endothelial cell leukocyte adhesion molecule-1) that is involved in the rolling/tethering process during an inflammatory response (Berg et al. 1991; Phillips et al. 1990). In light of the chemical verification for the existence of sLe^{X} on the surface of tumor cells (Hakomori 1989) and its key role in the adhesion of human tumor cells to endothelium (Takada et al. 1993), we chose sLe^{X} as our first target (Fig. 2). Human mAbs specific for sLe^{X} should be useful in general cancer therapy. Both sLe^{X} and its derivative **2** have been chemoenzymatically prepared by other workers and reproduced in our laboratory (Wittmann et al. 1998). We also prepared conjugates with various proteins, such as bovine serum albumin (BSA).

Single-chain antibody fragments (scFvs) (Bird et al. 1988; Huston et al. 1988) are unable to induce either CDC or ADCC, yet they have several distinct advantages over whole IgG in regards to cancer immunotherapy. First, due to their relatively small size (27 kDa), scFvs will clear plasma at an accelerated rate and can also penetrate more rapidly and deeper into tissue than a whole IgG molecule (150 kDa) (Thirion et al. 1996). For example, an iodine-123-labeled anti-CEA (carcinoembryonic antigen) scFv showed tumor localization superior to that of whole IgG (Chester and Hawkins 1995; Yokota et al. 1992). Second, potential side effects will be reduced, since lack of the constant region ensures that scFvs will not be retained in organs such as the liver and/or kidney. Third, an scFv can be constructed in an expedient manner using the polymerase chain reaction (PCR) and prepared as a phage library for direct selection. Finally, the scFv can be re-engineered into several different formats such as a diabody (Poljak 1994), triabody (Pei et al. 1997), or bispecific antibody (Mallender and Voss 1994).

We chose a combinatorial approach to directly construct scFv libraries from the peripheral blood lymphocytes (PBLs) of cancer patients. The cDNA was generated by reverse transcription from RNA isolated from PBLs of 20 patients. The cancer phenotypes and the corresponding numbers of patients were: breast (5), multiple myeloma (3), testicular (2), colon (2), non-Hodgkin's lymphoma (2), lung (1), stomach (1), malignant paraganglioma (1), cholangiocarcinoma (1), hairy-cell leukemia (1), and acute leukemia (1). Our libraries will constantly be upgraded upon acquisition of blood with new phenotypes. The PCR-amplified fragments of the cDNA were cloned into a phagemid vector, pCGMT, previously designed and constructed in our group (Gao et al. 1997b). The vector allows human scFv to be expressed as a fusion protein that can be assembled onto the surface of phage. The presence of antibody on the surface of phage enables selection using immunoconjugates and various panning procedures. Using conventional panning procedures, the sLe^x-BSA conjugate was immobilized on an immuno-tube. Four rounds of panning were performed that culminated in the isolation of four human scFv antibodies (S6, S7, S8, S10) all capable of recognizing the sLe^x antigen. Using the sLe^x derivative **2**, we were able to covalently attach the compound directly to modified microtiter plates. This experiment revealed that the four scFvs indeed recognized the sLe^x epitope. Specificity was confirmed through competitive inhibition experiments using free sLe^x, sLe^x-BSA, and an anti-sLe^x murine mAb (Sasaki et al. 1993).

To produce soluble scFvs, *Escherichia coli* HB2151 cells (Pharmacia Biotech) were infected with phage obtained after four rounds of pan-

Table 1. Kinetics and thermodynamics of binding of sLe^x and Le^x conjugates to scFvs[a]

	sLe^x-biotin			Le^x-BSA		
scFv	k_{on} ($\times 10^4\ M^{-1}\ s^{-1}$)	k_{off} ($\times 10^{-3}\ s^{-1}$)	K_d ($\times 10^{-7}\ M$)	k_{on} ($\times 10^4\ M^{-1}\ s^{-1}$)	k_{off} ($\times 10^{-3}\ s^{-1}$)	K_d ($\times 10^{-7}\ M$)
S6	4.3	4.7	1.1	1.3	8.3	6.2
S7	1.1	5.3	4.9	2.6	7.3	2.9
S8	5.1	9.1	1.8	2.3	6.5	2.9
S10	1.2	6.7	5.7	2.1	4.6	2.2

[a]The kinetic constants of scFvs were measured using the BIAcore biosensor: k_{on} association rate constant, k_{off} dissociation rate constant, K_d calculated ($K_d = k_{off}/k_{on}$)

ning. Recognition of the amber stop codon between the scFv gene and M13 gene III protein resulted in the production of soluble scFv by this strain. Affinity chromatographic purification of each antibody allowed binding data to be garnered. Since sLe^x and Le^x share a common biosynthetic pathway and have related carbohydrate structures, the fine specificities of the four scFvs were analyzed by testing the binding to Le^x-BSA by ELISA. It was found that the four scFv antibodies bound sLe^x as well as Le^x; therefore, sialic acid was not a recognition element for these antibodies. Notably, Le^x also occurs on a wide range of tumor cells, so that our antibodies will be valuable in this regard. The binding specificities and kinetic parameters of purified scFvs were further determined using surface plasmon resonance (SPR) (BIAcore; Pharmacia). It was shown that the scFvs bound to immobilized sLe^x or Le^x conjugates. Binding to the sensor chips was inhibited by these antigens. The dissociation equilibrium constants (K_d) of the scFvs were calculated from association and dissociation rate constants (Table 1).

Analysis of the kinetic data revealed that the scFvs, although derived from two different V_H germlines, showed similar affinities for sLe^x and Le^x. The K_d values ranged from 1.1 to 6.2×10^{-7} M. The affinities are the highest that have been observed for anti-oligosaccharide antibodies directly isolated from phage-antibody libraries. They are also comparable to the affinities of mAbs derived from the murine secondary immune response. The scFvs were also analyzed by fluorescence-activated cell sorting (FACS) for binding to human pancreatic adenocarcinoma SW1990 cells to confirm binding to cells overexpressing sLe^x (Fig. 3).

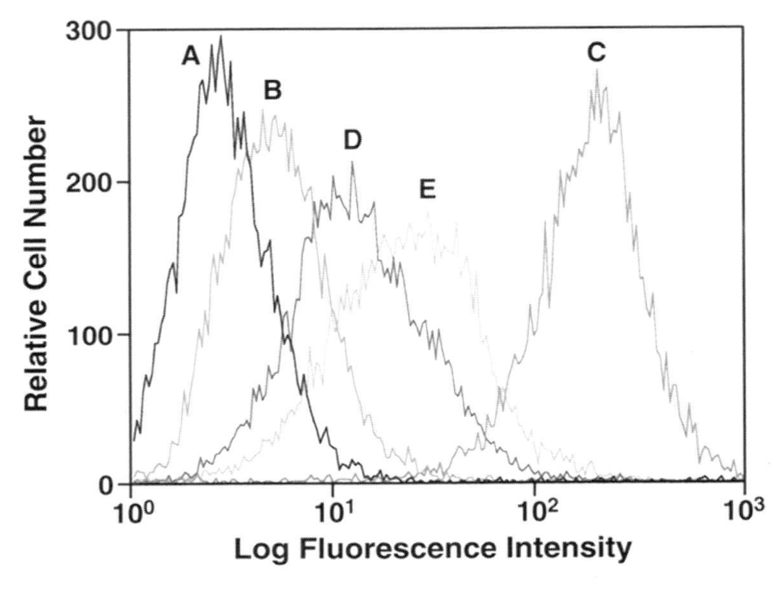

Fig. 3. Binding activity of human scFvs on SW1990 cells using FACS (*A* BSA control, *B* scFv S6, *C* scFv S7, *D* scFv S8, *E* scFv S10)

All of the scFv fragments bound to SW1990 cells. With this assay, clone S7 showed the best binding. Given the similar thermodynamic and kinetic behavior of the scFvs in previous assays, the differences in mean fluorescence intensity (MFI) in binding the cellular epitope likely reflected differences in the presentation of the sLe^x structure and different domains of sLe^x recognized by the scFvs. Indeed, we provided evidence for this hypothesis by performing a two-site BIAcore binding experiment. It was observed in all cases examined that different pairs of scFvs bound concurrently to the sLe^x antigen (Table 2). We have recently published our results in full detail (Mao et al. 1999).

Besides having therapeutic potential in their natural form, antibodies can carry radionuclides, enzymes, genes, drugs, or toxins to target cells (Reisfeld and Gillies 1996; Pietersz et al. 1994). Our aim is to utilize scFv conjugates to directly promote cytotoxicity. We will construct scFv-drug or scFv-radionuclide conjugates that would bring about cell death as traditionally studied in the past. Now, however, in accordance

Table 2. A BIAcore two-site binding assay of scFvs and sLe^x (*ND* not determined)

	Immobilized scFv[a]		
Second scFv	S7	S8	S10
S6	+	+	+
S7	-	ND	ND
S8	+	-	+
S10	+	ND	-

[a]Immobilized scFv was used to trap sLe^x. The second scFv was then tested for binding to the scFv-sLe^x complex. Plus signs (+) indicate scFvs that bound concurrently (typically 100–500 RU) and hence recognized different regions of sLe^x. Minus signs (-) indicate scFvs that did not bind concurrently (typically 0–20 RU).

with our proposal, we will be able to target tumors via the carbohydrate antigens using human antibodies.

We have pursued a multidisciplinary approach to the selection of human anti-glycosyl antibodies and their application in the targeting of tumor-associated antigens. In our laboratory, we have developed and continue to expand on the molecular biology and immunochemistry required to procure human scFvs. The glycosyl panning reagents analogous to the peripheral carbohydrate motifs of tumor cell-surface GSLs and glycoproteins are accessible based on methodologies developed by other workers. In addition, in cases where synthetic reagents are not feasible, cloned or isolated glycoproteins can be utilized within the context of our human antibody libraries. Either way, we have the ability to obtain human mAbs that bind tumor carbohydrate antigens with high affinity and specificity. Our studies encompass the use of our isolated scFvs for the construction of human monoclonal whole IgG, as well as scFv immunopharmacological reagents including drug and radionuclide conjugates for tumor targeting. Both scFv constructs and scFv reagents should be applicable to passive immunotherapeutic paradigms and rapid entry into clinical trials.

12.2 Cocaine Abuse Therapy

Cocaine abuse continues to be prevalent. Its escalation into a major medical and social problem has degraded the fabric of society in many countries the world over. Recent surveys for the United States indicated that more than 23 million people have tried cocaine, that nearly 400,000 use it daily, and that 5000 new users are added each day (Blaine and Ling 1992; National Institute on Drug Abuse Division of Epidemiology and Prevention 1991). Although abuse appears to be stabilizing, as much as 0.3% of the population may be dependent on the drug (US Department of Health and Human Services 1993). A myriad of medical problems, including death, often accompany cocaine use, and the association of the drug with the spread of AIDS is of concern (Klonoff et al. 1989; Cregler and Herbert 1986). Furthermore, the detrimental effects are especially tragic for pregnant women, where "crack" is the most abused illicit drug (Lee and Bennett 1991; Zuckerman et al. 1989; Frank et al. 1988). There is no doubt that cocaine abuse has insidious and far-reaching consequences on health, the legal system, and the family unit. Clearly, its eradication will have an enormous impact on improving the quality of our lives.

Despite intensive efforts, the development of effective therapies for cocaine craving and addiction remain elusive. In the absence of a "magic bullet" drug, available pharmacological agents must be part of a comprehensive approach toward treatment. A number of biopsychosocial models have been proposed and evaluated to address addiction and relapse prevention (Hoffman et al. 1994; Carroll 1993; Wallace 1992). Unquestionably, an improved pharmacotherapy would increase the effectiveness of such programs. At present, there is a compelling need for alternatives to the existing strategies if progress is to be made in the control of the cocaine problem. One alternative might rely on immunological reagents and the immune system. Early work demonstrated that antibodies specific for haptenic drugs were feasible and that they were useful in the attenuation of their effects (Spector et al. 1973; Berkowitz and Spector 1972). For any new drug or therapy, it is important to have reliable animal models for assessing safety and effectiveness. We have proposed to implement a number of paradigms that provide an excellent correlation with the human condition of cocaine abuse.

Fig. 4. Structures of cocaine and benzoylecgonine

The immune-mediated binding of cocaine would impede its passage into the central nervous system and would result in a suppression of its characteristic actions. The described assays provide a sensitive and quantitative tool for evaluating the effectiveness of anti-cocaine antibodies or catalytic antibodies in blocking the psychostimulant effects of cocaine. The advent of new technologies for monoclonal antibody production and for creating highly specific human antibodies can bring these proteins into the realm of clinical evaluation. In addition, there is the exciting opportunity for a vaccine against the drug's toxic and addictive properties. Immunopharmacotherapy offers a new avenue in the challenging battle against cocaine.

Since cocaine is a small, haptenic molecule it requires coupling to a carrier protein to elicit an immune response. The design and preparation of a cocaine immunogen require special regard for the stability of free cocaine in solution and particularly as a haptenic determinant. It is instructive to compare the structures of cocaine **3** and benzoylecgonine **4** (Fig. 4).

Cocaine degrades spontaneously in vitro and in vivo, largely through pH- and temperature-dependent hydrolysis of the methyl ester to produce the nonpsychoactive compound benzoylecgonine (Garrett and Seyda 1983; Stewart et al. 1979; Cunningham and Lakoski 1990). Nonspecific esterases are also known to contribute to the in vivo degradation of cocaine and have been shown to cleave both the methyl and the benzoate esters (Stewart et al. 1979; Brzezinski et al. 1994; Boyer and Petersen 1992; Dean et al. 1991; Matsubara et al. 1984; Liu et al. 1982). While the latter process is significantly diminished for an immunoconjugate, spontaneous hydrolysis is not affected. Conjugates that have epitopes structurally similar to those of metabolites, especially benzoylecgonine, would compromise the avidity and specificity of an anti-

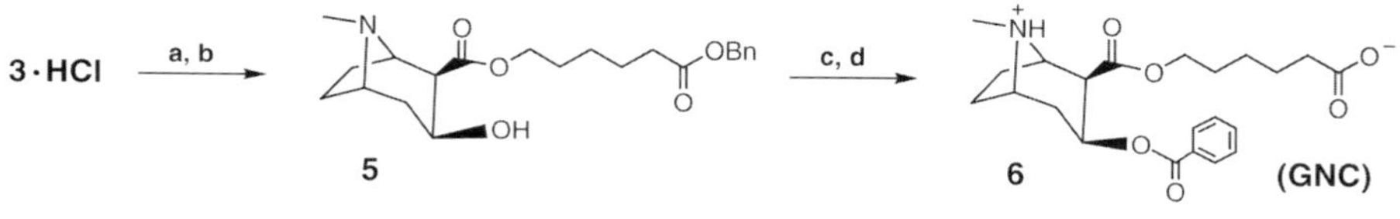

3·HCl →(a, b) 5 →(c, d) 6 (GNC)

a) 1.25 M HCl, reflux; b) $Br(CH_2)_5CO_2Bn$, NaOH, pyridine; c) benzoyl chloride, NEt_3, DMAP; d) H_2, Pd/C, MeOH

Fig. 5. Synthesis of the GNC hapten used in the preparation of an anti-cocaine vaccine

cocaine immune response as a result of antigenic competition (Taussig 1973). Also, an appreciable benzoylecgonine titer would be exceptionally detrimental because the antiserum would be inadequate to neutralize cocaine, particularly in the presence of rapidly formed and stable metabolites. Both cocaine and benzoylecgonine present the phenyl ring as a major recognition element, but the neutrality of cocaine contrasts with the negatively charged benzoylecgonine that is a factor in antibody binding. Hence, it is possible to maximize the affinity and selectivity for cocaine of antibodies generated in an immune response.

Although the chemistry of cocaine has been studied for over 100 years, it was not until we modified some of the methodology of Gray (1925) that the compounds needed for immunological studies became known (unpublished data; patent pending, The Scripps Research Institute). Recently, a different sequence of reactions affording similar structures appeared in the literature (Lewin et al. 1992). The requisite chemistry necessitates the introduction of a linker group onto the cocaine framework. Significantly, the linkage must be made at the position of the methyl ester. By joining the carrier protein to the cocaine framework in this way, any minor decomposition of the linked hapten results not in a benzoylecgonine response, but primarily in nonhaptenic recognition. Coupling by way of other functionalities (e.g., nitrogen, phenyl ring, bicyclic ring) increases the likelihood of interference from the cocaine metabolite. Attention to such aspects of the immunochemistry must be emphasized, in view of an unsuccessful report of a potential cocaine prophylactic (Bagasra et al. 1992; Gallacher 1994).

The compound **6** (hapten code: **GNC**) was synthesized in four steps starting from (-)-cocaine (Fig. 5). The key reaction, alkylation of (-)-ecgonine, introduced the required tether. The configuration remained intact at C-2 of the tropane nucleus. This ester linker mimics the alkyl character of the methyl ester of cocaine that is important for recognition of this part of the molecule.

Coupling of **6** to keyhole limpet hemocyanin (KLH) and bovine serum albumin (BSA) afforded the conjugate **6**-KLH (**GNC**-KLH) for immunization and **6**-BSA (**GNC**-BSA) for enzyme-linked immunosorbent assay (ELISA) analyses. Our work has demonstrated that **GNC**-KLH can be used to immunize mice in a highly consistent fashion and thereby produce quality monoclonal antibodies (mAbs). Significantly, we have demonstrated that these antibodies have excellent affinity and specificity for cocaine compared with benzoylecgonine and other cocaine metabolites.

Perhaps the ultimate goal in an immunological approach aimed at the abatement of cocaine abuse is the de novo design of a vaccine. In principle, vaccination would impart to the host's immune system the ability to defend itself against the acute psychostimulant and toxic effects of cocaine and its powerful reinforcing properties.

One mechanism for a vaccine is that circulating cocaine can stimulate the memory B cells, induced by an appropriate immunogen, to differentiate and produce antibodies. However, this may not be a tenable hypothesis, since it is generally accepted that B cells must collaborate with T cells to culminate activation and progression to a cycling, immunoglobulin-producing cell (Möller 1974). Even though the secondary immune response is much less T-cell dependent and requires less antigen than the intimately linked T-helper cell and lymphokine-activated primary response, it is unlikely that a free hapten can provide adequate recognition to initiate T-cell-dependent interactions (Vitetta et al. 1991; Kehrl et al. 1984). Yet it cannot be completely ruled out, given that secondary B cells are able to undergo stimulation by hapten-carrier complexes bearing relatively nonimmunogenic carriers, in the absence of T cells, albeit to a reduced extent (Klinman and Doughty 1973). Many aspects of B- and T-cell properties remain ambiguous. The details of immunologic memory and the mechanisms underlying the origin and generation of memory B cells have not been determined (Vitetta et al. 1991).

While the advantages of a sustained memory effect derived from the maturation and refinement of the immune system upon repeated use of cocaine would be favorable, this is not anticipated in light of our current understanding. A more realistic expectation of a "first-generation" vaccine would rely on antibody production established through immunoconjugate memory along with a process of repeated immunization. It is known that memory B cells express high-affinity, antigen-specific surface receptors generated via the process of somatic hypermutation (Vitetta et al. 1991; Pike et al. 1987). With some vaccines, subsequent immunity can last for months, years, or the lifetime of the host. In addition, terminally differentiated IgG-secreting plasma cells can have life spans of up to 2 months following a post-boost vaccination (MacLennan and Gray 1986).

The scavenging and neutralization of cocaine via such a mechanism would involve a fixed number of binding sites. Since the in vivo breakdown of cocaine produces benzoylecgonine that can compete for these sites, the available surface and soluble immunoglobulins must have a high specificity for cocaine. This reemphasizes the importance of the immunoconjugates we are developing that establish the necessary characteristics of the immune response. A vaccine that operates in this way would require periodic boosters as part of a self-compliant rehabilitation and therapeutic program. Importantly, an active immunization against cocaine offers a means of blocking the actions of the drug by preventing it from entering the central nervous system and should have fewer side effects than treatments based on manipulation of central neurotransmitter function. Thus, immunopharmacotherapy may offer a nontoxic, substance-specific strategy that should not affect normal neurochemical physiology, presenting a solid scientific approach for cocaine abuse treatment.

We brought together immunochemistry and a well-defined behavioral model and demonstrated the suppression of the psychoactive effects of cocaine (Carrera et al. 1995). An experimental group of male Wistar rats was immunized with **GNC**-KLH and a control group inoculated with a control modified-carrier protein (propionate-KLH). When the two groups were challenged with cocaine, the ambulatory response (crossovers) measured in photocell cages was significantly reduced in the experimental animals (Fig. 6). Stereotyped behavior was also suppressed in the experimental animals consistent with the decreases in locomotor activity, and this effect was significant across all challenges.

Analysis of serum immunoglobulin dissociation kinetics, antigen-binding capacity, and the effect of antigen dilution (Minden and Farr

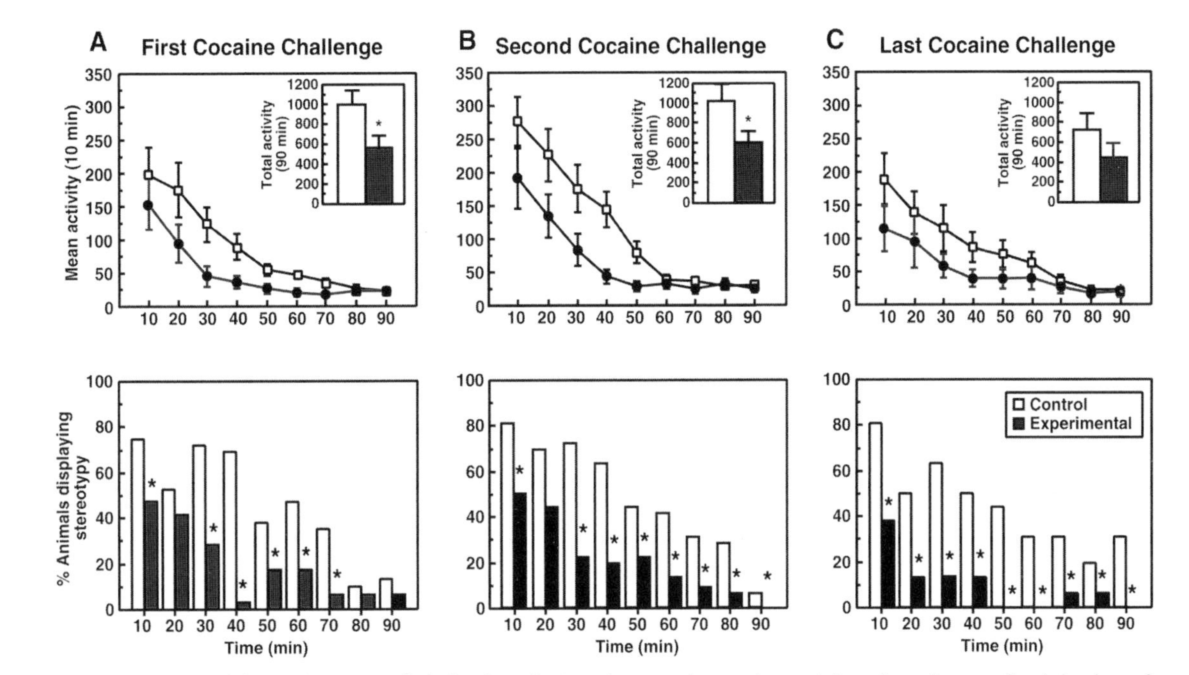

Fig. 6. Locomotor activity and stereotypic behavior of control rats and experimental (vaccinated) rats after injection of cocaine

1967; Steward and Steensgaard 1983; Eisen 1964; Hill et al. 1973) were consistent with an in vivo antibody excess having a micromolar average binding constant. In addition, comparison with one of our purified anti-cocaine murine mAbs (GNC92H2) indicated that 1 ml undiluted serum from an immunized rat could bind the same amount of cocaine as 4 mg ml^{-1} mAb. To complement these experiments, plasma cocaine levels were determined by HPLC following a 15 mg kg^{-1} intraperitoneal injection in a separate group of animals. A measured peak concentration of 5.56± 0.31 μM occurred at 5 min that decayed with a half-life of 25 min. Taken together, the data suggested that at equilibrium at least 50% of the cocaine present in the blood under physiological conditions is likely to be bound.

The basis for the blunted behavioral response to cocaine in the animals immunized with **GNC**-KLH was substantiated by measurement of cocaine concentrations in the brain. The levels of cocaine were found to be 52% lower in the striatal tissue (Fig. 7a) and 77% lower in the cerebellar tissue (Fig. 7b) of the experimental animals compared with the controls. This difference correlated well with the estimated levels of bound cocaine in the bloodstream. These results are important in demonstrating that modulation of cocaine levels in the circulation by antibody binding directly influence behavior.

The antibody titer was the same before the first challenge and 8 days after the last challenge, a span of 20 days. Hence, given enough time between cocaine injections, saturation of the immune mechanism may be avoided and protection sustained. The results warrant further investigation of the potential for immunotherapy to address the relapse and "binge-like" patterns of the human condition in the context of drug-dependence models.

We have now generated preliminary evidence in this direction. In order to test the hypothesis that this immunotherapy is effective in preventing cocaine self-administration reinstatement, an animal model of priming-induced reinstatement was established. Male Wistar rats (n=10) were trained to press a lever for food under a fixed schedule of 1 (FR-1) (General Methods, see below). Rats were prepared with intrajugular catheters and trained to self-administer cocaine in daily 1-h sessions preceded by a single noncontingent cocaine prime (0.75 mg/kg/infusion). After baseline criteria were met, animals were initiated in the immunization protocol (n=5, **GNC**-KLH; n=5, KLH)

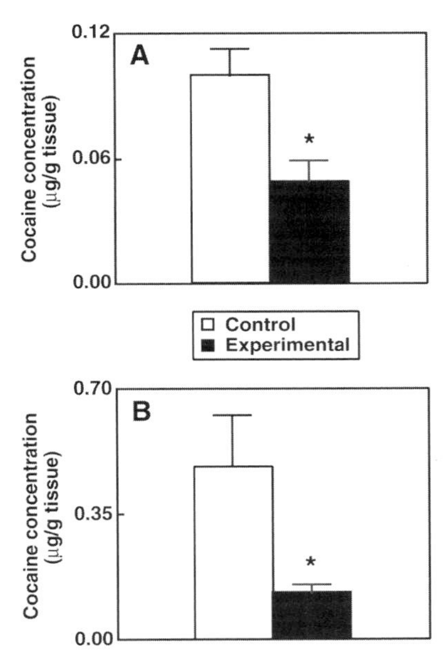

Fig. 7. Levels of cocaine in the brains of control and experimental (vaccinated) rats after injection of cocaine

and exposed to a period of extinction by replacing drug with saline. Once immunization was completed and extinction-like responding was stable, animals were again primed with cocaine and allowed to press the lever, but their responses were not rewarded. After 2 days, lever-pressing was reinforced as follows: one contingent infusion on day 3; two contingent infusions on day 4; free access to cocaine on days 5, 6, 7, and 8. Reinstatement of self-administration behavior was observed in all vehicle (KLH)-treated rats, where baseline rates of responding values were reacquired. After contingent reinforcement and free access to cocaine, their rate of responding remained within baseline criteria (Table 3). Conversely, single cocaine primes did not result in reinstatement of responding in the immunized (**GNC**-KLH) group. These results indicate that active immunization with **GNC**-KLH effectively prevents reinstatement of cocaine self-administration after a single prime injection in rats. Upon reestablishment of contingent reinforcement condi-

Table 3. Mean number of cocaine self-administration responses ±SEM per 1-h session after cocaine infusion (0.75 mg/kg) prior to and following immunization, in rats

	GNC-KLH	KLH
Baseline	15.86±3.04	13.46±2.91
Extinction	8.66±1.21	5.93±0.96
Reinstatement		
Day 1 (1 prime)	9.60±1.88[a]	15.00±2.14
Day 2 (1 prime)	7.00±1.50[a]	12.20±2.17
Day 3 (1 prime + 1 infusion)	15.40±5.01	13.20±1.43
Day 4 (1 prime + 2 infusions)	13.80±2.51	17.20±1.72
Free cocaine	28.40±3.63[a]	14.70±2.63

[a]Significantly different from control group (KLH).

tions, three of the five animals reinstated responding only in the immunized group. Also, free cocaine access produced a 180% increase in the rate of responding (Table 3). Thus, surmountability of antibody titers by increasing amounts of ingested cocaine elicited reinstatement in some animals and doubled the values of cocaine intake as compared with baseline in immunized animals but not in controls.

We are in the process of accumulating large amounts of our best anti-cocaine mAbs elicited with GNC-KLH for in vivo experimentation. We have already produced several grams of the murine mAb GNC92H2 and have done exploratory studies by injecting several rats. The antibody was introduced to afford an in vivo concentration of approximately 1–1.5 mg/ml. There were no apparent physiological or behavioral anomalies, and preliminary evidence suggested cocaine was effectively bound in the bloodstream. Our previous data using rats indicated a peak plasma concentration of 5 μ*M* cocaine following a 15 mg/kg i.p. injection (see above). This is similar to the concentration that occurs in human beings following a 100-mg intravenous dose (Barnett et al. 1981). Consequently, it is anticipated that circulating antibody at 2 mg/ml (~25 μ*M* in binding sites) in the human bloodstream would have significant therapeutic value. This concentration would provide an antibody excess wherein free cocaine would be rapidly and completely bound at equilibrium. Intravenous application of mAbs at 2 mg/ml is considered within the usual limits of antibody therapy (Imbach 1991).

For such experiments to achieve adequate success, it is again important to underscore the necessity of high antibody selectivity for cocaine versus cocaine metabolites, a criterion fulfilled by our mAbs.

We believe the passive administration of anti-cocaine antibodies may prove beneficial in overdose cases to reduce serum levels and attenuate toxic effects, or as (bi)weekly pharmacotherapy during rehabilitation. The latter could entail self-injection of mAb to maintain a high circulating level of antibody. In a more novel approach, it may even be possible to establish passive mucosal immunity protocols through the use of aerosolized immunoglobulin (Crowe et al. 1994). These approaches may be effective for cocaine abuse by keeping cocaine below a threshold concentration required for reinforcement. Our murine mAbs could be "humanized" via several techniques (LoBuglio et al. 1989; Verhoeyen et al. 1988; Riechmann et al. 1988; Padlan 1991). In addition, our methodology for selecting antibodies of desired specificity from combinatorial libraries would make human mAbs directly available. Both approaches make clinical application quite feasible.

We have investigated and continue to develop an immunopharmacological treatment for cocaine abuse. Our program incorporates both active and passive immunization protocols to establish a sound scientific approach to the suppression of the psychoactive effects and detoxification of cocaine. Animal models have demonstrated the efficacy of our strategy. Application of vaccination and antibody administration to the human condition, together with solid psychosocial counseling, should aid in reducing the cocaine problem in our society.

References

Apostolopoulos V, McKenzie IFC (1994) Cellular mucins: targets for immunotherapy. Crit Rev Immunol 14:293–309

Bagasra O,Forman LJ, Howeedy A, Whittle P (1992) A potential vaccine for cocaine abuse prophylaxis. Immunopharmacology 23:173–179

Barbas CF III (1994) The combinatorial approach to human antibodies. In: Rosenberg M, Moore GP (eds) The pharmacology of monoclonal antibodies. Springer, Berlin Heidelberg New York, pp 242–266 (Handbook of experimental pharmacology, vol 113)

Barbas CF III, Kang AS, Lerner RA, Benkovic SJ (1991) Assembly of combinatorial antibody libraries on phage surfaces: the gene III site. Proc Natl Acad Sci U S A 88:7978–7982

Barbas CF III, Bain JD, Hoekstra DM, Lerner RA (1992) Semisynthetic combinatorial antibody libraries: a chemical solution to the diversity problem. Proc Natl Acad Sci U S A 89:4457–4461

Barnett G, Hawks R, Resnick R (1981) Cocaine pharmacokinetics in humans. J Ethnopharmacol 3:353–366

Bender E, Pilkington GR, Burton DR (1994) Human monoclonal Fab fragments from a combinatorial library prepared from an individual with a low serum titer to a virus. Human Antibod Hybrid 5:3–8

Berg EL, Robinson MK, Mansson O, Butcher EC, Magnani JL (1991) A carbohydrate domain common to both sialyl Le^{a} and sialyl Le^{x} is recognized by the endothelial cell leukocyte adhesion molecule ELAM-1. J Biol Chem 266:14869–14872

Berkowitz B, Spector S (1972) Evidence for active immunity to morphine in mice. Science 178:1290–1292

Bhavanandan VP (1991) Cancer-associated mucin and mucin-type glycoprotein. Glycobiology 1:493–503

Bird RE, Hardman KD, Jacobson JW, Johnson S, Kaufman BM, Lee SM, Lee T, Pope SH, Riordan GS, Whitlow M (1988) Single-chain antigen-binding proteins. Science 242:423–426

Blaine JD, Ling W (1992) Psychopharmacologic treatment of cocaine dependence. Psychopharmacol Bull 28:11–14

Boyer CS, Petersen DR (1992) Enzymatic basis for the transesterification of cocaine in the presence of ethanol: evidence for the participation of microsomal carboxylesterases. J Pharmacol Exp Ther 260:939–946

Brown SL, Miller RA, Horning SJ, Czerwinski D, Hart SM, McElderry R, Basham T, Warnke RA, Merigan TC, Levy R (1989) Treatment of B-cell lymphomas with anti-idiotype antibodies alone and in combination with alpha interferon. Blood 73:651–661

Brzezinski MR, Abraham TL, Stone CL, Dean RA, Bosron WF (1994) Purification and characterization of a human liver cocaine carboxylesterase that catalyzes the cocaine. Biochem Pharmacol 48:1747–1755

Burton DR, Barbas CF III (1993a) Human antibodies to HIV-1 by recombinant DNA methods. Chem Immunol 56:112–126

Burton DR, Barbas CF III (1993b) Human antibodies from combinatorial libraries. In: Clark M (ed) Protein engineering of antibody molecules for prophylactic and therapeutic applications in man. Academic Titles, Nottingham, pp 65–82

Burton DR, Barbas CF III (1994) Human antibodies from combinatorial libraries. Adv Immunol 57:191–280

Burton DR, Barbas CF III, Persson MAA, Koenig S, Chanock RM, Lerner RA (1991) A large array of human monoclonal antibodies to type 1 human immunodeficiency virus from combinatorial libraries of asymptomatic seropositive individuals. Proc Natl Acad Sci U S A 88:10134–10137

Carrera MRA, Ashley JA, Parsons LH, Wirsching P, Koob GF, Janda KD (1995) Suppression of psychoactive effects of cocaine by active immunization. Nature 378:727–730

Carroll KM (1993) Psychotherapeutic treatment of cocaine abuse: Models for its evaluation alone and in combination with pharmacotherapy. NIDA Res Monogr 135:116–132

Castro-Palomino JC, Ritter G, Fortunato SR, Reinhardt S, Old LJ, Schmidt RR (1997) Efficient synthesis of ganglioside GM2 for use in cancer vaccines. Angew Chem Int Ed Engl 36:1998–2001

Chaudhary VK, Queen C, Junghans RP, Waldmann TA, FitzGerald DJ, Pastan I (1989) A recombinant immunotoxin consisting of two antibody variable domains fused to Pseudomonas exotoxin. Nature 339:394–397

Chester KA, Hawkins RE (1995) Clinical issues in antibody design. Trends Biotechnol 13:294–300

Cheung NK, Lazarus H, Miraldi FD, Abramowsky CR, Kallick S, Saarinen UM, Spitzer T, Strandjord SE, Coccia PF, Berger NA (1987) Ganglioside GD2 specific monoclonal antibody 3F8: a phase I study in patients with neuroblastoma and malignant melanoma. J Clin Oncol 5:1430–1440

Cheung NK, Kushner BH, Cheung IY, Kramer K, Canete A, Gerald W, Bonilla MA, Finn R, Yeh SJ, Larson SM (1998a) Anti-G(D2) antibody treatment of minimal residual stage 4 neuroblastoma diagnosed at more than 1 year of age. J Clin Oncol 16:3053–3060

Cheung NK, Kushner BH, Yeh SDJ, Larson SM (1998b) 3F8 monoclonal antibody treatment of patients with stage 4 neuroblastoma: a phase II study. Int J Oncol 12:1299–1306

Co MS, Deschamps M, Whitley RJ, Queen C (1991) Humanized antibodies for antiviral therapy. Proc Natl Acad Sci U S A 88:2869–2873

Co MS, Avdalovic NM, Caron PC, Avdalovic MV, Scheinberg DA, Queen C (1992) Chimeric and humanized antibodies with specificity for the CD33 antigen. J Immunol 148:1149–1154

Co MS, Baker J, Bednarik K, Janzek E, Neruda W, Mayer P, Plot R, Stumper B, Vasquez M, Queen C, Loibner H (1996) Humanized anti-Lewis Y antibodies: in vitro properties and pharmacokinetics in rhesus monkeys. Cancer Res 56:1118–1125

Coons T (1995) Monoclonal antibodies: the promise and the reality. Radiol Technol 67:39–60

Crameri A, Raillard SA, Bermudez E, Stemmer WP (1998) DNA shuffling of a family of genes from diverse species accelerates directed evolution. Nature 391:288–291

Cregler LL, Herbert M (1986) Medical complications of cocaine abuse. N Engl J Med 315:1495–1500

Crowe JE jr, Murphy BR, Chanock RM, Williamson RA, Barbas CF III, Burton DR (1994) Recombinant human respiratory syncytial virus (RSV) monoclonal antibody Fab is effective therapeutically when introduced directly into the lungs of RSV-infected mice. Proc Natl Acad Sci U S A 91:1386–1390

Cunningham KA, Lakoski JM (1990) The interaction of cocaine serotonin dorsal raphe neurons. Single-unit extracellular recording studies. Neuropsychopharmacology 3:41–50

Dean RA, Christian CD, Sample B, Bosron WF (1991) Human liver cocaine esterases: ethanol-mediated formation of ethylcocaine. FASEB J 5:2735–2739

DeNardo SJ, DeNardo GL, O'Grady LF, Hu E, Sytsma VM, Mills SL, Levy NB, Macey DJ, Miller CH, Epstein AL (1988) Treatment of B cell malignancies with ^{13}1I Lym-1 monoclonal antibodies. Int J Cancer [Suppl] 3:96–101

Des Jarlais DC, Friedland SR (1989) AIDS and IV drug use. Science 245:578–579

Dickman S (1998) Antibodies stage a comeback in cancer treatment. Science 280:1196–1197

Dillman RO, Beauregard J, Shawler DL, Halpern SE, Markman M, Ryan KP, Baird SM, Clutter M (1986) Continuous infusion of T101 monoclonal antibody in chronic lymphocytic leukemia and cutaneous T-cell lymphoma. J Biol Response Mod 5:394–410

Dillman RO, Shawler DL, Dillman JB, Royston I (1984) Therapy of chronic lymphocytic leukemia and cutaneous T-cell lymphoma with T101 monoclonal antibody. J Clin Oncol 2:881–891

Dillman RO, Beauregard JC, Halpern SE, Clutter M (1986) Toxicities and side effects associated with intravenous infusions of murine monoclonal antibodies. J Biol Response Mod 5:73–84

Dyer MJ, Hale G, Hayhoe FG, Waldmann H (1989) Effects of CAMPATH-1 antibodies in vivo in patients with lymphoid malignancies: influence of antibody isotype. Blood 73:1431–1439

Eisen HN (1964) Equilibrium dialysis for measurement of antibody-hapten affinities. In: Eisen HN (ed) Methods in medical research. Yearbook Medical Publishers, Chicago, pp 106–114

Foon KA, Schroff RW, Bunn PA, Mayer D, Abrams PG, Fer M, Ochs J, Bottino GC, Sherwin SA, Carlo DJ, et al (1984) Effects of monoclonal antibody therapy in patients with chronic lymphocytic leukemia. Blood 64:1085–1093

Frank DA, Zuckerman B, Amaro H, Aboagye K, Bauchner H, Cabral H, Freid L, Hingson R, Kayne H, Levenson SM, Parker S, Reece H, Vinci R (1988) Cocaine use during pregnancy: prevalence and correlates. Pediatrics 82:888–895

Frost JD, Hank JA, Reaman GH, Frierdich S, Seeger RC, Gan J, Anderson PM, Ettinger LJ, Cairo MS, Blazar BR, Krailo MD, Matthay KK, Reisfeld RA, Sondel PM (1997) A phase I/IB trial of murine monoclonal anti-GD2 antibody 14, G2a plus interleukin-2 in children with refractory neuroblastoma: a report of the Children's Cancer Group. Cancer 80:317–333

Fukushima K, Hirota M, Terasaki PI, Wakisaka A, Togashi H, Chia D, Suyama N, Fukushi Y, Nudelman E, Hakomori S (1984) Characterization of sialosylated Lewis x as a new tumor-associated antigen. Cancer Res 44:5279–5285

Furukawa K, Yamaguchi H, Oettgen HF, Old LJ, Lloyd KO (1989) Two human monoclonal antibodies reacting with the major gangliosides of human melanomas and comparison with corresponding mouse monoclonal antibodies. Cancer Res 49:191–196

Gallacher G (1994) A potential vaccine for cocaine abuse prophylaxis? Immunopharmacology 27:79–81

Gao C, Lavey BJ, Lo CHL, Datta A, Wentworth P, Janda KD (1997a) Direct selection for catalysis from combinatorial antibody libraries using a boronic acid probe: primary amide bond hydrolysis. J Am Chem Soc 120:2211–2217

Gao C, Lin C-H, Lo C-HL, Mao S, Wirsching P, Lerner RA, Janda KD (1997b) Making chemistry selectable by linking it to infectivity. Proc Natl Acad Sci U S A 94:11777–11782

Garrett ER, Seyda K (1983) Prediction of stability in pharmaceutical preparations XX: stability evaluation and bioanalysis of cocaine and benzoylecgonine by high-performance liquid chromatography. J Pharm Sci 72:258–271

Gijsen HJM, Qiao L, Fitz W, Wong C-H (1996) Recent advances in the chemoenzymatic synthesis of carbohydrates and carbohydrate mimetics. Chem Rev 96:443–473

Goldenberg DM, Horowitz JA, Sharkey RM, Hall TC, Murthy S, Goldenberg H, Lee RE, Stein R, Siegel JA, Izon DO, et al (1991) Targeting dosimetry and radioimmunotherapy of B-cell lymphomas with iodine-131-labeled LL2 monoclonal antibody (see comments). J Clin Oncol 9:548–564

Gray WH (1925) Aromatic esters of acylecgonines. J Chem Soc 128:1150–1158

Grossbard ML, Press OW, Appelbaum FR, Bernstein ID, Nadler LM (1992) Monoclonal antibody-based therapies of leukemia and lymphoma. Blood 80:863–878

Grossbard ML, Freedman AS, Ritz J, Coral F, Goldmacher VS, Eliseo L, Spector N, Dear K, Lambert JM, Blattler WA, et al (1992) Serotherapy of B-cell neoplasms with anti-B4-blocked ricin: a phase I trial of daily bolus infusion. Blood 79:576–585

Hakomori S (1986) Glycosphingolipids. Sci Am 254:44–53

Hakomori S (1989) Aberrant glycosylation in tumors and tumor-associated carbohydrate antigens. Adv Cancer Res 52:257–331

Hakomori S (1996) Tumor malignancy defined by aberrant glycosylation and sphingo(glyco)lipid metabolism. Cancer Res 56:5309–5318

Hakomori S, Zhang Y (1997) Glycosphingolipid antigens and cancer therapy. Chem Biol 4:97–104

Hill JH, Wainer BH, Fitch FW, Rothberg RM (1973) The interaction of ^{14}C-morphine with sera from immunized rabbits and from patients addicted to heroin. Clin Exp Immunol 15:213–224

Hoffman JA, Caudill BD, Koman JJ III, Luckey JW, Flynn PM, Hubbard RL (1994) Comparative cocaine abuse treatment strategies: Enhancing client retention and treatment exposure. J. Addict Dis 13:115–128

Hounsell EF, Young M, Davies MJ (1997) Glycoprotein changes in tumours: a renaissance in clinical applications. Clin Sci 93:87–293

Huston JS, Levinson D, Mudgett-Hunter M, Tai MS, Novotny J, Margolies MN, Ridge RJ, Bruccoleri RE, Haber E, Crea R, Oppermann H (1988) Protein engineering of antibody binding sites: recovery of specific activity in an anti-digoxin single-chain Fv analogue produced in Escherichia coli. Proc Natl Acad Sci U S A 85:5879–5883

Imbach P (ed) (1991) Immunotherapy with intravenous immunoglobulins. Academic, San Diego

Irie RF, Matsuki T, Morton DL (1989) Human monoclonal antibody to ganglioside GM2 for melanoma treatment. Lancet 1:786

Ito M, Suzuki E, Naiki M, Sendo F, Arai S (1984) Carbohydrates as tumor-associated antigens. Int J Cancer 34:689–697

Jacobson JM (1998) Passive immunization for the treatment of HIV infection. Mt Sinai J Med 65:22–26

James K (1994) Human monoclonal antibodies. In: Rosenberg M, Moore GP (eds) The Pharmacology of Monoclonal Antibodies. Springer, Berlin Heidelberg New York, pp 3–19 (Handbook of experimental pharmacology, vol 113)

Janda KD, Lo CH, Li T, Barbas CF III, Wirsching P, Lerner RA (1994) Direct selection for a catalytic mechanism from combinatorial antibody libraries. Proc Natl Acad Sci U S A 91:2532–2536

Janda KD, Lo LC, Lo CHL, Sim MM, Wang R, Wong CH, Lerner RA (1997) Chemical selection for catalysis in combinatorial antibody libraries. Science 275:945–948

Jones PT, Dear PH, Foote J, Neuberger MS, Winter G (1986) Replacing the complementarity-determining regions in a human antibody with those from a mouse. Nature 321:522–525

Kang AS, Barbas CF III, Janda KD, Benkovic SJ, Lerner RA (1991) Linkage of recognition and replication functions by assembling combinatiorial antibody Fab libraries along phage surfaces. Proc Natl Acad Sci U S A 88:4363–4366

Kehrl JH, Muraguchi A, Butler JL, Falkoff RJM, Fauci AS (1984) Human B cell activation proliferation and differentiation. Immunol Rev 78:75–96

Khazaeli MB, Saleh MN, Wheeler RH, Huster WJ, Holden H, Carrano R, LoBuglio AF (1988) Phase I trial of multiple large doses of murine monoclonal antibody CO17–1A. II: Pharmacokinetics and immune response. J Natl Cancer Inst 80:937–942

Kim YS, Gum J jr, Brockhausen I (1996) Mucin glycoproteins in neoplasia. Glycoconj J 13:693–707

Klinman NR, Doughty RA (1973) Hapten-specific stimulation of secondary B cells independent of T cells. J Exp Med 138:473–478

Klonoff DC, Andres BT, Obana WG (1989) Stroke associated with cocaine use. Arch Neurol 46:989–993

Köhler G, Milstein C (1975) Continuous cultures of fused cells secreting antibody of predefined specificity. Nature 256:495–497

Kudryashov V, Kim HM, Ragupathi G, Danishefsky SJ, Livingston PO, Lloyd KO (1998) Immunogenicity of synthetic conjugates of Lewis(y) oligosaccharide with proteins in mice: towards the design of anticancer vaccines. Cancer Immunol Immunother 45:281–286

Lee JH, Bennett G (1991) Substance abuse in adulthood. In: Bennett G, Woolf D (eds) Substance abuse: pharmacologic developmental and clinical perspectives. Delman, Albany, New York, pp 157–170

Lerner RA, Kang AS, Bain JD, Barbas CF III (1992) Antibodies without immunization. Science 258:1313–1314

Lewin AH, Gao Y, Abraham P, Boja JW, Kuhar MJ, Carroll FI (1992) 2bb-substituted analogues of cocaine. Synthesis and inhibition of binding to the cocaine receptor. J Med Chem 35:135–140

Liu Y, Budd RD, Griesemer EC (1982) Study of the stability of cocaine and benzoylecgonine its major metabolite in blood samples. J Chromatogr 248:318-320

Lloyd KO (1993) Tumor antigens known to be immunogenic in man. Ann NY Acad Sci 690:50–58

LoBuglio AF, Saleh MN (1992a) Advances in monoclonal antibody therapy of cancer. Am J Med Sci 304:214–224
LoBuglio AF, Saleh MN (1992b) Monoclonal antibody therapy of cancer. Crit Rev Oncol Hematol 13:271–282
LoBuglio AF, Wheeler RH, Trang J, Haynes A, Rogers K, Harvey EB, Sun L, Ghrayeb J, Khazaeli MB (1989) Mouse/human chimeric monoclonal antibody in man: kinetics and immune response. Proc Natl Acad Sci U S A 86:4220–4224
MacLennan ICM, Gray D (1986) Antigen-driven selection of virgin and memory B cells. Immunol Rev 91:61–85
Mallender WD, Voss EW jr (1994) Construction expression and activity of a bivalent bispecific single-chain antibody. J Biol Chem 269:199–206
Maloney DG, Grillo-Lopez AJ, White CA, Bodkin D, Schilder RJ, Neidhart JA, Janakiraman N, Foon KA, Liles TM, Dallaire BK, Wey K, Royston I, Davis T, Levy R (1997) IDEC-C2B8 (Rituximab) anti-CD20 monoclonal antibody therapy in patients with relapsed low-grade non-Hodgkin's lymphoma. Blood 90:2188–2195
Mao S, Gao C, Lo C-HL, Wirsching P, Wong C-H, Janda KD (1999) Phage-display library selection of high affinity human single-chain antibodies to tumor-associated carbohydrate antigens sialyl Lewis^{x} and Lewis^{x}. Proc Natl Acad Sci U S A 96:6953–6958
Mao S, Gao C, Lo C-HL, Janda KD (2000) A human Fab against cholera toxin B from a combinatorial library. (in preparation)
Marks C, Marks JD (1996) Phage libraries – a new route to clinically useful antibodies. N Engl J Med 335:730–733
Marwick C (1997) Monoclonal antibody to treat lymphoma. JAMA 278:616–618
Matsubara K, Kagawa M, Fukui Y (1984) In vivo and in vitro studies on cocaine metabolism: Ecgonine methyl ester as a major metabolite of cocaine. Forensic Sci Intl 26:169–180
Miller RA, Oseroff AR, Stratte PT, Levy R (1983) Monoclonal antibody therapeutic trials in seven patients with T-cell lymphoma. Blood 62:988–995
Minden P, Farr RS (1967) The ammonium sulphate method to measure antigen-binding capacity. In: Weir DM (ed) Handbook of experimental immunology, chap 13. Davis, Philadelphia, pp 463–492
Möller E (1974) Specificity of hapten-reactive T and B mouse lymphocytes. Affinity and avidity of T- and B-cell receptors and anti-hapten antibodies as factors of dose and time after immunization. Scand J Immunol 3:339–355
Morrison SL (1985) Transfectomas provide novel chimeric antibodies. Science 229:1202–1207
Muramatsu T (1993) Carbohydrate signals in metastasis and prognosis of human carcinomas. Glycobiology 3:294–296

Muroi K, Suda T, Nojiri H, Ema H, Amemiya Y, Miura Y, Nakauchi H, Singhal A, Hakomori S (1992) Reactivity profiles of leukemic myeloblasts with monoclonal antibodies directed to sialosyl-Le(x) and other lacto-series type 2 chain antigens: absence of reactivity with normal hematopoietic progenitor cells. Blood 79:713–719

National Institute on Drug Abuse Division of Epidemiology and Prevention (1991) Research National Household Survey on Drug Abuse: population estimates 1990. US Department of Health and Human Services, Rockville MD

Oettgen HF, Old LJ (1991) In: DeVita V, Hellman S, Rosenberg SA (eds) Biologic therapy of cancer. Lippincott, Philadelphia

Oldham RK (1991) Custom-tailored drug immunoconjugates in cancer therapy. Mol Biother 3:148–162

Padlan EA (1991) A possible procedure for reducing the immunogenicity of antibody variable domains while preserving their ligand-binding properties. Mol Immunol 28:489–498

Pegram MD, Lipton A, Hayes DF, Weber BL, Baselga JM, Tripathy D, Baly D, Baughman SA, Twaddell T, Glaspy JA, Slamon DJ (1998) Phase II study of receptor-enhanced chemosensitivity using recombinant humanized anti-p185HER2/neu monoclonal antibody plus cisplatin in patients with HER2/neu-overexpressing metastatic breast cancer refractory to chemotherapy treatment. J Clin Oncol 16:2659–2671

Pei XY, Holliger P, Murzin AG, Williams RL (1997) The 2,0-Å resolution crystal structure of a trimeric antibody fragment with noncognate VH-VL domain pairs shows a rearrangement of VH CDR3. Proc Natl Acad Sci U S A 94:9637–9642

Phillips ML, Nudelman E, Gaeta FC, Perez M, Singhal AK, Hakomori S, Paulson JC (1990) ELAM-1 mediates cell adhesion by recognition of a carbohydrate ligand sialyl-Le^{x}. Science 250:1130–1132

Pietersz GA, Krauer K, McKenzie IFC (1994) The use of monoclonal antibody immunoconjugates in cancer therapy. Adv Exp Med Biol 353:169–179

Pike BL, Alderson MR, Nossal GJV (1987) T-independent activation of single B cells: an orderly analysis of overlapping stages in the activation pathway. Immunol Rev 99:120–152

Poljak RJ (1994) Production and structure of diabodies. Structure 2:1121–1123

Press OW, Eary JF, Badger CC, Martin PJ, Appelbaum FR, Levy R, Miller R, Brown S, Nelp WB, Krohn KA, et al (1989) Treatment of refractory non-Hodgkin's lymphoma with radiolabeled MB-1 (anti-CD37) antibody. J Clin Oncol 7:1027–1038

Ragupathi G, Park TK, Zhang S, Kim IJ, Graber L, Adluri S, Lloyd KO, Danishefsky SJ, Livingston PO (1997) Immunization of mice with a fully synthetic globo H antigen results in antibodies against human cancer cells: a combined chemical-immunological approach to the fashioning of an anti-cancer vaccine. Angew Chem Int Ed Engl 36:125–128

Reisfeld RA, Gillies SD (1996) Recombinant antibody fusion proteins for cancer immunotherapy. Curr Top Mircrobiol Immunol 213:27–53

Riechmann L, Clark M, Waldmann H, Winter G (1988) Reshaping human antibodies for therapy. Nature 332:323–327

Riethmüller G, Schneider-Gadicke E, Johnson JP (1993) Monoclonal antibodies in cancer therapy. Curr Opin Immunol 5:732–739

Riethmüller G, Schneider-Gädicke E, Schlimok G, Schmiegel W, Raab R, Hoffken K, Gruber R, Pichlmaier H, Hirche H, Pichlmayr R, Buggisch P, Witte J (1994) Randomised trial of monoclonal antibody for adjuvant therapy of resected Dukes' C colorectal carcinom. Lancet 343:1177–1183

Ritz J, Pesando JM, Sallan SE, Clavell LA, Notis-McConarty J, Rosenthal P, Schlossman SF (1981) Serotherapy of acute lymphoblastic leukemia with monoclonal antibody. Blood 58:141–152

Robertson D (1998) Genentech's anticancer Mab expected by November. Nat Biotechnol 16:615

Rosen ST, Zimmer AM, Goldman-Leikin R, Gordon LI, Kazikiewicz JM, Kaplan EH, Variakojis D, Marder RJ, Dykewicz MS, Piergies A, et al (1987) Radioimmunodetection and radioimmunotherapy of cutaneous T cell lymphomas using an ^{13}1I-labeled monoclonal antibody: an Illinois Cancer Council Study. J Clin Oncol 5:562–573

Sames D, Chen X-T, Danishefsky SJ (1997) Convergent total synthesis of a tumour-associated mucin motif. Nature 389:587–590

Sasaki K, Watanabe E, Kawashima K, Sekine S, Dohi T, Oshima M, Hanai N, Nishi T, Hasegawa M (1993) Expression cloning of a novel Galbb(1–3/1–4)GlcNAc aa-23- sialyltransferase using lectin resistance selection. J Biol Chem 268:22782–22787

Sawyer RG, McGory RW, Gaffey MJ, McCullough CC, Shephard BL, Houlgrave CW, Ryan TS, Kuhns M, McNamara A, Caldwell SH, Abdulkareem A, Pruett TL (1998) Improved clinical outcomes with liver transplantation for hepatitis B-induced chronic liver failure using passive immunization. Ann Surg 227:841–850

Schneider-Gädicke E, Riethmüller G (1995) Prevention of manifest metastasis with monoclonal antibodies: a novel approach to immunotherapy of solid tumours. Eur J Cancer 31A:1326–1330

Scott AM, Welt S (1997) Antibody-based immunological therapies. Curr Opin Immunol 9:717–722

Seeberger PH, Bilodeau MT, Danishefsky SJ (1997) Synthesis of biologically important oligosaccharides and other glycoconjugates by the glycal assembly method. Aldrichim Acta 30:75–92

Spector S, Berkowitz B, Flynn EJ, Peskar B (1973) Antibodies to morphine barbiturates and serotonin. Pharmacol Rev 25:281–291

Stemmer WP (1994) Rapid evolution of a protein in vitro by DNA shuffling. Nature 370:389–391

Sterk C (1988) Cocaine and HIV seropositivity. Lancet 2:1052–1053

Steward MW, Steensgaard J (1983) Antibody affinity: thermodynamic aspects and biological significance. CRC Press, Boca Raton

Stewart DJ, Inaba T, Lucassen M, Kalow W (1979) Cocaine metabolism: cocaine and norcocaine hydrolysis by liver and serum esterases. Clin Pharmacol Ther 25:464–468

Tai T, Paulson JC, Cahan LD, Irie RF (1983) Ganglioside GM2 as a human tumor antigen (OFA-I-1). Proc Natl Acad Sci U S A 80:5392–5396

Takada A, Ohmori K, Yoneda T, Tsuyuoka K, Hasegawa A, Kiso M, Kannagi R (1993) Contribution of carbohydrate antigens sialyl Lewis A and sialyl Lewis X to adhesion of human cancer cells to vascular endothelium. Cancer Res 53:354–361

Takayama S, Wong C-H (1997) Chemo-enzymatic approach to carbohydrate recognition. Curr Org Chem 1:109–126

Taussig MJ (1973) Antigenic competition. Curr Top Microbiol Immunol 60:125–174

Thirion S, Motmans K, Heyligen H, Janssens J, Raus J, Vandevyver C (1996) Mono- and bispecific single-chain antibody fragments for cancer therapy. Eur J Cancer Prev 5:507–511

Tonegawa S (1983) Somatic generation of antibody diversity. Nature 302:575

US Department of Health and Human Services (1992) Preliminary estimates from the 1992 Household Survey on Drug Abuse (June 1993) Substance Abuse and Mental Health Service Administration Office of Applied Studies. Washington DC, US Department of Health and Human Services Public Health Services Advance Report, no 3

Vaswani SK, Hamilton RG (1998) Humanized antibodies as potential therapeutic drugs. Ann Allergy Asthma Immunol 81:105–115

Verhoeyen M, Milstein C, Winter G (1988) Reshaping human antibodies: grafting an antilysozyme activity. Science 239:1534–1536

Vitetta ES, Berton MT, Burger C, Kepron M, Lee WT, Yin X-M (1991) Memory B and T cells. Annu Rev Immunol 9:193–217

Von Behring E, Kitasato S (1890) Ueber das Zustandekommen der Diphentherie-Immunität und der Tetanus-Immunität bei Thieren. Dtsch Med Wochenschr 16:1113

Waldmann TA, Strober W (1969) Metabolism of immunoglobulins. Prog Allergy 13:1–110

Wallace BC (1992) Treating crack cocaine dependence: the critical role of relapse prevention. J. Psychoactive Drugs 24:213–222

Williamson RA, Burioni R, Sanna PP, Partridge LJ, Barbas CF III, Burton DR (1993) Human monoclonal antibodies against a plethora of viral pathogens from single combinatorial libraries. Proc Natl Acad Sci U S A 90:4141–4145

Wittmann V, Takayama S, Gong K, Weitz-Schmidt G, Wong C (1998) Ligand recognition by E- and P-selectin: chemoenzymatic synthesis and inhibitory activity of bivalent sialyl Lewis x derivatives and sialyl Lewis carboxylic acids. J Org Chem 63:5137–5143

Wong C-H, Halcomb RL, Ichikawa Y, Kajimoto T (1995a) Enzymes in organic synthesis: application to the problems of carbohydrate recognition (part 1). Angew Chem Int Ed Engl 34:412–432

Wong C-H, Halcomb RL, Ichikawa Y, Kajimoto T (1995b) Enzymes in organic synthesis: application to the problems of carbohydrate recognition (part 2). Angew Chem Int Ed Engl 34:521–546

Yamaguchi H, Furukawa K, Fortunato SR, Livingston PO, Lloyd KO, Oettgen HF, Old LJ (1990) Human monoclonal antibody with dual GM2/GD2 specificity derived from an immunized melanoma patient. Proc Natl Acad Sci U S A 87:3333–3337

Yang WP, Green K, Pinz-Sweeney S, Briones AT, Burton DR, Barbas CF III (1995) CDR walking mutagenesis for the affinity maturation of a potent human anti-HIV-1 antibody into the picomolar range. J Mol Biol 254:392–403

Yokota T, Milenic DE, Whitlow M, Schlom J (1992) Rapid tumor penetration of a single-chain Fv and comparison with other immunoglobulin forms. Cancer Res 52:3402–3408

Yu AL, Uttenreuther-Fischer MM, Huang CS, Tsui CC, Gillies SD, Reisfeld RA, Kung FH (1998) Phase I trial of a human-mouse chimeric anti-disialoganglioside monoclonal antibody ch14,18 in patients with refractory neuroblastoma and osteosarcoma. J Clin Oncol 16:2169–2180

Zuckerman B, Frank DA, Hingson R, Amaro H, Levenson SM, Kayne H, Parker S, Vinci R, Aboagye K, Fried LE (1989) Effects of maternal marijuana and cocaine use on fetal growth. N Engl J Med 320:762–768

Subject Index

Ernst Schering Research Foundation Workshop

Editors: Günter Stock
Monika Lessl

Vol. 1 *(1991)*: Bioscience ⇆ Society – Workshop Report
Editors: D. J. Roy, B. E. Wynne, R. W. Old

Vol. 2 (1991): Round Table Discussion on Bioscience ⇆ Society
Editor: J. J. Cherfas

Vol. 3 (1991): Excitatory Amino Acids and Second Messenger Systems
Editors: V. I. Teichberg, L. Turski

Vol. 4 (1992): Spermatogenesis – Fertilization – Contraception
Editors: E. Nieschlag, U.-F. Habenicht

Vol. 5 (1992): Sex Steroids and the Cardiovascular System
Editors: P. Ramwell, G. Rubanyi, E. Schillinger

Vol. 6 (1993): Transgenic Animals as Model Systems for Human Diseases
Editors: E. F. Wagner, F. Theuring

Vol. 7 (1993): Basic Mechanisms Controlling Term and Preterm Birth
Editors: K. Chwalisz, R. E. Garfield

Vol. 8 (1994): Health Care 2010
Editors: C. Bezold, K. Knabner

Vol. 9 (1994): Sex Steroids and Bone
Editors: R. Ziegler, J. Pfeilschifter, M. Bräutigam

Vol. 10 (1994): Nongenotoxic Carcinogenesis
Editors: A. Cockburn, L. Smith

Vol. 11 (1994): Cell Culture in Pharmaceutical Research
Editors: N. E. Fusenig, H. Graf

Vol. 12 (1994): Interactions Between Adjuvants, Agrochemical and Target Organisms
Editors: P. J. Holloway, R. T. Rees, D. Stock

Vol. 13 (1994): Assessment of the Use of Single Cytochrome P450 Enzymes in Drug Research
Editors: M. R. Waterman, M. Hildebrand

Vol. 14 (1995): Apoptosis in Hormone-Dependent Cancers
Editors: M. Tenniswood, H. Michna

Vol. 15 (1995): Computer Aided Drug Design in Industrial Research
Editors: E. C. Herrmann, R. Franke

Vol. 16 (1995): Organ-Selective Actions of Steroid Hormones
Editors: D. T. Baird, G. Schütz, R. Krattenmacher

Vol. 17 (1996): Alzheimer's Disease
Editors: J.D. Turner, K. Beyreuther, F. Theuring

Vol. 18 (1997): The Endometrium as a Target for Contraception
Editors: H.M. Beier, M.J.K. Harper, K. Chwalisz

Vol. 19 (1997): EGF Receptor in Tumor Growth and Progression
Editors: R. B. Lichtner, R. N. Harkins

Vol. 20 (1997): Cellular Therapy
Editors: H. Wekerle, H. Graf, J.D. Turner

Vol. 21 (1997): Nitric Oxide, Cytochromes P 450,
and Sexual Steroid Hormones
Editors: J.R. Lancaster, J.F. Parkinson

Vol. 22 (1997): Impact of Molecular Biology
and New Technical Developments in Diagnostic Imaging
Editors: W. Semmler, M. Schwaiger

Vol. 23 (1998): Excitatory Amino Acids
Editors: P.H. Seeburg, I. Bresink, L. Turski

Vol. 24 (1998): Molecular Basis of Sex Hormone Receptor Function
Editors: H. Gronemeyer, U. Fuhrmann, K. Parczyk

Vol. 25 (1998): Novel Approaches to Treatment of Osteoporosis
Editors: R.G.G. Russell, T.M. Skerry, U. Kollenkirchen

Vol. 26 (1998): Recent Trends in Molecular Recognition
Editors: F. Diederich, H. Künzer

Vol. 27 (1998): Gene Therapy
Editors: R.E. Sobol, K.J. Scanlon, E. Nestaas, T. Strohmeyer

Vol. 28 (1999): Therapeutic Angiogenesis
Editors: J.A. Dormandy, W.P. Dole, G.M. Rubanyi

Vol. 29 (2000): Of Fish, Fly, Worm and Man
Editors: C. Nüsslein-Volhard, J. Krätzschmar

Vol. 30 (2000): Therapeutic Vaccination Therapy
Editors: P. Walden, W. Sterry, H. Hennekes

Vol. 31 (2000): Advances in Eicosanoid Research
Editors: C.N. Serhan, H.D. Perez

Vol. 32 (2000): The Role of Natural Products in Drug Discovery
Editors: J. Mulzer, R. Bohlmann

Supplement 1 (1994): Molecular and Cellular Endocrinology of the Testis
Editors: G. Verhoeven, U.-F. Habenicht

Supplement 2 (1997): Signal Transduction in Testicular Cells
Editors: V. Hansson, F. O. Levy, K. Taskén

Supplement 3 (1998): Testicular Function:
From Gene Expression to Genetic Manipulation
Editors: M. Stefanini, C. Boitani, M. Galdieri, R. Geremia, F. Palombi

Supplement 4 (2000): Hormone Replacement Therapy
and Osteoporosis
Editors: J. Kato, H. Minaguchi, Y. Nishino

Supplement 5 (1999): Interferon:
The Dawn of Recombinant Protein Drugs
Editors: J. Lindenmann, W.D. Schleuning

Supplement 6 (2000): Testis, Epididymis and Technologies
in the Year 2000
Editors: B. Jégou, C. Pineau, J. Saez

This series will be available on request from
Ernst Schering Research Foundation, 13342 Berlin, Germany